Reproduction in the
Dog and Cat

GW00670520

Reproduction in the Dog and Cat

Ib J. Christiansen

Assistant Professor, Institute for Animal Reproduction
Royal Veterinary and Agricultural University
Copenhagen, Denmark

Baillière Tindall

LONDON   PHILADELPHIA   TORONTO   MEXICO CITY
RIO DE JANEIRO   SYDNEY   TOKYO   HONG KONG

Baillière Tindall   1 St Anne's Road
W. B. Saunders   Eastbourne, East Sussex BN21 3UN, England

West Washington Square
Philadelphia, PA 19105, USA

1 Goldthorne Avenue
Toronto, Ontario M8Z 5T9, Canada

Apartado 26370—Cedro 512
Mexico 4, DF Mexico

Rua Evaristo da Veiga 55, 20° andar
Rio de Janeiro—RJ, Brazil

ABP Australia Ltd, 44–50 Waterloo Road
North Ryde, NSW 2113, Australia

Ichibancho Central Building, 22–1 Ichibancho
Chiyoda-ku, Tokyo 102, Japan

10/fl, Inter-Continental Plaza, 94 Granville Road
Tsim Sha Tsui East, Kowloon, Hong Kong

First published 1984

Typeset by Preface Ltd, Salisbury
Printed in Great Britain by Butler and Tanner Ltd, Frome and London

British Library Cataloguing in Publication Data

Christiansen, Ib J.
 Reproduction in the dog and cat.
 1. Dogs—Reproduction   2. Cats—
 Reproduction
 I. Title
 636.7′08926   SF427.2

ISBN 0-7020-0918-0

Contents

(Cont. overleaf)

Part 3   Appendices and Index

Preface

For centuries dogs and cats have been closely related to man, each species in its own way. The dog has been treated as a pet, used for racing and sport, as a herdsman's dog, a guard and for many other purposes that suited the characteristics of particular breeds. The cat, aloof and self-sufficient, has been primarily regarded as a pet or a nearly equal partner, but on its own conditions. Both species have contributed to the solution of scientific, medical and surgical problems of the greatest importance to mankind. In spite of all this, it is striking how little was known about the reproductive processes of these animals until very recently. It has gradually become apparent that not only do the dog and the cat differ between themselves, but both species also differ markedly from farm animals and from human beings in their physiology and endocrinology.

This book draws widely on published literature throughout the world to present a concise, scientific account of reproduction in the dog and cat. The structure of the text is based on my teaching course for students at Copenhagen, and it is hoped that it will be of particular help to veterinary students and also to breeders and owners with a scientific interest in their animals. Small animal clinicians wishing to study particular topics at greater depth will be assisted by the full citation of references.

The book is divided into two parts, the first devoted to the dog, the second to the cat, and both parts use the same chapter headings to cover normal as well as pathological events in the female and the male separately. The length at which the subjects are treated varies. My purpose has been to survey the reproductive patterns, sexual behaviour, gynaecological and andrological examinations, breeding, pregnancy and parturition, but subjects such as basic physiology and endocrinology, nutrition and infection are treated very briefly, since they are well covered in existing texts. Aspects of reproductive behaviour that are fundamentally the same in both species are described at length in the chapters on the dog, and are not repeated in detail in the cat section.

Disease symptoms and diagnoses are included. The therapeutic advice includes indications for the use of newer drugs as well as those long established.

Ib J. Christiansen
March 1984

Acknowledgements

I am grateful to N.O. Rasbech, Professor, the Institute for Animal Reproduction, Royal Veterinary and Agricultural University, Copenhagen, and J.A. Laing, Professor of Animal Husbandry and Hygiene, Royal Veterinary College, University of London, for having urged and encouraged me to write this book and for their advice during its preparation. Professor Laing has followed the development of this book closely and has given valuable comments, evaluations and suggestions, for all of which I am deeply thankful.

Tables, illustrations and photographs have been generously provided by J. Archibald, L.M. Cobb, Å. Hedhammer, E. Holm Nielsen and Mette Schmidt. Others have given permission to reproduce tables and figures from their published works, and here I want to thank G.H. Arthur, D.J. Bartlett, J.S. Boyd, T.J. Burke, D.W. Christie, E.D. Colby, P.W. Concannon, A.B. Dawson, H.E. Evans, J. Grandage, M. Haemmerli, W. van der Holst, H. Hurni, K. Jensen, W. Jöchle, E. Klug, E. Larsen, A. Lyngset, S. McKeever, R.T. Mowrer, C. Naaktgeboren, M.H. Pineda, C. Platz, C.W. Prescott, N.O. Rasbech, R. Robinson, I.W. Rowlands, P. P. Scott, N. Sojka, J.B. Tedor, K. Tiedemann, H.G. Verhage and R.G. Wales.

For the reproduction of illustrations from previously published works my gratitude is extended to the editors of the following journals: *American Journal of Veterinary Research*, *Australian Veterinary Journal*, *Biology of Reproduction*, *Dansk Veterinærtidsskrift*, *Deutsche tierärztliche Wochenschrift*, *Feline Practice*, *Journal of Reproduction and Fertility*, *Journal of Small Animal Practice*, *Laboratory Animal Care*, *Nordisk Veterinærmedicin*, *Theriogenology*, *Tijdschrift voor Diergeneeskunde*, *Veterinary Medicine/Small Animal Clinic*, *Veterinary Record*, *Zeitschrift für Versuchstierkunde* and *Zentralblatt für Veterinärmedizin*, *Reihe C*, and to the following publishing houses: Baillière Tindall, Heffers Ltd, Lea & Febiger, Medical Book Company, Merck & Company Inc, Pergamon Press Ltd, and John Wiley & Sons Inc.

The typewriting of the manuscript was undertaken by Mrs B. Bengtsson, I. Kühnel and B. Sandgreen and I express my greatest thanks for their meticulous attention to the text. I would also particularly like to thank Mrs Aa. Vestergaard Hansen who, with unfailing enthusiasm and patience, has carefully co-ordinated the practical work of typewriting, preparation of figures and tables and cross-checking of these and the references. My gratitude is also extended to my colleagues at the Institute for Animal Reproduction, Royal Veterinary and Agricultural University, Copenhagen, for their valuable support during the progress of the book.

Particular thanks are directed to Dr J. Hasholt, the Department of Small Domestic Animals, Royal Veterinary and Agricultural University, Copenhagen, and the librarians of the National Veterinary and Agricultural Library, Copenhagen, for their assistance in providing me with articles and books necessary for the work.

Special thanks should also be given to all dogs and cats that have supplied the wealth of information on which this book is based.

Finally I wish to express my very best thanks to my wife and daughters for their patience and tolerance during the long periods of gestation and delivery which have been necessary for the appearance of this book.

Part 1
Reproduction in the Dog

Chapter 1
Gynaecology of the Normal Female

ANATOMY AND PHYSIOLOGY

The ovaries

The ovaries of the bitch are located in the dorsal part of the abdominal cavity caudal to the kidneys at about the level of the 3rd or 4th lumbar vertebra. They are supported by a fold of peritoneum, the mesovarium, which contains the nerve and blood supply. The length of the ovarian attachment varies individually and is also influenced by age and pregnancy. Each ovary is surrounded by a bursa with fat-covered walls which has an opening 0.2–1.8 cm long. The ovary is oval, 2 cm in length, 1.5 cm in diameter, similar in shape to a lima bean, although the exact shape depends on the presence of follicles and corpora lutea.

The relative weight of the ovaries in the neonate is about ten times greater than in the adult bitch.[55] In most bitches the left ovary is heavier than the right, and the number of preovulatory follicles is significantly greater there. At birth the ovary is globular in shape; in beagles, for example, it is 4.0 × 6.0 mm in size.[3] In 15-day-old puppies true primary follicles are present,[75] and at 6½ months Graafian follicles have been demonstrated,[3] but until about 8 months the majority of the follicles undergo degeneration. At the time of puberty maturing Graafian follicles are present in the ovary. In the adult beagle the average dimension of the ovary is 1.7 × 0.9 × 0.4 cm. In older bitches, cysts may be present which vary in size and contain a watery fluid. There is a correlation of 0.77 between the ovarian weight and the bodyweight of adult bitches.

The size of the ovary increases in pro-oestrum, reaching its maximum weight and size at about the time of ovulation, after which it decreases until anoestrum.[92] The ovaries of a newborn bitch contain an estimated 700 000 ova, declining to 350 000 at maturity. At the age of 5 years the number is 33 000 and at 10 years 500.[81] During oestrus the size of the follicles may reach a diameter of 6 mm.

The uterine tube and the uterus

The uterine tube is 4–10 cm in length and 1–2 mm in diameter; it has a funnel-like opening at the ovarian end and is of smaller diameter

3

towards the uterus; it connects the ovary with the corresponding uterine horn. The uterine horns, of elliptical shape in cross-section, are long and narrow and unite caudally, forming the uterine body. The size and the weight of the uterus increase as the bitch matures and enters pro-oestrum and oestrus, attaining maximum size during early metoestrum. It then decreases when anoestrum starts, although not returning to the size of that of the mature bitch.[92] The peak values for thickness and width are reached 7–9 weeks after the onset of oestrus.[14] During metoestrum a characteristic morphological alteration termed corkscrewing occurs. This happens shortly after ovulation and is associated with maximum activity of the corpus luteum and maximum influence of endogenous progesterone (P).[92]

The cervix and vagina

The cervix appears as an oval-shaped mass separating the uterus from the vagina. The vagina extends from the cervix to the poorly defined hymen, beyond which the vestibule extends to the vulva.

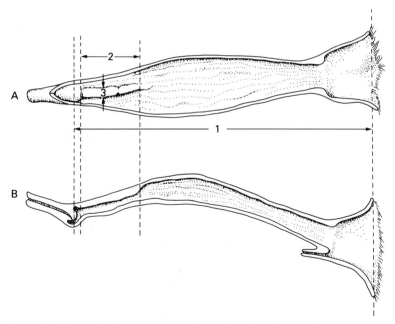

Figure 1.1 Schematic drawing of the canine vagina. A = Ventral view of the dorsal wall. B = Lateral view of a sagittal section. 1 = Length of vagina plus vulva; 2 = length of dorsal median postcervical fold; and 3 = width of dorsal median postcervical fold. (From Pineda *et al.*[73])

In the vagina there is a dorsomedial postcervical fold extending from the vaginal portion of the cervix and terminating caudally by passing into lesser dorsal longitudinal folds (Figure 1.1). The postcervical fold varies in size during the cycle, being at its maximum during oestrus and pregnancy. The vestibule and vagina increase in width during the oestrous cycle, and the genital tract becomes firm and turgid.[3] In pro-oestrum and anoestrum the cervix and the vagina are large, thick and oedematous, and the myometrical thickness is increased. In anoestrum the cervix and the vagina are relatively quiescent.

Hormones

Three main groups of hormones, the releasing hormones, the gonadotropic hormones, and the steroid hormones, control the reproductive cycle. The releasing hormones originate from the hypothalamus and control the synthesis and the release of the gonadotropic hormones from the anterior part of the pituitary gland. The gonadotropic hormones – follicle-stimulating hormone (FSH), luteinizing hormone (LH), and prolactin (PRL) – control the maturation of the germ cells and the production of the steroid hormones: oestrogen and progesterone in the female, testosterone in the male (Table 1.1).

The steroid hormones also act, via a feedback mechanism, to stimulate or inhibit the secretion of releasing and pituitary hormones.

PUBERTY

Puberty is normally reached at an age of 7–12 months (range 6–18 months), in other words 2–3 months after the bitch has achieved adult bodyweight.[3] Small breeds of dogs consequently experience puberty earlier than larger breeds, as the adult bodyweight is achieved at an earlier age. The bitch enters maturity a few months before the male dog. In beagle bitches the age at first oestrus averages 343.4 ± 15.7 days (range 216–696).[89]

Various factors may influence the start of puberty. The sire may influence the time of his daughters' first oestrus.[77] Semi-feral and domesticated dogs allowed to roam free become sexually mature earlier than kennelled dogs.[43] Crossbreeding may also play a part (Table 1.2).

A distinct climacteric does not occur in the dog, but the intervals between oestrus gradually lengthen with increasing age. In aged dogs that still have oestrus the fertility may not be seriously affected,[1] and oestrous cycles have been reported to continue regularly up to 20 years of age.[65]

Table 1.1   Hypothalamic and hypophyseal hormones involved in the reproductive cycle.

Source	Hormone	Main functions
Hypothalamus	Gonadotropic releasing hormone (GnRH)	Release of luteinizing hormone (LH) and follicle-stimulating hormone (FSH)
	Thyrotropic releasing hormone (TRH)	Release of thyrotropic hormone (TSH)
	Prolactin releasing factor (PRF)	Release of prolactin (PRL)
	Prolactin inhibiting factor (PIF)	Inhibits release of prolactin
Pituitary gland	Luteinizing hormone (LH)	Ovulation Formation of corpus luteum Secretion of progesterone Secretion of oestrogens Secretion of androgens
	Follicle-stimulating hormone (FSH)	Follicle growth Stimulation of Sertoli cells Spermatogenesis
	Prolactin (PRL)	Secretion of progesterone Lactation Secretion of testosterone Stimulation of male accessory sexual organs
	Oxytocin	Parturition, contraction of uterine muscles Milk letdown Transportation of ova Transportation of spermatozoa

Table 1.2   Mean age (with standard error) of different breeds at onset of sexual maturity. Difference between purebred and crossbred animals = 5.5 ± 1.1 months. F × A = Foxhound–airedale cross. (From Rowlands[80])

Breed	Number of bitches	Mean age at first oestrus (months)
Airedale	21	20 ± 0.9
Mongrel	8	13 ± 0.6 ⎫
F × A	8	16 ± 0.8 ⎭ 14.5 ± 0.6

BREEDING SEASON

It was believed that 2 breeding seasons a year exist in dogs, but methodical study has failed to show a bimodal distribution of the oestrous periods.[4,30,90,94] Some investigators have found that oestrus occurs all the year round, although more cycles occur during the first

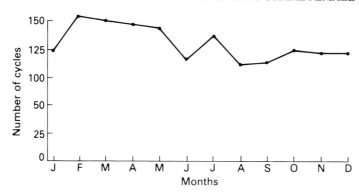

Figure 1.2 Frequency of the time of onset of pro-oestrum during the months of the year in the breeding bitch. (From Christie and Bell[13])

half as compared with the latter half of the year (Figure 1.2). Others have found certain peak months (May, July and October) of oestrous activity.[89]

The ambient temperature and other environmental factors may have some influence upon the oestrous activity.[45] Latitude and photoperiodicity seem to have no effect,[13] nor do natural or artificial lighting,[92] but reproduction is reported to be affected by both inanition and adiposity.[60]

OESTROUS CYCLE

The bitch is mono-oestrous, that is, she has 1 oestrus per breeding season, and the widespread opinion that all bitches have 2 oestrous cycles per year is, as mentioned above, incorrect. The feral dog has only 1 oestrous cycle per year, as can be seen in the dingo and wolf × dog cross. Meischner,[63] examining only a few bitches, found that oestrus occurred twice in 65.2% with an average interval of 182.4 days (range 156.7–219.0 days). In 26.1% oestrus occurred once and in 8.7% three times a year. In another investigation studying a greater number of animals it was shown that most do not have 2 oestrous cycles per year and that the duration of the interoestrus period varies with breed from 22 to 47 weeks (average 31 weeks) (Table 1.3). This corresponds fairly well with other observations that the duration of the interoestrous period is 7.7 months.[80] In the greyhound, for example, interoestrum averages 8.1 months (244 days), but it may vary from 100 to 410 days.[74]

The basenji is known to have only one season per year, usually in the autumn;[35] the reason for this is a single recessive gene, and cross-

Table 1.3   Median length and variation for the oestrus interval in various breeds. (After Christie and Bell[13])

Breed	Interval (weeks)	
	Median	Variation
Alsatian	26	22–30
Beagle	34	30–40
Boxer	34	30.5–42.5
Cairn terrier	29	27–31
Cavalier King Charles spaniel	34	30–36
Chihuahua	31	—
Cocker spaniel	32	26–39
Dachshund, miniature	34	30–42
Dachshund, standard	33.5	29–41
Golden retriever	33	27–39.5
Labrador retriever	29	25–34
Pekinese	29	26–32
Pembroke Corgi	31	—
Pomeranian	27	—
Poodle, miniature	27	24–32
Rhodesian ridgeback	29	25.5–31.5
Rough collie	36.5	31–47
Scottish terrier	28	26–31
Shetland sheepdog	32.5	29–39
Yorkshire terrier	32	31–34
Pregnant bitches	32	28–39
Non-pregnant bitches	29	25–34
Mated non-pregnant bitches	29	25–32
Total study	31	26–36

breeding results in offspring some of which have 1 breeding season and others 2 per year.

Significant breed differences are not the rule, but some are of interest. Thus, the interoestrous period is, on average, 26 weeks in alsatians and 33.5 weeks in standard dachshunds;[35] the alsatian has a significantly shorter interval (149 ± 28.5 days) between oestrous periods than other breeds.[91,93] Bodyweight has no influence; thus the interoestrous period lasts 29 weeks in the average pekingese, cairn terrier, labrador and rhodesian ridgeback in spite of variations in body size.[13] Crossbreeding shortens the interval between oestrous periods (Table 1.4).

Pregnancy prolongs the period until the next oestrus; thus the interoestrous period in beagles is 230 ± 3 days in pregnant and 202 ± 5 days in non-pregnant bitches.[19] These variations, according to whether the bitch is pregnant, non-pregnant, or mated but non-pregnant, are well substantiated (see Table 1.3).

Table 1.4   Length of interoestrous period in purebred and crossbred bitches. Purebred bitches include 16 airedales and 2 foxhounds, crossbreds comprise 12 mongrels. Breed difference and standard error = 0.7 ± 0.3 months. (From Rowlands[80])

Length of interoestrous period (months)	Number of interoestrous periods		
	Purebred	Crossbred	Total
4	1	0	1
5	0	4	4
6	4	6	10
7	13	13	26
8	17	7	24
9	7	3	10
10	5	0	5
11	3	3	6
Mean interoestrous period (months)	8.0 ± 0.2	7.3 ± 0.3	7.7 ± 0.16

Most bitches show a slight, though progressive, increase in the length of the interoestrous period up to 4 years of age;[3] this seems to have no deleterious effect on the conception rate.[94] The aged bitch commonly has irregular oestrous cycles and offers an extended anoestrum period.[1,71]

The duration of the oestrous cycle is about 10 times longer in the bitch than in farm animals. Each of the components of the cycle is prolonged, but anoestrum contributes most to the oestrous cycle. Oestrus continues a few days after ovulation in spite of the high plasma level of progesterone. Furthermore, the bitch differs in that there is no luteolysis if a mating does not result in fertilization.

The fact that the bitch, in contrast to large animal species, remains sexually receptive for a number of days after ovulation, and after the formation of and during the initial function of the corpora lutea, has contributed to the problems of terminology in the sequence of the oestrous cycle (Table 1.5). According to the original terminology, which is also used in this text, the oestrous cycle comprises pro-oestrum, oestrus, metoestrum (limited by the last days of acceptance of the male dog and the regression of corpus luteum), and anoestrum. The term metoestrum is also used for the period starting from the last acceptance of the male and lasting until the endometrium has completely regenerated, a duration of 80–90 days[31] or 140–155 days.[2] The term dioestrum has been suggested for the luteal phase of the cycle, starting with a distinct decline in the content of cornified cells accompanied by an increasing number of non-cornified cells in vaginal smears about 3 days before the end of oestrus.[46] Others have described metoestrum as that part of oestrus with luteal activity, and

Table 1.5 Terminology of the sequence of the oestrous cycle.

					Author
Pro-oestrum	Oestrus	Metoestrum		Anoestrum	[1] Evans & Cole (1931)
Pro-oestrum	Oestrus	Metoestrum		Anoestrum	[2] Andersen & Simpson (1973)
Pro-oestrum	Oestrus	Di-oestrum		Anoestrum	Holst & Phemister (1975)
Pro-oestrum	Oestrus	Metoestrum	Di-oestrum	Anoestrum	Stabenfeldt & Shille (1977)
Pro-oestrum	Oestrus	Metoestrum	Pseudopregnancy or pregnancy	Anoestrum	McDonald (1975)
	a	b	c	d	

a. Ovulation
b. Arrest of accept
c. Arrest of the function of C.l.
d. Regeneration finished

[1] Duration of metoestrum 80–90 days
[2] Duration of metoestrum 140–155 days

Table 1.6 Parameters of the oestrous cycle in the bitch.

	Pro-oestrum	Oestrus	Metoestrum	Anoestrum
Duration (days) { Average	9	9	75	125
{ Range	3–16	4–12	60–90	15–265
The vulva	Oedematous	Oedematous	Folded	Folded
Vaginal discharge	Blood-tinged	Yellow/straw-coloured	None	None
Vaginoscopy, introduction of the vaginoscope	Easy	Some resistance	Smooth	Smooth
Mucosa	Rose pink, oedematous, deep folds	Pale pink to anaemic, hypoplastic	Pale, longitudinal folds	Pale, longitudinal folds
Male dog	Attracted, non-acceptance	Attracted, acceptance	Not attracted	Not attracted
The vaginal smear				
Epithelial cells nucleated	Decreasing number	Disappear	Appear in increasing numbers	Present
nucleus absent or pyknotic	Absent	Pyknotic	Absent	Maybe a few
cornified	Increasing numbers towards the end	Numerous	Disappear rapidly	
Erythrocytes	Great number, decreasing with increasing time	Maybe a few		None
Leukocytes	Decreasing towards the end	None in the beginning, but reappear after the ovulation	Increasing number the first 10 days, then decreasing, and none present after day 20	A few
Hormones				
Oestrogen	Concentration increases with maximum 1–2 days prior to LH maximum	Decreasing concentration	Not measurable	Slow increase towards the end
LH	Increases with maximum the day before or the day with standing oestrus	Ovulation at the maximum concentration or 1–2 days later and then decreasing concentration	Low concentration	
P	Very low concentration	Abrupt increase	Maximum 10 days after LH maximum, then decreasing concentration	

use the term dioestrum for the last part of the luteal phase beginning with the loss of sexual receptivity.[94] The luteal period has also been divided into metoestrum, defined as above, and a period of pregnancy or pseudopregnancy.[62]

The sequences of the normal oestrous cycle and the variations that may occur are shown in Table 1.6.

Pro-oestrum

The duration of pro-oestrum averages 9 days (range 3–16), starting from the first signs of vulval oedema and blood-tinged vaginal discharge and ending when the male is accepted. A duration of 27 days has been recorded.[97]

A few days prior to the visible signs of pro-oestrum, most bitches become listless and somewhat indifferent to their surroundings. The maiden bitch may even refuse food. A few females have been observed to have convulsions preceding signs of oestrus, without apparent cause, but these cease as soon as external signs of pro-oestrum become evident.[3]

Pro-oestrum is characterized by a change in the behaviour of the bitch; she becomes restless, excitable and does not obey orders; some bitches may mount other bitches and make coital movements like male dogs. She drinks increasing volumes of water and therefore urinates frequently. There are pheromones, which attract male dogs but not bitches, in the vaginal secretion, and thus also in the urine.[27] One pheromone has been identified as methyl-p-hydroxybenzoate.[37] Attempts have been made to feed the bitch with chlorophyll, 8 mg/kg, in order to attract fewer male dogs, but this does not interfere with breeding and fertility.[36]

The bitch is not interested in male dogs at this period and may even be aggressive towards them.[14] In the beginning of pro-oestrum the bitch acts as though anoestrous, and if a male approaches will turn around, show her teeth, and maybe even bite. Within a few days this behaviour changes – she becomes calmer and may try to escape if a male dog is near; alternatively she may act absolutely passively, or even permit the male to mount and, rarely, allow intromission and ejaculation.

The visible genitalia are enlarged and oedematous, and within 2–4 days a bloody vaginal discharge appears.

Oestrus

The duration of oestrus is similar to that of pro-oestrum, about 9 days (range 4–12 days), but a duration of 27 days[34,69,97] or 30 days[34] has been described. The period is limited by the first and the last days of acceptance of the male dog. At puberty the duration is somewhat longer than average.[65]

The bitch shows interest in the male and tries to attract him. She turns the hindquarters towards him, lowers the back and raises the pelvic region, displays the perineal region, waves the tail to one side and opens the vulva. Vulval oedema is still present but the vaginal discharge changes from being red and bloody to clear and colour-free or straw-yellow.

Tsutsui and Shimizu[97] describe an average duration of pro-oestrum plus oestrus of 19.4 days (range 11–35 days), and of vulval bleeding of 20.5 days (range 4–37 days).

Ovulation

According to the literature there are great variations in the duration of the period in which ovulations take place, from only a few hours [22] to 12–72 h.[38]

Ovulation occurs from 2 days before to 7 days after the initiation of oestrus,[61,72] but most frequently it happens 1–3 days after the first acceptance of the male dog.[49,62] Using laparoscopy to follow the development of the follicles and the ovulations, it has been found that most ovulations occur from days 1 to 7 in oestrus.[99] From observations based on laparotomies, sections and histological investigations in single dogs at different times of oestrus, it is found that on average the ovulations take place on the third day of oestrus and that the age of the bitch influences the ovulations; thus young bitches ovulate early in oestrus whereas older bitches ovulate a little later.[2] A similar investigation concludes that in most bitches ovulation starts 48 h after acceptance of the male and terminates 60 h after acceptance and that 4 to 6 follicles ovulate at each ovulation.[95]

The ovulated eggs are not ready for fertilization but have to go through a meiotic cleavage of 2–5 days' duration. This, combined with the fact that spermatozoa may survive for a long time in the genital tract of the bitch, may explain why it is difficult to estimate exactly the duration of the pregnancy. Motile spermatozoa have been found up to 268 h after mating.[25]

It is, therefore, possible that a bitch bred on the first day of oestrus may have fertile spermatozoa in the uterus through most of the oestrous period. The survival of the spermatozoa over this long period also explains how superfecundation occurs.

Metoestrum

The average length of this period is 75 days (range 60–90 days), limited by the last acceptance of the male and the regression of the corpus luteum. The vulval oedema diminishes rather quickly and disappears, and only a limited amount of vaginal discharge may be present. The bitch becomes calm and relaxed and the attraction of male dogs soon decreases.

Anoestrum

This period of ovarian inactivity averages 125 days (range 15–265 days), during which there is no vaginal discharge.

GYNAECOLOGICAL EXAMINATION

This may be required because of the need for contraception, in attempts to find the optimal time for breeding, or for obtaining diagnostic specimens in pathological conditions to find the cause of infertility. Information on previous use should be given, and the specific gynaecological examination should be accompanied by a general clinical examination. In some bitches further investigations may be required, such as haematological examinations and hormone assays (Table 1.7).

History

The owner should first fill in a questionnaire as accurately as possible (Table 1.8). This should include information on age; state of health; vaccinations; feeding and management; date and duration of the last oestrous period; mating behaviour and information on whether more than one male dog has been used; previous pregnancies and parturitions; date and result of the last parturition; number of stillborn and live puppies born and number at weaning; fertility of relatives, especially if the bitch has not yet bred; medical treatment, especially use of hormones for contraception; prevention of implantation or interruption of pregnancy.

Clinical examination

This should include the current state of health and physical condition, as well as behaviour and measurable variables such as pulse, respiration rate, temperature and weight.

A specific gynaecological examination should include a careful examination of all parts of the genital tract and organs.

The vulva

The examination includes inspection and palpation. The vulva changes its appearance and consistency during the cycle and also during specific pathological conditions.

Pro-oestrum
The vulva increases in size due to oedema in the vulval region, accompanied by the presence of droplets of secretion in the ventral commissure, which result in moistening and adhesion of the hairs round the vulva and on the tail.

Oestrus
Oedema is very obvious, decreasing in the latter part of the period; 1–2 weeks after oestrus the vulva is of normal size. When palpating the perineal and vulval regions the vulva is found to be elevated, the perineal region elongated and the tail waved aside.

Exactly the same appearance, and pronounced oedema of the vulval region, may accompany hormonal disturbances and vaginal tumours.

Metoestrum–anoestrum
The vulva is relatively small, furrowed, and with a good tonus.

Discharge
Vaginal discharge, of varying appearances, occurs normally, not only during the oestrous period but also accompanying rupture of the fetal membranes, parturition and in the puerperium. Pathological discharge is seen in bitches with pyometra, metritis and other pathological conditions of the uterus or vagina.

Pro-oestrum
Bloody or blood-tinged vaginal discharge occurs normally in pro-oestrum and the very beginning of oestrus. Bloody vaginal discharge may occur at other stages of the oestrous cycle and in various pathological conditions, such as tumours in the uterus or vagina, vaginal ulceration, cystitis, ovarian cysts, and in bitches with sub-involution or placental separation during pregnancy.

Oestrus
The colour of the vaginal discharge normally changes gradually due to a decreasing amount of blood, altering from red to yellow, and becoming straw-coloured in proper oestrus.

Metoestrum–anoestrum
There is normally no vaginal discharge in these periods.

Parturition
Due to the pigmentation of the placenta the rupture of the fetal membranes is accompanied by a vaginal discharge which is dark green in colour with a slippery consistency.

Table 1.7 The stages in gynaecological examination.

History/region/sample	Method	Parameter	Reason
History	Interview Questionnaire	Previous breeding	Stage in cycle Because infertile Reproduction control
Vulva	Inspection Palpation	Size Consistency Discharge	Stage in cycle Pathological conditions, e.g. anatomical abnormalities
Vagina	Inspection	Colour Folds Structure Discharge	Stage in cycle Optimal time for breeding Pathological conditions anatomical abnormalities
Vaginal smear	Cytological examination Microbiological examination Biochemical examination	Cell content Pathogenic organisms Anions and cations Glucose	tumours infections

Sample	Method	Parameter	Indication
Uterus	Palpation X-ray examination Ultrasound Laparoscopy Laparotomy	Dimension Shape Adherences Content	Insemination Pregnancy diagnosis Pathological conditions anatomical abnormalities tumours infections
Oviducts	X-ray examination Laparoscopy Laparotomy	Anatomy Adherences	Insemination Pathological abnormalities, e.g. anatomical abnormalities
Ovaries	Laparoscopy Laparotomy	Anatomy Content Adherences	Stage in cycle Optimal time for breeding Pathological conditions, e.g. anatomical abnormalities, cysts
Blood/plasma/serum	Hormone assay Haematological examination Chromosome analysis Agglutination test	Hormone content Erythrocytes Leukocytes Blood plates Cell volume Haemoglobin percentage Haematocrit Sedimentation rate Chromosome picture *Brucella* titre	Stage in cycle Reproduction control Pathological conditions, e.g. pyometra Hereditary diseases Sterility Abortion, stillbirth

Table 1.8 Questionnaire for gynaecological examination.

Owner	

Identification of the bitch–breed	Date of birth:

Oestrus season number	1	2	3	etc.

Date of onset of pro-oestrous bleeding

Date for end of pro-oestrous bleeding

Date for first acceptance

Breeding–kind: tie (T), art. ins. (AI), not bred (NB)

       date

       date

       date

Date for the first refusal

Identification of the male–breed

Has the male sired other litters?

Was the bitch pregnant at 28 days?

Abortion – date

Number of aborted fetuses

Parturition date

Number of puppies born alive

       dead

Parturition: normal (NP), assistance (AP), caesarian section (CS)

Postparturient problems – kind

Hormonal treatment: indication

       drug

       date

Check for *Br. canis*   bitch?

       male?

Comments

Metritis and pyometra
Accompanying these conditions there may be a malodorous red or red-brown purulent discharge, if the uterine cervix is more or less dilated.

Vaginitis
In juvenile vaginitis the discharge is sticky, grey to yellow-green in colour, and of a creamy consistency; in adult vaginitis, it is purulent.

Pregnancy
Normally there is no discharge, but a mucopurulent or even bloody vaginal discharge may occur for a period of up to 4 weeks before parturition without being a symptom of pathological events.

The vagina
Vaginal examination includes inspection, cytological and micro-biological examinations, possibly followed by a sensitivity test.

Vaginal inspection
After careful cleaning of the vulva it is possible, by the use of a vaginoscope and adequate light source, to follow the change in colour of the mucous membrane during the oestrous cycle, and to demonstrate the presence of normal secretion and also any pathological conditions such as tumours, vestibulo-vaginal stenosis, well-developed hymenal tissue or pathological exudate. Introducing the vaginoscope also gives information on the patency of the vagina. Due to the anatomy of the vestibule and vagina the instrument should be directed cranially and dorsally through the vestibule and the caudal part of vagina, and then cranially and horizontally.

The caudal part of the dorsal postcervical fold of the vagina and the constriction of the lateral and ventral vaginal wall may be misinterpreted as the vaginal part of the cervix, with a ventral fissure simulating the external uterine ostium. In beagles and crossbreeds the true vaginal part of the cervix is about 26 mm cranial to this pseudo-cervix.[73]

Pro-oestrum   At the beginning of pro-oestrum the vaginoscope is easily slipped into the vagina, but later there is a certain resistance. The mucosa is rose pink in colour and oedematous, there are longitudinal folds, and a large amount of blood-tinged discharge is visible. The cervix is only very slightly opened.

Oestrus   There is some resistance against the introduction of the vaginoscope, felt as adhesiveness. The mucosa is hyperplastic and pale pink to anaemic in colour. The folds are ill-marked and the cervix partly open.

Metoestrum–anoestrum   The vaginoscope can be introduced very smoothly; the mucosa at this stage is pale in colour with longitudinal folds.

Cytological examination of the vaginal smear[17,82–84,101]
This examination is one of the most important in connection with the gynaecological investigation of the bitch. Microscopical changes occur in addition to the visible changes in the vaginal mucosa during the oestrous cycle. The cell content of the vaginal secretion alters in response to the level of oestrogen in the blood which causes proliferation of the endometrium and diapedetic bleeding and cornification of the vaginal epithelial cells.

Analysis of the epithelial cell content of the vaginal secretion is used to demonstrate the stage of the cycle and the optimum time for mating or for artificial insemination. It is also very useful in the diagnosis of various pathological conditions. However, it should be stressed that this examination must be supported by a careful history and clinical examination, and that several successive smears should be examined at 24 h intervals, as the cytological content may be influenced by various factors such as management and feeding.[32] For sample collection the bitch should be placed on a table, an assistant holding the head with one hand and the tail with the other, thereby reducing the angle of the vestibule so that the lumens of the vestibule and vagina become almost straight and parallel with the long axis of the body. The vulva is cleaned and opened with two fingers, and the smear may be collected with a flattened stick,[15] a sterile cotton-tipped applicator[26,85,86] moistened with isotonic saline solution, the broad end of a Bard–Parker scalpel,[101] or a sterile pipette filled with isotonic saline solution.[32]

The collecting instrument should be introduced at the dorsal commissure so as to avoid the clitoral fossa. Using a vaginoscope may help to ensure there is no contamination of the instrument with vestibular smears. A pipette should be introduced deep to the cervix, withdrawn a little and a few drops of saline fluid emptied into the lumen, and resucked before the pipette is withdrawn. Cells collected with a cotton-tipped applicator are transferred to a slide by rolling the cotton along the length of the slide. A specimen collected with a pipette is placed on a slide, spread by another slide and then placed in fixative.

To avoid misinterpretation of the content in vaginal smears it is of vital importance that the collection of the specimen is performed correctly, and that the specimen is fixed immediately after being placed upon the slide. Improper and incorrect collection without the use of a vaginoscope may result in contamination with fungal spores, urine crystals, crystals from talcum powder, or erythrocytes from injuries, which may lead to a wrong diagnosis. A delay of fixation may result in an initial drying of the cells that subsequently, in trichrome staining, may appear as swollen, pink-to-orange homogeneous masses, which may confuse the interpretation of the smear.

Staining The vaginal smear may be stained in various ways (Table 1.9) and examined under an ordinary microscope or, if not stained, examined with a phase-contrast microscope.[68]

A modified method may be used in which the nuclei are stained with haematoxylin prior to Shorr's single staining[26] or trichrome staining.[78] Both these staining techniques may be modified by the use of spray fixation and reduction in the number of staining steps, resulting in a shorter time required for staining.[58,78,79]

Trichrome staining is very useful for hormonal cytology as the cytoplasm of epithelial cells changes under the influence of circulating oestrogen from blue to orange, because of the development of the

Table 1.9 Examples of staining techniques which can be used for staining vaginal smears.

Staining technique	Staining step	Fluid	Time
Shorr staining[58]	1	Fixative: 95% ethanol and ether in a 1:1 ratio	30 min
	2	95% ethanol	Repeated dippings
	3	70% ethanol	
	4	50% ethanol	
	5	Distilled water	
	6	Harris' haematoxylin	6 min
	7	Running tap water	
	8	3% ethanol–ammonia	1 min
	9	Shorr S III solution	2 min
	10	70% ethanol	Repeated dippings
	11	80% ethanol	
	12	95% ethanol	
	13	Xylol	Washing
	14	Mounting medium	
Rapid Shorr staining[58]	1	Fixative: Merckofix spray	3–5 min
	2	Distilled water	Repeated dippings
	3	70% ethanol	
	4	Harris' haematoxylin	4–6 min
	5	Running tap water	
	6	3% ethanol–ammonia	1 min
	7	Shorr S III solution	2 min
	8	70% ethanol	Repeated dippings
	9	Xylol	Washing
	10	Mounting medium	

Table 1.9 *contd.*

Staining technique	Staining step	Fluid	Time
Sano trichrome staining[78]	1	Fixative: 95% ethanol, or ether and 95% ethanol in a 1:1 ratio	30 min to 1 year
	2	80% ethanol	1½ min
	3	70% ethanol	1½ min
	4	50% ethanol	1½ min
	5	Distilled water	1½ min
	6	Mayer's haematoxylin	3 min
	7	Running tap water	6 min
	8	Modified Pollak's trichrome	3 min
	9	Acetic acid (0.2%)	1 min
	10	95% ethanol	1½ min
	11	100% ethanol	1½ min
	12	100% ethanol and xylol (1:1)	1½ min
	13	Xylol	1½ min
	14	Xylol	1½ min
	15	Xylol	1½ min
	16	Mounting medium	
Rapid trichrome staining[78]	1	DeLaunay's fixative: 50 ml acetone, 50 ml 100% methanol, 0.2 g trichloracetic acid crystals	10 s to 10 days
	2	70% ethanol	5 s
	3	Distilled water	10 s
	4	Mayer's haematoxylin	5 s
	5	Running tap water	15 s
	6	Modified Pollak's trichrome	6 s
	7	70% ethanol	5 s
	8	95% ethanol	5 s
	9	100% ethanol	10 s
	10	Xylol	10 s
	11	Mounting medium	
Leishman staining	1	Fixative: 100% ethanol and ether 1:1	30 min
	2	Leishman's solution, undiluted	1 min
	3	Dilute with equal volume of distilled water	10–15 min
	4	Wash and differentiate in distilled water	
	5	Dry and mount	
Leishman staining without prior fixation	1	Leishman's solution, undiluted (12–15 drops)	2 min
	2	Dilute with twice the volume of distilled water (24–30 drops)	10–15 min
	3	Wash and differentiate in distilled water	av. 30 s
	4	Dry and mount	

keratin precursor.[79] Thus oestrus is easily recognized because smears comprise almost exclusively large epithelial cells with orange cytoplasm. For routine examination Leishman's staining with or without a specific fixation is adequate. For rapid but impermanent staining toluidine blue can be used without fixation by adding one drop of 0.1% toluidine solution to the smear after it has been transferred to a slide. After 5 min of staining, the smear can be examined.[67,87] The same goes for the use of slides prepared commercially with stain, where the collected sample is transferred to the slide and the microscopical examination can be performed a few seconds later.

Cells   Various types of cell, erythrocytes, leukocytes and epithelial cells of several forms and sizes, may be present depending on the time of collection.

Epithelial cells may be classified as follows (see Figure 1.3):

1. Anuclear cells, squames, are 30–60 μm keratinous superficial cells without a nucleus, with acidophilic or basophilic staining.

2. Superficial cells are 30–60 μm polygonal cells with a pyknotic nucleus less than 6 μm in diameter. The cytoplasm may be acidophilic due to the presence of the precursor of keratin, keratohyalin.

3. Large intermediate cells are 30–60 μm polygonal cells with a vesicular nucleus 7–11 μm in diameter.

4. Small intermediate cells are 20–30 μm, round to oval in shape with a relatively large vesicular nucleus surrounded by a clear non-keratinized cytoplasm. In the vesicular nucleus of the intermediate cell the female chromosome marker, the Barr body, may be evident.

5. Parabasal cells are 15–25 μm, varying from round to oval in shape with a 9–13 μm large vesicular central nucleus.

6. 'Foam' cells are parabasal cells with cytoplasmic vacuoles.

7. 'Metoestrum' cells are parabasal cells with infiltration of a varying number of leukocytes.[82] This cell type may be present at periods other than metoestrum in smears from bitches with vaginitis.

8. Basal cells are small elongated cells, 10–20 μm in size with a basal nucleus.

It must be stressed that a definite correlation is not always found between the stage of the cycle and the cytological findings. In most such cases the mixture of cells in the smear is 1–3 days ahead of the clinical symptoms. In other bitches the vaginal smear may contain cells typical of metoestrum even though the animals, judged from examination of uterus and the ovaries, are in pro-oestrum.[33]

No information can be deduced from either the absence or the presence of erythrocytes in the smear. Thus erythrocytes may be

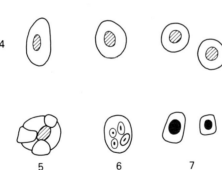

Figure 1.3 Classification of exfoliate cells in vaginal smears. 1, Anuclear superficial cells; 2, superficial cells; 3, intermediate cells; 4, parabasal cells; 5, foam cell; 6, 'metoestrum' cell; 7, basal cells. (From Schmidt, 1981)

absent in pro-oestrum,[8] but may also be present for a long period after oestrus has ended, even as a visible bloody discharge.[80] Erythrocytes have been found in vaginal smears during oestrus and the first 18 days of metoestrum.[8] In some bitches there are no cellular changes in the vaginal smear.[96]

Cell content of the vaginal smear in normal conditions

Pro-oestrum   This period is characterized by the presence of epithelial cells with a nucleus. Initially parabasal and intermediate cells are present in large numbers together with leukocytes and erythrocytes. The latter have been found in smears as early as 10 days before haemorrhage becomes visible in the vaginal discharge. Later in the period the number of parabasal cells decreases, superficial cells appear, the number of erythrocytes and leukocytes decreases; the staining is acidophilic.

Oestrus   Initially there are some large intermediate cells and a great number of either acidophilic or basophilic superficial cells. Anuclear cells and a few erythrocytes may also be present. Later in the period the picture is dominated by solitary acidophilic superficial cells, anucleated cells and cell debris. When acceptance of males decreases, leukocytes and nucleated epithelial cells reappear.

Metoestrum   This is characterized by the presence of superficial cells and an increasing number of parabasal and intermediate cells. During the first 10 days of the period the number of leukocytes increases, then decreases, and after 20 days leukocytes are normally absent. 'Foam' cells are numerous and 'metoestrum' cells are present in great numbers.

Anoestrum   This is a period during which only a few cells – basal, parabasal and intermediate – and a few leukocytes are present. Foam cells may also be present.

Pregnancy   Examination of the cell content in vaginal smears cannot be used to confirm pregnancy, as there is no difference in the smears from pregnant and non-pregnant bitches at corresponding times after oestrus.

The postpartum period   For up to 3 weeks post partum the smears contain copious cellular debris, neutrophils, erythrocytes and endometrial cells and, occasionally, trophoblastic syncytium.

Senility   In a vaginal smear from a senile bitch, dyskaryotic and non-typical cells may be present, even if there are no clinical or pathological signs of genital disorder.[59] Cells from the endocervix and the endometrium may be present as well as an increased number of leukocytes.

Cell content of the vaginal smear in pathological conditions
The cell content of the vaginal smear in various pathological conditions, as described by Dreier,[28] is summarized below.

Subfertility or infertility   When due to infectious conditions these often result in the presence of micro-organisms in the vaginal mucus in varying number.

Subinvolution of the placental site   The smear has the same content as one from a bitch in the normal postpartum period.

Fetal maceration   Examination of vaginal smears from infertile bitches after mating may reveal the same types of cell as in postpartum smears, but there may also be cells filled with melanin fragments from striated muscles, as well as hair and squames as evidence of fetal maceration.[79]

Acute purulent endometritis   This is followed by vaginal mucus containing pyknotic cells with vacuoles and enlarged nuclei. There may be an increased content of leukocytes. Staining is partly acidophilic and partly basophilic.

Closed pyometra   This is characterized by a small amount of vaginal mucus with only a very limited number of cell types (basophilically stained basal, parabasal and intermediate cells and leukocytes), whereas involutional pyometra is accompanied by increased numbers of leukocytes.

Vaginitis   The smear contains neutrophils and atypical desquamated vaginal epithelial cells with enlarged nuclei and often binucleation or multinucleation arranged in clusters. Spherical structures, histiocytes filled with cell debris, are often present.

Microbiological examination of the vagina
This is indicated in cases of abnormal vaginal discharge, abortion or stillbirth and may be extended to include sensitivity tests. A specimen of vaginal mucus may be obtained after careful cleansing of the vulva in the same way as when taking specimens for cytological examination.

Electrical resistance
Measurement of the electrical resistance of vaginal smears has been tried for the detection of oestrus in farm animals and is used in foxes, but it cannot be used in bitches – presumably because of a much more voluminous secretion during oestrus and the presence of a rather large amount of blood.[56]

Bodyweight
There are clear alterations in the bodyweight during oestrus and metoestrum, although the extent of the fluctuations is relatively small in relation to total bodyweight. This parameter cannot therefore be

used for diagnostic purposes in bitches maintained under normal domestic conditions.

During pro-oestrum and the beginning of oestrus the bodyweight increases, decreasing during the later part of oestrus and early part of metoestrum followed by an increase later in metoestrum.[7]

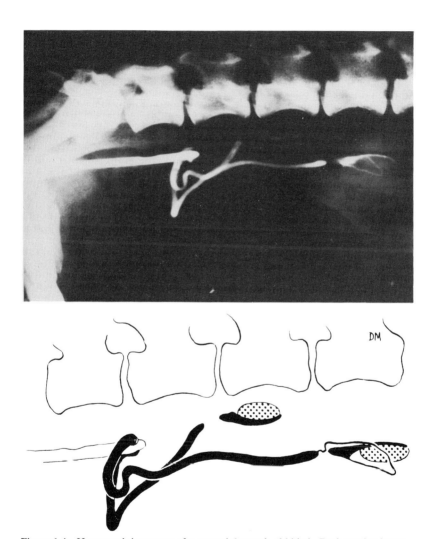

Figure 1.4   Hysterosalpingogram of a normal 6-month-old bitch. Both uterine horns, the right oviduct and both ovarian bursae are outlined. The cervix and uterine body are displaced forward by the catheter. (From Cobb[18])

Rectal temperature
This cannot be used for diagnosing oestrus. It has been shown that most bitches show a thermal nadir during oestrus followed by a period of higher temperature without any correlation with either the vaginal cytology or the progesterone content of the plasma.[12]

Haematological examination
This should include a complete blood count and determination of the *Brucella canis* titre.

X-ray examination
This may be used for the diagnosis of pregnancy and, with the aid of contrast material, for the diagnosis of such pathological conditions as infections and occlusions or anomalies in the tubular tract (Figures 1.4 to 1.7).[18,51]

Laparotomy or laparoscopy
These may give information not obtainable by X-ray examination. The procedure is described in Chapter 8, page 166.

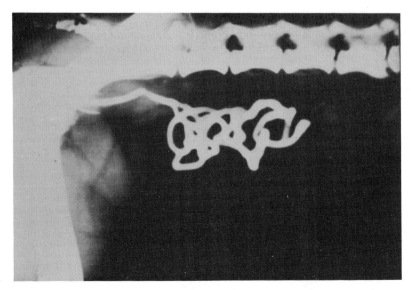

Figure 1.5 Hysterosalpingogram taken during 'heat'. Both horns and one oviduct are outlined. While the uterus is greatly enlarged at oestrus the normal coiling is exaggerated in this radiograph, because the exposure was made with the abdominal muscles tensed. (From Cobb[18])

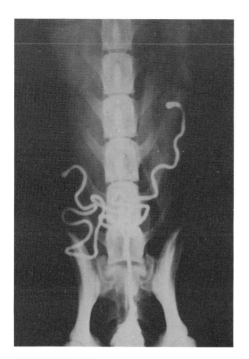

Figure 1.6 Hysterogram taken at oestrus. Both horns and the cranial half of the uterine body are outlined. Note the terminal dilation of the uterine horns; this is found in approximately 50% of hysterograms. (From Cobb[18])

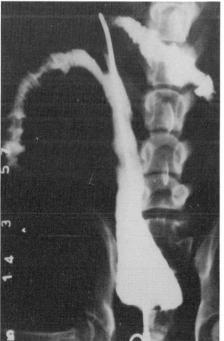

Figure 1.7 Hysterogram taken at 3 days postpartum. Six denuded regions are seen from which maternal placenta was removed together with the fetal placenta at parturition. The tip of the catheter is seen pressing against the wall of one horn of the uterus. (From Cobb[18])

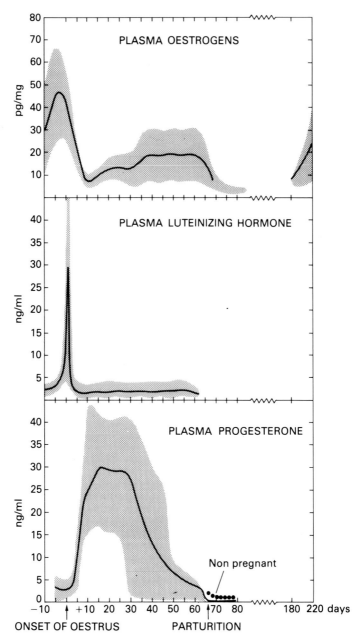

Figure 1.8 Plasma hormone profiles during the reproductive period of the bitch. Mean hormone values are represented by solid lines and the range by shadowing; note absence of data during portions of metoestrum and anoestrum. (From Jöchle and Andersen[49])

Hormone assay

References differ on the content of hormones in the blood of the bitch (Figure 1.8). This may be explained by differences in the samples investigated, whether plasma, serum or whole blood; by different preparations, with or without chromatographic separation; or by differences in biochemical methods (radioimmunoassay or protein-binding). Variations may also be attributed to breed differences, the number of ova shed per animal, to the age of the animals, or to the fact that the serum hormone concentration is elevated during the second cycle compared with the first.[11]

The concentrations of LH, oestrogen and P in pro-oestrum and oestrus and the corresponding behaviour of the bitch, the character of the vaginal discharge and the time of mating are shown in Figure 1.9.

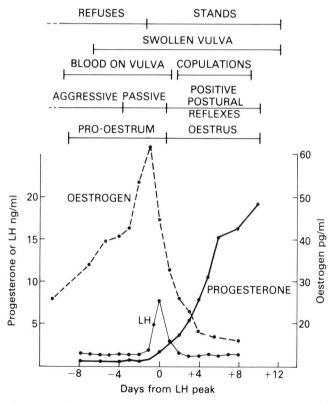

Figure 1.9 Mean plasma levels of oestrogen, LH and progesterone during pro-oestrum and oestrus in beagle bitches. The vertical bars represent the mean time of onset and termination of the parameter indicated. (From Concannon et al.[19])

Luteinizing hormone (LH)

The LH concentration increases from a mean base level of 1.4 ng/ml during pro-oestrum to a maximum between 3 days before and as late as 9 days after the onset of oestrus.[99] The release of LH may be a result of the withdrawal of the negative feedback effect of the rising oestrogen levels and this may be facilitated by the preovulatory increase of P.[22] The LH peak occurs 24–48 h prior to ovulation.[19,21,52,54,61,64,66,72,100] The average maximum value varies from 7.3 ± 1.0 ng/ml[21] to 35.5 ± 10.0 ng/ml.[66] The concentration then decreases and remains at a low level of 2–4 ng/ml.[88]

Age variations occur, and the peak preovulatory LH has been found to be 23.1 ± 4.2 ng/ml in the first oestrous cycle and 51.6 ± 6.9 ng/ml in the second cycle.[11]

Within the first 48 h after the LH peak, 40.2% of the follicles ovulate, and 93.5% of the follicles are ovulated 96 h after the peak.[99] The greatest number of ovulations (77.2%) occur in the period from 24 to 72 h after the maximum concentration of LH. There does not seem to be any correlation between the LH maximum or ovulation time on the one hand, and any particular day in the pro-oestrum or the oestrous period on the other, and ovulations are seen as late as on day 9 of oestrus.[99]

In ovariectomized bitches the concentration of LH is elevated to values similar to the peak levels during the preovulatory surge in intact bitches.[10,22]

Progesterone (P)

The concentration of P is very low in pro-oestrum with 0.2 ng/ml at the beginning[41] and 0.6 ng/ml towards the end, followed by a slow increase in oestrus and an average peak ranging from 22.9 ± 2.7 ng/ml on day 25[19] to 47 ± 3.1 ng/ml 20–25 days after the LH peak.[88]

Behavioural oestrus in the bitch is potentiated by exposure to elevated levels of oestrogen, and subsequently initiated by a decrease in the oestrogen : progesterone ratio, effected by either a fall in oestrogen and/or a rise in progesterone.[23]

No differences in the mean maximum P levels, or in the time when these occur, have been found during the luteal phase between pregnant and non-pregnant bitches. After this maximum level the concentration decreases during metoestrum or pregnancy.[5,16,19,29,39,41,42,50,52,54,66,70,88,98] A shorter period of P secretion has been found in the non-pregnant bitch, whereas the time at which the concentration falls below 1 ng/ml is more variable.[52,53] In non-pregnant bitches the concentration falls below 1 ng/ml between days 51 and 82, and in pregnant bitches between days 61 and 65.[19]

In the pregnant bitch a prepartum decrease occurs during the last 2 days before parturition.[19,29,53,88] In the non-pregnant bitch a minimum concentration of P of 420 pg/ml is found on day 145, coinciding with the regeneration of the endometrium.[44]

Oestrogen
The concentration of oestrogen in plasma increases during pro-oestrum, a peak concentration is reached just prior to or at the start of oestrus, and is followed by a decrease in concentration.[9] The maximum concentration of oestrogen has been measured as varying from 25.3 ± 4.8 pg/ml[41] to 669 ± 91 pg/ml,[72] probably because of different assay techniques. The oestradiol concentration peaks prior to that of oestrone, but both oestrogens decline coincidentally with the LH surge.[100] The concentration of oestradiol-17β is 7–13 pg/ml in the later part of anoestrum, 4.5 weeks before pro-oestrum,[29] and a peak varying from 17.2 pg/ml[52] to 97.7 pg/ml[5] is found at the beginning of or just prior to pro-oestrum. The preovulatory surge of oestradiol-17β is probably responsible for triggering the LH release, supported by the rising oestrone concentration. The oestradiol-17β : oestrone ratio is 2.5 during pro-oestrum and oestrus, but the ratio drops to 1.0 within 30–50 days after the beginning of metoestrum.[29]

During pro-oestrum and oestrus there is a significant correlation between the urine and plasma oestrogen levels, and the highest oestrogen values occur late in pro-oestrum and in early oestrus.[6,48]

A pro-oestrous oestrogen concentration of 26 pg/ml, a slow increase to 43 pg/ml 3 days before the LH peak and a maximum concentration of 62 pg/ml 1 day before the LH peak followed by an abrupt decrease have been observed.[19] On the fourth day of oestrus the concentration, 18 pg/ml, is below that of pro-oestrum. In non-pregnant bitches the concentration remains fairly constant at 9–15 pg/ml, whereas in pregnant bitches the concentration increases to 27 pg/ml on day 36 and remains at this level until 2 days before parturition, where it suddenly decreases to such levels (11 pg/ml) as are found in non-pregnant bitches. Both higher and lower oestrogen levels have been found in plasma during the last trimester of pregnancy in comparison with the non-pregnant dog.[19,42,66]

In some investigations a peak is found in metoestrum with a maximum oestrogen concentration of 10–400 pg/ml and that for oestradiol-17β of 10–15 pg/ml. There are some variations in the time of this peak – if taken from the start of oestrus, the oestrogen peak is reached in the period from 13–15 to 36–60 days, and that of oestradiol-17β from 16 to 69 days.

The oestradiol-17β peak is 28.0 ± 7.0 pg/ml in the first and 54.0 ± 8.0 pg/ml in the second oestrous cycle, but no significant difference is found in the concentration of oestrogen.[11]

In pregnant bitches following implantation, the blood haematocrit levels (packed cell volume) are lower than in non-pregnant bitches, decrease continuously during pregnancy and reach a minimum just prior to parturition.[20] Mean haematocrit levels at days 60–62 were 44.8 ± 1.2 in non-pregnant and 30.6 ± 0.8 in pregnant bitches. During pregnancy the bodyweight increases. These changes imply a large increase in plasma volume which dilutes the circulating hormones so that the concentrations found do not truly reflect the changes in secretion rate of hormones during pregnancy. Recalculation using the mean haematocrit levels suggests the existence of definite post-implantation increases in both progestin and oestrogen secretion specific to pregnancy.

Follicle-stimulating hormone (FSH)
The concentration of FSH is lowest during pro-oestrum (56.3 ± 8.7 ng/ml), increasing simultaneously with LH to a maximum of 167.0 ± 36.9 ng/ml on the first day of oestrus, then slowly decreasing until day 6 (69.2 ± 14.7 ng/ml). Later in the period a new increase occurs to 254.8 ± 27.8 ng/ml in the pregnant bitch and to 107.5 ± 22.2 ng/ml in the pseudopregnant animal on days 55–58.[76] The FSH peak may occur 48 h after the LH peak.[24]

Prolactin (PRL)
The PRL concentrations in pregnant and non-pregnant bitches have been recorded as averaging 38.8 ± 2.2 ng/ml and 23.6 ± 4.0 ng/ml, respectively, on days 28–30, and 52.6 ± 13.6 ng/ml and 35.0 ± 6.5 ng/ml on days 55–58.[76] Others have reported lower concentrations – presumably because of differences in assay methods.[39,57] The rise during the later part of the pregnancy is continued in the beginning of the lactation period, and 84 days after mating the concentration begins to decrease and continues to do so until lactation ceases. In the latter half of pregnancy there is a negative correlation between the concentrations of P and PRL.[38,39] An abrupt increase in PRL concentration is observed just prior to spontaneous abortion.[57]

In a limited number of bitches examined no significant difference was found between plasma from lactating and from non-lactating pseudopregnant bitches.[76]

REFERENCES

1. Andersen, A.C. (1957) Puppy production in the weaning age. *J. Amer. vet. med. Ass.*, **130**, 151–158.
2. Andersen, A.C. and Simpson, M.E. (1973) *The Ovary and Reproductive Cycle of the Dog (Beagle)*. Geron-X Inc., Los Altos, California.
3. Andersen, A.C. and Wooten, E. (1959) The estrous cycle of the dog. In: *Reproduction in Domestic Animals*. eds Cole, H.H. and Cupps, P.T., 1st ed., Chapter 11.
4. Asdell, S.A. (1964) Dog. In: *Patterns of Mammalian Reproduction*. 2nd ed., pp. 426–433. Comstock Publ. Ass., Cornell University Press. Ithaca, NY.

5. Austad, R., Lunde, Astrid & Sjaastad, Ø.V. (1976) Peripheral plasma levels of oestradiol-17β and progesterone in the bitch during the oestrous cycle, in normal pregnancy and after dexamethasone treatment. *J. Reprod. Fertil.*, **46**, 129–136.

6. Batchelor, A., Bell, E.T. and Christie, D.W. (1972) Urinary oestrogen excretion in the beagle bitch. *Brit. vet. J.*, **128**, 560–566.

7. Bell, E.T. and Christie, D.W. (1971) Bodyweight changes during the canine oestrous cycle. *Brit. vet. J.*, **127**, 460–465.

8. Bell, E. T. and Christie, D.W. (1971) Erythrocytes and leucocytes in the vaginal smear of the Beagle bitch. *Vet. Rec.*, **88**, 546–549.

9. Bell, E.T., Christie, D.W. and Younglai, E.V. (1971) Plasma oestrogen levels during the canine oestrous cycle *J. Endocrinol.*, **51**, 225–226.

10. Boyns, A. R., Jones, G.E., Bell, E.T., Christie, D.W. and Parkes, M.F. (1972) Development of a radioimmunoassay for canine luteinizing hormone. *J. Endocrinol.*, **55**, 279–291.

11. Chakraborty, P.K., Panko, W.B. and Fletcher, W.S. (1980) Serum hormone concentrations and their relationships to sexual behavior at the first and second estrous cycles of the Labrador bitch. *Biol. Reprod.*, **22**, 227–232.

12. Christie, D. W. and Bell, E.T. (1970) Changes in the rectal temperature during the normal oestrous cycle in the beagle bitch. *Brit. vet. J.*, **127**, 93–98.

13. Christie, D.W. and Bell, E.T. (1971) Some observations on the seasonal incidence and frequency of oestrus in breeding bitches in Britain. *J. small Anim. Pract.*, **12**, 159–167.

14. Christie, D.W. and Bell, E.T. (1972) Studies on canine reproductive behaviour during the normal oestrous cycle. *Anim. Behav.*, **20**, 621–631.

15. Christie, D.W., Bailey, J.B. and Bell, E.T. (1970) The collection of vaginal smears from the Beagle bitch. *Vet. Rec.*, **87**, 265.

16. Christie, D.W., Bell, E.T., Horth, C.E. and Palmer, R.F. (1971) Peripheral plasma progesterone levels during the canine oestrous cycle. *Acta endocrinol. (Kbh.)*, **68**, 543–550.

17. Christie, D.W., Bailey, J.B. and Bell, E.T. (1972) Classification of cell types in vaginal smears during the canine oestrous cycle. *Brit. vet. J.*, **128**, 301–310.

18. Cobb. L.M. (1959) The radiographic outline of the genital system of the bitch. *Vet. Rec.*, **71**, 66–68.

19. Concannon, P.W., Hansel, W. and Visek, W.J. (1975) The ovarian cycle of the bitch: plasma estrogen, LH and progesterone. *Biol. Reprod.*, **13**, 112–121.

20. Concannon, P.W., Powers, M.E., Holder, W. and Hansel, W. (1977) Pregnancy and parturition in the bitch. *Biol. Reprod.*, **16**, 517–526.

21. Concannon, P., Hansel, W. and McEntee, K. (1977) Changes in LH, progesterone and sexual behaviour associated with preovulatory luteinization in the bitch. *Biol. Reprod.*, **17**, 604–613.

22. Concannon, P., Cowan, R. and Hansel, W. (1979) LH release in ovariectomized dogs in response to estrogen withdrawal and its facilitation by progesterone. *Biol. Reprod.*, **20**, 523–531.

23. Concannon, P. W., Weigand, N., Wilson, S. and Hansel, W. (1979) Sexual behavior in ovariectomized bitches in response to estrogen and progesterone treatments. *Biol. Reprod.*, **20**, 799–809.

24. de Coster, R., Beckers, J.F., Wouters-Ballman, P. and Ectors, F. (1979) Variations de la LH, de la FSH, du 17β oestradiol et de la progestérone au cours de cycle oestral de la chienne. *Ann. Méd. vét.*, **123**, 177–184.

25. Doak, R.l., Hall, A. and Dale, H.E. (1967) Longevity of spermatozoa in the reproductive tract of the bitch. *J. Reprod. Fertil.*, **13**, 51–58.

26. Dore, M.A. (1978) The value of vaginal smears in determining ovulation and optimum breeding times in the bitch. *Irish vet. J.*, **32**, 54–60.

27. Doty, R.L. and Dunbar, I. (1974) Attraction of beagles to conspecific urine, vaginal and anal sac secretion odors. *Physiol. Behav.*, **12**, 825–833.

28. Dreier, H.K. (1975) Diagnostische Möglichkeiten mit der Vaginalsmearuntersuchung bei der Hündin. *Kleintier-Prax.*, **20**, 48–54.
29. Edqvist, L.-E., Johansson, E.D.B., Kasström, H., Olsson, S.-E. and Richkind, M. (1975) Blood plasma levels of progesterone and oestradiol in the dog during the oestrous cycle and pregnancy. *Acta endocrinol. (Kbh.)*, **78**, 554–564.
30. Engle, E.T. (1946) No seasonal breeding cycle in dogs. *J. Mammal.*, **27**, 79–81.
31. Evans, H.M. and Cole, H.H. (1931) An introduction to the study of the oestrous cycle in the dog. *Mem. Univ., Calif.*, **9**, 2, 65–103 (cited in Jöchle and Andersen, 1977[49]).
32. Evans, J.M. and Savage, T.J. (1970) The collection of vaginal smears from bitches. *Vet. Rec.*, **87**, 598–599.
33. Fowler, E.H., Feldman, M.K. and Loeb, W.F. (1971) Comparison of histologic features of ovarian and uterine tissues with vaginal smears of the bitch. *Amer. J. vet. Res.*, **32**, 327–334.
34. Frost, R.C. (1963) Observations concerning ovarian and related conditions in bitches kept as domestic pets. *Vet. Rec.*, **75**, 653–654.
35. Fuller, J.L. (1956) *J. Hered.*, **47**, 179.
36. Gier, H.T. (1954) Effect of chlorophyll derivatives on mating odors of dogs. *Vet. Med.*, **49**, 377–391.
37. Goodwin, M., Gooding, K.M. and Regnier, F. (1979) Sex pheromone in the dog. *Science*, **203**, 559–561.
38. Gräf, K.-J. (1978) Serum oestrogen, progesterone and prolactin concentrations in cyclic, pregnant and lactating beagle dogs. *J. Reprod. Fertil.*, **52**, 9–14.
39. Gräf, K.-J., Friedreich, E., Matthes, S. and Hasan S. H. (1977) Homologous radioimmunoassay for canine prolactin and its application in various physiological states. *J. Endocrinol.*, **75**, 93–103.
40. Griffiths, W.F.B. and Amoroso, E.C. (1939) Pro-oestrus, oestrus, ovulation and mating in the greyhound bitch. *Vet. Rec.*, **51**, 1279–1284.
41. Hadley, J.C. (1975) Total unconjugated oestrogen and progesterone concentrations in peripheral blood during the oestrous cycle of the dog. *J. Reprod. Fertil.*, **44**, 445–451.
42. Hadley, J.C. (1975) Total unconjugated oestrogen and progesterone concentrations in peripheral blood during pregnancy in the dog. *J. Reprod. Fertil.*, **44**, 453–460.
43. Hancock, J.L. and Rowlands, I.W. (1949) The physiology of reproduction in the dog. *Vet. Rec.*, **61**, 771–779.
44. Hansel, W., Concannon, P.W. and McEntee, K. (1976) Plasma hormone profiles and pathological observations in the medroxyprogesterone acetate treated beagle bitch. Symposium, *Pharmacology of Steroid Contraceptive Drugs*. Milan, Italy, Raven Press, New York (cited in Jöchle and Andersen, 1977[49]).
45. Harrop, A.E. (1960) *Reproduction in the Dog*. London: Baillière, Tindall and Cox.
46. Holst, P.A. and Phemister, R.D. (1974) Onset of diestrus in the beagle bitch: definition and significance. *Amer. J. vet. Res.*, **35**, 401–406.
47. Holst, P.A. and Phemister, R.D. (1975) Temporal sequence of events in the estrous cycle of the bitch. *Amer. J. vet. Res.*, **36**, 705–706.
48. Horst, H.-J., Grunert, E. and Stoye, M. (1973) Harnöstrogenbestimmungen beim Hund wärhend der Brunst und Trächtigkeit. *Zbl. Vet.-Med. A*, **20**, 77–83.
49. Jöchle, W. and Andersen, A.C. (1977) The estrous cycle in the dog: a review. *Theriogenology*, **7**, 113–140.
50. Jöchle, W., Tomlinson, R.V. and Andersen, A.C. (1973) Prostaglandin effects on plasma progesterone levels in the pregnant and cycling dog (beagle). *Prostaglandins*, **3**, 209–217.
51. Johnston, S.D. (1980) Diagnostic and therapeutic approach to infertility in the bitch. *J. Amer. vet. med. Assoc.*, **176**, 1335–1338.

52. Jones, G.E., Boyns, A.R., Bell, E.T., Christie, D.W. and Parkes, M.F. (1973) Immunoreactive luteinizing hormone and progesterone during pregnancy and following gonadotrophin administration in beagle bitches. *Acta endocrinol. (Kbh.)*, **72**, 573–581.

53. Jones, G.E., Boyns, A.R., Cameron, E.H.D., Bell, E.T., Christie, D.W. and Parkes, M.F. (1973) Plasma oestradiol, luteinizing hormone and progesterone during pregnancy in the beagle bitch. *J. Reprod. Fertil.*, **35**, 187–189.

54. Jones, G.E., Boyns, A.R., Cameron, E.H.D., Bell, E.T., Christie, D.W. and Parkes, M.F. (1973) Plasma oestradiol, luteinizing hormone and progesterone during the oestrous cycle in the beagle bitch. *J. Endocrinol.*, **57**, 331–332.

55. Kaiser, G. (1977) The significance of the ovaries in relation to body size for the reproductive performance of the dog. *Zool. Anzeiger*, **198**, 3/4, 203–244.

56. Klötzer, I. (1974) Untersuchungen über den elektrischen Widerstand des Vaginalschleimes der Hündin. *Kleintier-Prax.*, **19**, 125–133.

57. Knight, P.J., Hamilton, J.M. and Hiddleston, W.A. (1977) Serum prolactin during pregnancy and lactation in the beagle bitch. *Vet. Rec.*, **101**, 202–203.

58. Kubicek, J. (1978) Vereinfachte Aufbereitung von Vaginalabstrichen zur hormonalen Zytodiagnostik bei der Hündin. *Kleintier-Prax.*, **23**, 259–262.

59. Kubicek, J. (1978) Atypische und scheidenfremde Zellen in Vaginalabstrichen von seneszenten und senilen Hündinnen. *Kleintier-Prax.*, **23**, 325–327.

60. Laing, J.A. (1955) *Fertility and Infertility in the Domestic Animals, Aetiology, Diagnosis and Treatment.* Baillière, Tindall and Cox, London.

61. Masken, J.F. (1973) Circulating hormone levels in the cycling beagle. *22nd Gaines vet. Symp.*, Gaines Dog Research Center, White Plains, New York. pp. 33–34.

62. McDonald, L.E. (1975) Reproductive patterns of dogs. In: *Veterinary Endocrinology and Reproduction*, 2nd ed., Chapter 17. Philadelphia: Lea and Febiger.

63. Mesichner, W. (1966) Zur Brunstperiodik von Haushündinnen. *Wiss. Z. der Karl Marx-Universität, Leipzig*, **15**, 481–482.

64. Mellin, T.N., Orczyk, G.P., Hichens, M. and Behrman, H.R. (1976) Serum profiles of luteinizing hormone, progesterone and total estrogens during the canine estrous cycle. *Theriogenology*, **5**, 175–187.

65. Mulligan, R.M. (1942) Histological studies on the canine female genital tract. *J. Morphol.*, **71**, 431–448.

66. Nett, T.M., Akbar, A.M., Phemister, R.D., Holst, P.A., Reichert, Jr., L.E. and Niswender, G.D. (1975) Levels of luteinizing hormone, estradiol and progesterone in serum during the estrous cycle and pregnancy in the beagle bitch. *Proc. Soc. exp. Biol. (NY)*, **148**, 134–139.

67. Newberry, W.E. and Gier, H.T. (1952) Determination of breeding time in the bitch from vaginal smears. *Vet. Med.*, **47**, 390–392.

68. Northway, R.B. (1972) Use of phase microscopy with vaginal smears to determine ovulation in the bitch. *Vet. Med./small Anim. Clin.*, **67**, 538–541.

69. Oettel, M. (1975) Der Einsatz von Seksualhormonen beim Kleintier. In: *Veterinärmedizinische Endokrinologie.* pp. 631–651. VEB Gustav Fischer Verlag, Jena.

70. Parkes, M.F., Bell, E.T. and Christie, D.W. (1972) Plasma progesterone levels during pregnancy in the beagle bitch. *Brit. vet. J.*, **128**, 15–16.

71. Pearson, M. and Pearson, K. (1930–31) On the relation of the duration of pregnancy to size of litter and other characters in bitches. *Biometrika*, **22**, 309–323.

72. Phemister, R.D., Holst, P.A., Spano, J.S. and Hopwood, M.L. (1973) Time of ovulation in the beagle bitch. *Biol. Reprod.*, **8**, 74–82.

73. Pineda, M.H., Kainer, R.A. and Faulkner, L.C. (1973) Dorsal median postcervical fold in the canine vagina. *Amer. J. vet. Res.*, **34**, 1487–1491.

74. Prole, J.H.B. (1973) Some observations in the physiology of reproduction in the greyhound bitch. *J. small Anim. Pract.*, **14**, 781–784.

75. Raps, G. (1948) The development of the dog ovary from birth to 6 months of age. *Amer. J. vet. Res., 9*, 61–64.
76. Reimers, T.J., Phemister, R.D. and Niswender, G.D. (1978) Radioimmunological measurement of follicle stimulating hormone and prolactin in the dog. *Biol. Reprod., 19*, 673–679.
77. Rogers, A.L., Tempelton, J.W. and Stewart, A.P. (1970) Preliminary observations of estrous cycles in large, colony raised laboratory dogs. *Lab. Anim. Care, 20*, 1133–1136.
78. Roszel, J.F. (1975) Genital cytology of the bitch. *Vet. Scope, 19*, 2–15.
79. Roszel, J.F. (1977) Normal canine vaginal cytology. *Vet. Clin. N. Amer., 7*, 667–681.
80. Rowlands, I.W. (1950) Some observations on the breeding of dogs. *Proc. Conf. Soc. Study Fertil., London, 2*, 40–55.
81. Schotterer, A. (1928) Beitrag zur Feststellung der Eianzahl in verschiedenen Altersperioden bei der Hündin. *Anat. Anzeiger, 65*, 11/13, 177–192.
82. Schutte, A.P. (1967) Canine vaginal cytology, I. Technique and cytological morphology. *J. small Anim. Pract., 8*, 301–306.
83. Schutte, A.P. (1967) Canine vaginal cytology, II. Cyclic changes *J. small Anim. Pract., 8*, 307–311.
84. Schutte, A.P. (1967) Canine vaginal cytology, III. Compilation and evaluation of cellular indices. *J. small Anim. Pract., 8*, 313–317.
85. Settergren, I. (1971) Examination of the canine genital system. *Vet. Clin. N. Amer., 1*, 103–118.
86. Settergren, I. (1971) Vaginalcytologien hos tik och dess användning brunstdiagnostiken. *Svensk Vet.-Tidn., 23*, 79–81.
87. Simmons, J. (1970) The vaginal smear and its practical application. *Vet. Med./small Anim. Clin., 65*, 369–373.
88. Smith, S.M. and McDonald, L.E. (1974) Serum levels of luteinizing hormone and progesterone during the estrous cycle, pseudopregnancy and pregnancy in the dog. *Endocrinology, 94*, 404–412.
89. Smith, W.C. and Reese, Jr., W.C. (1968) Characteristics of a beagle colony. I. Estrous cycle. *Lab. Anim. Care, 18*, 602–606.
90. Sokolowski, J.H. (1973) Reproductive features and patterns in the bitch. *J. Amer. Anim. Hosp. Assoc., 9*, 71–81.
91. Sokolowski, J.H. (1977) Reproductive patterns in the bitch. *Vet. Clin. N. Amer., 7*, 653–666.
92. Sokolowski, J.H., Zimbelman, R.G. and Goyings, L.S. (1973) Canine reproduction: reproductive organs and related structures of the nonparous, parous, and postpartum bitch. *Amer. J. vet. Res., 34*, 1001–1013.
93. Sokolowski, J.H., Stover, D.G. and VanRavenswaay, F. (1977) Seasonal incidence of estrus and interestrous interval for bitches of seven breeds. *J. Amer. vet. med. Assoc., 171*, 271–273.
94. Stabenfeldt, G.H. and Shille, V.M. (1977) Reproduction in the dog and cat. In: *Reproduction in Domestic Animals*, eds. Cole, H. H. and Cupps, P. T., 3rd ed., pp. 499–527. Academic Press, New York, San Francisco, London.
95. Tsutsui, T. (1973) Studies on the physiology of reproduction in the dog, II. Observation on the time of ovulation. *Jap. J. Anim. Reprod., 18*, 137–142.
96. Tsutsui, T. (1975) Studies on the reproduction in the dog, III. Observations of vaginal smear in estrous cycle. *Jap. J. Anim. Reprod., 21*, 37–42.
97. Tsutsui, T. and Shimizu, T. (1973) Studies on the physiology of reproduction in the dog, I. Duration of estrus. *Jap. J. Anim. Reprod., 18*, 132–136.
98. Vogel, F. (1973) *Some Biochemical and Clinical Aspects of the Oestrous Cycle and the Action of Oestrogen Injections in the Bitch.* Proefschrift, Faculteit der Diergeneeskunde Rijksuniversiteit, Utrecht, 101 pp.
99. Wildt, D.E., Chakraborty, P.K., Panko, W.B. and Seager, S.W.J. (1978) Rela-

tionship of reproductive behaviour, serum luteinizing hormone and time of ovulation in the bitch. *Biol. Reprod.*, **18**, 561–570.

100. Wildt, D.E., Panko, W.B., Chakraborty, P.K. and Seager, S.W.J. (1979) Relationship of serum estrone, estradiol-17β and progesterone to LH, sexual behavior and time of ovulation in the bitch. *Biol. Reprod.*, **20**, 648–658.

101. Witiak, E. (1967) The use of vaginal smears to determine ovulation in the bitch. *Vet. Med/small Anim. Clin.*, **62**, 869–878.

ADDITIONAL READING

Abel, Jr., J.H., Tietz, Jr., W.J. and Verhage, H.G. (1974) Development and regression of the corpus luteum in the bitch. *Anat. Rec.*, **178**, 296.

Andersen, A.C., McKelvie, D.H. and Phemister, R. (1962) Reproductive fitness of the female beagle. *J. Amer. vet. med. Assoc.*, **141**, 1451–1454.

Barrau, M.D., Abel, Jr., J.H., Verhage, H.G. and Tietz, Jr., W.J. (1975) Development of the endometrium during the estrous cycle in the bitch. *Amer. J. Anat.*, **142**, 47–65.

Butler, W.F. and Wright, A.I. (1981) Hair growth in the greyhound. *J. small Anim. Pract.*, **22**, 655–661.

Del Campo, C.H. and Ginther, O.J. (1974) Arteries and veins of uterus and ovaries in dogs and cats. *Amer. vet. Res.*, **35**, 409–415.

Christie, D.W. and Bell, E.T. (1971) Endocrinology of oestrous cycle in the bitch. *J. small Anim. Pract.*, **12**, 383–389.

Concannon, P. (1980) Effects of hypophysectomy and of LH administration on luteal phase plasma progesterone levels in the beagle bitch. *J. Reprod. Fertil.*, **58**, 407–410.

Frost, R.C. (1963) Observations concerning ovarian and related conditions in bitches kept as domestic pets. *Vet. Rec.*, **75**, 653–654.

Hale, P. A. (1982) Periodic hair shedding by a normal bitch. *J. small Anim. Pract.*, **23**, 345–350.

Janiak, M.I. (1975) Das Sexualleben des Hundes. *Dtsch. tierärztl. Wochenschr.*, **82**, 499–503.

Jöchle, W. (1976) Neuere Erkenntnisse über die Fortpflanzungsbiologie von Hund und Katze: Konsequenzen für die Östruskontrolle, Konzeptionsverhütung, Abortauslösung und Therapie. *Dtsch. tierärztl. Wochenschr.*, **83**, 564–569.

Jöchle, W. (1980) Reproduction in small animals: advances in biology, pet population control and hormonal therapy. *XI. Int. Congr. Diseases of Cattle. Satellite Symposium Diseases of Small Animals.* Tel-Aviv, 7–30.

Kuwabara, S., Yoshida, H., Tanaka, S. and Murasugi, E. (1973) Studies on the diagnosis of the estrus cycle in the bitch. *J. Tokyo Soc. vet. zootech. Sci.*, **19/20**, 120–127.

Linde, C. (1978) Transport of radiopaque fluid into the uterus after vaginal deposition in the oestrous bitch. *Acta vet. scand.*, **19**, 463–465.

Lindsay, F.E.F. (1981) Detection of cyclic changes in the bitch. *Proc. Spring Meet. Netherl. small Anim. vet. Assoc.* Amsterdam, pp. 18–20.

Lunaas, T. (1978) Ovarialsyklus og drektighet hos hund: Endokrine omstillinger. *Norsk Vet.-T.*, **90**, 397–404.

McCandlish, I.A.P., Munro, C.D., Breeze, R.G. and Nash, A.S. (1979) Hormone producing ovarian tumours in the dog. *Vet. Rec.*, **105**, 9–11.

McCann, J.P. and Concannon, P.W. (1983) Effects of sex, ovarian cycles, pregnancy and lactation on insulin and glucose response to exogenous glucose and glucagon in dogs. *Biol. Reprod.*, **28**, Suppl. 1, 41.

Noakes, D.E. (1980) Pathology of reproduction in the dog. *9th Int. Congr. Anim. Reprod.*, Madrid, vol. 1, 245–260.

Oettel, M. (1979) Reproduktionsbiologie der Hündin. *Mh. Vet.-Med.*, **34**, 937–942.

Olson, P.N., Bowen, R.A., Behrendt, M.D., Olson, J.D. and Nett, T.M. (1982) Concentrations of reproductive hormones in canine serum throughout late anestrus, proestrus and estrus. *Biol. Reprod.*, **27**, 1196–1206.

Olson, P.N., Bowen, R. A. and Nett, T.M. (1982) Hormonal changes in the bitch during the onset of proestrus. *Biol. Reprod.*, **26**, Suppl. 1, 82A.

Reimers, T.J. and Lein, D.H. (1980) Patterns of reproductive hormones in the bitch. *XI. Int. Congr. Diseases of Cattle. Satellite Symposium Diseases of Small Animals.* Tel-Aviv, 31–36.

Reimers, T.J., Mummery, L.K. and Cowan, R.G. (1983) Effects of reproductive state on thyroid and adrenal function in dogs. *Biol. Reprod.*, **28**, Suppl. 1, 41.

Shille, V.M. (1974) Clinical approach to small animal reproductive problems. *Amer. Anim. Hosp. Assoc., 41st Ann. Meet.*, 501–509.

Shille, V.M. (Consultant Editor) (1980) Canine. In: *Current Therapy in Theriogenology Diagnosis, Treatment and Prevention of Reproductive Diseases in Animals*, Chapter VI, 565–579, ed. Morrow, D.A. W.B. Saunders, Philadelphia, London, Toronto.

Shille, V.M. and Stabenfeldt, G.H. (1980) Current concepts in reproduction of the dog and cat. In: *Advanc. Vet. Sci. Comp. Med.*, **24**, 211–243, ed. Brandley, C.A. and Cornelius, C.E. Academic Press, New York and London.

Stabenfeldt, G.H. and Shille, V. (1974) Canine reproductive biology. *Amer. Anim. Hosp. Assoc., 41st Ann. Meet.*, 461–462.

Tesoriero, J.V. (1979) The morphology of the developing ovarian follicle in the dog. *Anat. Rec.*, **193**, 702.

Vandaele, W. (1977) Progrès récents dans la connaissance de cycle sexual des chiennes. Précautions à prendre lors de l'emploi de progestagènes. *Ann. Méd. vét.*, **121**, 369–381.

Chapter 2
Breeding and Mating

BREEDING

Several factors are of importance for optimum breeding. The age of the dogs, the frequency with which the male is used, and the time interval between pregnancies in the bitch should be considered before dogs are used for breeding (Table 2.1).

The bitch should be rested if the litter size is decreasing, if the physical condition after lactation is below normal, or if the interval between cycles is less than 6 months. The male dog has to be given some rest if he has been used frequently. It is advisable to check the semen of stud animals regularly if they are still wanted for breeding.

Optimum time for breeding

The ovulation time varies from bitch to bitch. Newly ovulated eggs must go through a meiotic cleavage of 2–5 days' duration and maturation before fertilization is possible. Furthermore, spermatozoa may be motile for quite a long time in the female genital tract, and therefore it is possible that a bitch bred on the first day of oestrus may have fertile spermatozoa in the uterus through most of the oestrous period. The survival of the spermatozoa during this long period may also explain the possibility of superfecundation.

Table 2.1 Factors influencing breeding results.

Factor	Bitch	Male dog
Age when breeding can begin	11–18 months 30 months at the latest	17–19 months
Frequency of breeding or mating	One or two cycles up to two or three cycles, depending upon: litter size, condition after rearing, cycle interval	One bitch per 2–3 weeks; more frequent use implies periods of sexual rest
Last litter or mating	Depending upon: ease of whelping, average age of the breed, mothering behaviour	Depending upon: previous use, fertility judged by semen examination
Time for breeding	See Table 2.2	

41

Table 2.2 The suitability of various parameters for finding the optimal time for breeding in the bitch.

Parameter	Optimum time for breeding in relation to parameter
Behaviour of the bitch	
first interest in the male dog	3–5 days after
first acceptance of the male dog	2–3 days after
willingness to wave the tail to the side	No relationship
degree of waving the tail to the side	No relationship
Clinical findings	
appearance of bloody vaginal discharge	10–14 days after
appearance of straw-coloured vaginal discharge	2–3 days after
degree of vulval oedema	No relationship
colour of the vaginal mucosa	No relationship
degree of contraction of vagina	No relationship
Vaginal discharge	
reaction of glucose	Second negative reaction
electrical resistance	No relationship
Vaginal smears	
anuclear and superficial cells	100%
leukocytes	Reappearance
cornification index (CI)	80% and second maximum
eosinophilic index (EI)	60%
superficial cell index (SCI)	No relationship
karyopyknotic index (KPI)	No relationship

Under natural conditions, when the bitch and male dog are left alone, there are seldom problems with mating and fertilization. In bitches being bred artificially, knowledge of the optimum time for insemination is essential. Various tests can be performed to determine the best time for breeding (Table 2.2), and finding the most suitable time is facilitated if several of these are combined with observation of the symptoms of oestrus.[10]

The behaviour of the bitch
Even though a bitch in pro-oestrum attracts the male, he is not accepted until late in pro-oestrum and mounting is not permitted until she is in true oestrus. This, together with the fact that ovulation often occurs on the second or third day of oestrus, suggests that the time for breeding is optimal 3–5 days after the bitch has first shown interest in the male dog, or 2–3 days after mounting is first allowed. The acceptance of mounting is the most reliable sign of oestrus.

Clinical findings
With approaching oestrus the bitch shows increasing willingness to

stand for the male and tends to hold the tail to one side when she is touched on the vulva or the perineal region. In oestrus the vulval oedema has reached a maximum, and the vaginal mucosa is pale in colour with longitudinal folds. Vaginal contractions may be provoked when the vagina is touched or insemination is carried out.

Vaginal discharge
In pro-oestrum, which has an average duration of 9 days, there is a bloody vaginal discharge, whereas in oestrus the discharge is straw-coloured. In most bitches the optimum time for breeding is 10–14 days after the first sign of bloody vaginal discharge or 2–3 days after the appearance of the straw-coloured discharge. As mentioned in Chapter 1, a few bitches may not show bloody vaginal discharge in pro-oestrum, and in others the bloody vaginal discharge may persist into real oestrus and even for some days into metoestrum.

Glucose content
The amount of glucose in the vaginal discharge varies during the oestrous cycle. Its presence can be detected by the use of TES tape

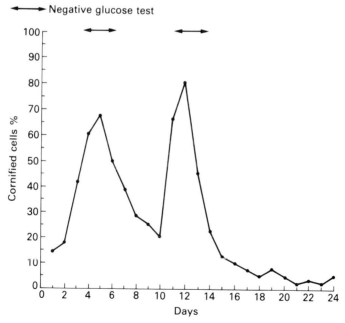

Figure 2.1 The percentage of cornified epithelial cells and the 2 periods in pro-oestrum and oestrus during which the glucose test is negative. (From van der Holst and Best[19])

Table 2.3 The glucose content of the vaginal smear at the time of breeding and subsequent pregnancy rate. (After van der Holst and Best[19])

Glucose in vaginal smear	Pregnancy rate %	Total number of bitches
+	39	141
−	69	143

placed in the vagina for 1 min.[19] The border of the tape becomes green if glucose is present whereas it remains yellow if there is no glucose. Blood may interfere with the evaluation.

It has been found that the glucose content of the vaginal discharge is a valuable indicator of the time for breeding and consequently also for optimum fertility.[19] A negative reaction corresponds to the 2 maxima of cornified cells (Figure 2.1).[2] It has also been found that the best fertility is obtained if breeding takes place when the glucose test is negative (Table 2.3), and furthermore that breeding should not occur before the second negative reaction. In one study[19] only 2 out of 23 dogs inseminated before this time became pregnant. In some investigations, however, it has been found that this method cannot be used for detection of oestrus because there is a negative reaction in some bitches while others show only a weak reaction, but not until after acceptance of the male dog,[1,14,23] or a continuous positive reaction during the whole oestrous period.[21]

Electrical resistance
The electrical resistance varies too much to be of use for finding the optimum time for breeding. It varies from 495 to 1216 ohm on the last day of pro-oestrum and from 250 to 700 ohm on the day before. Variations occur also in oestrus, especially at the beginning. In some bitches the resistance decreases and in others there is an increase compared with pro-oestrum, but the resistance decreases during the last part of oestrus in all bitches.[20]

Vaginal smears
The proportions of the different kinds of epithelial cell vary during the oestrous cycle (see Chapter 1 and Table 2.4). The number of leukocytes decreases during pro-oestrum, and they are absent in the early part of oestrus, but reappear after ovulation; therefore the best time for breeding is when the leukocytes reappear in the vaginal smear.[2] Daily examination of vaginal smears thus makes it possible to estimate the optimum time for breeding, i.e. after ovulation in the mid-stage of oestrus, when the cells comprise only anuclear and superficial cells, and leukocytes have reappeared.

Table 2.4 The number of epithelial cells in the vaginal smear during the oestrous cycle in the bitch. The figures indicate the distribution of epithelial and other types of cell (%)

| | Pro-oestrum | | Oestrus | | | | |
	Early	Late	Early	Mid	Late	Metoestrum	Anoestrum
Anuclear cells		10	50	90	30	10	
Superficial cells	10	40	40	10	20	10	
Intermediate cells	70	50	10		20	20	+
Parabasal cells	20				30	60	+
'Foam' cells						+	(+)
'Metoestrum' cells						+	
Basal cells	+						+
Erythrocytes	+	+	(+)				
Leukocytes	+	(+)		+	+	+	+

(An arrow labelled "Ovulation" runs vertically between the Early and Mid Oestrus columns, pointing downward.)

+ Presence of the particular type of cell.
(+) Final presence of the particular type of cell.

Theoretically the optimum time for breeding is the day after ovulation, when the epithelial cells in the smear are arranged in groups or clusters,[1,23] but the best result is said to be achieved by breeding when the cells are singly arranged (Table 2.5).

Various indices based on counting and differentiating at least 100 epithelial cells have been used for finding the optimum time for breeding.

Cornification index (CI)
In pro-oestrum and at the beginning of oestrus there are many epithelial cells in the vaginal smear, and as oestrus progresses the cells undergo cornification.

$$CI = \frac{\text{number of cornified cells} \times 100}{\text{total number of epithelial cells}}$$

Table 2.5 Percentage of cornified epithelial cells and their arrangement at the time of breeding compared with the pregnancy rate. (After van der Holst and Best[19])

| Cornified cells | | Pregnancy rate % | Total number of bitches |
%	Arrangement		
0–20	In groups	12	42
20–40	In groups and single	40	30
40–60	Single	63	43
60–80	Single	59	76
80–100	Single	67	92

Fertility improves with increasing numbers of cornified cells at the time of breeding (Table 2.5) and the optimum time is when the CI is more than 80%. It is possible to follow the number of cornified cells by daily examination of vaginal smears. The number increases during the oestrous period and, after a minor decrease, increases again (see Figure 2.1). It is characteristic that the male is not accepted at the first peak of cornified cells but at the second, when breeding normally results in pregnancy.[19]

Eosinophilic index (EI)
In stained smears there is an increasing number of eosinophilic cells with advancing oestrus.

$$EI = \frac{\text{number of eosinophilic cells} \times 100}{\text{total number of epithelial cells}}$$

The EI and CI are less than 10% during anoestrum and increase in pro-oestrum, reaching a maximum at the time of ovulation. The maximum EI of between 56 and 100% occurs 10–14 days after the beginning of pro-oestrum, and on average pregnancy was obtained in 9 out of 10 bitches mated at this time.[22] Pregnancy can follow breeding when the EI averages 60% but should be repeated if there is a further increase in EI within the following 48 h.

Superficial cell index (SCI)
The proportions of different cell types in the smear vary during the cycle. There is an increase in the number of cells from the superficial layers, i.e. superficial cells with or without a nucleus and large intermediate cells, whereas cells from the deeper layers, i.e. the small intermediate cells and parabasal and basal cells, decrease in number.

$$SCI = \frac{\text{number of cells from the superficial layers} \times 100}{\text{number of cells from the deeper layers}}$$

An SCI of up to 70% without a specific maximum is found in pro-oestrum and characteristically there is a decrease 24–48 h after the maximum of the EI. This method cannot be used to find the optimum time for breeding since it requires a more detailed cell study and there is no specific maximum.[22]

Karyopyknotic index (KPI)
During oestrus the nuclei of the superficial cells undergo pyknotic changes, and a karyopyknotic index may be calculated:

$$KPI = \frac{\text{number of superficial cells with pyknotic nucleus} \times 100}{\text{number of superficial cells with vesicular nucleus}}$$

This method is not satisfactory since the maximum values are not constant.[22]

MATING

Sexual behaviour

As early as 5–6 weeks of age sexual behaviour, with mounting, pelvic clasping and thrusting movements, but without erection, intromission and ejaculation, may normally occur in both male and female puppies.

Male puppies may mount other male puppies and, less frequently, female puppies mount females. This behaviour is normal, not a sign of homosexuality, and usually ends at puberty.

Normal mating and mating behaviour

The optimum condition for successful mating is to let it take place in surroundings familiar to the male dog, i.e. in his home.[12] It is essential that the floor is not slippery to avoid difficulties during intromission and ejaculation. In some instances it may be necessary to hold and support the hindlimbs of the male during the first stage of coitus. If either the bitch or the male is shy and nervous, another calm and experienced partner should be used for the first mating. Some believe that the bitch and the dog should be kept together for a longer period of time – at least during pro-oestrum – to allow the full range of behaviour to develop during courtship and copulation.[7] Others take the opposite view, because prolonged association with bitches at various stages of oestrus may subdue the dog's libido.[17]

Mating is initiated by a sort of courtship, the male being stimulated by the smell of pheromones from the bitch's vagina.[15] Pheromones produced in the anal gland of oestrous bitches are also found to attract male dogs.[8,9] Visual and auditory stimuli are of minor importance and blind as well as deaf dogs are able to breed. The male dog first sniffs and licks the vulval and perineal regions. Normally the oestrous bitch permits this investigation and sometimes even turns her hindquarters towards him. A bitch not in oestrus will turn away or try to escape.

Erection of the penis takes place in two phases; initially the arteries are dilated and the cavernous tissue is filled with blood, leading to a partial erection. The male mounts the bitch, which holds the tail to one side. The penis is introduced into the vagina by a few thrusting movements during which the prepuce is pushed backwards behind the bulbus glandis. Some assistance may be needed during intromission, especially in bitches whose vulval region is covered with long hair. During the thrusting movements the male makes paddling movements with the hindlegs.

After intromission the erection is completed, partly because of contraction of the muscles near the base of the penis and partly

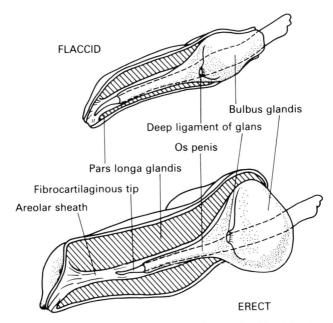

Figure 2.2   Schematic diagram showing the changing relationship of the glans and os penis during erection. (From Grandage[16])

through contraction of the sphincter muscles of the vulva which compress the dorsal veins of the penis. The male is thus fixed in position until ejaculation is completed, the so-called 'tie' or 'lock'.

Erection results in expansion of the glands (Figure 2.2) so that the bulbus glandis trebles its width to 6 cm or more and doubles its thickness to about 4 cm. The pars longa glandis elongates and increases in diameter.

The first semen fraction is ejaculated while the penis is partly erected, and the second fraction is ejaculated just after the thrusting movements cease. Shortly afterwards the male dog dismounts, turns round, lifting one hindleg over the bitch, and the animals stand with their hindquarters in contact and their heads facing opposite directions while the third fraction is ejaculated (Figure 2.3). This stage, the tie, may last from 5 to 30 min or even longer, and may be maintained by holding the tails of both dogs. During this stage the penis is bent through 180° at the middle of the body (Figures 2.4, 2.5(a) and (b)). The location of the bend at the level of the scrotal neck is dictated by the prepuce, the preputial muscles and superficial fascia.[16] Occasionally a bitch will try to escape just after the turn, so that the male is

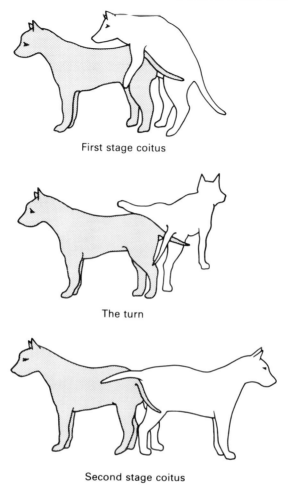

First stage coitus

The turn

Second stage coitus

Figure 2.3   Typical coital postures in the dog. (From Grandage[16])

dragged along or both animals fall over, but this does not interfere with ejaculation.

After ejaculation when the bulbus is relaxed, the animals separate and the male dog, and sometimes also the bitch, lick the partly erect penis without showing any sexual interest. After a period of rest the male's interest returns and he may mate up to 5 times in a day.[13]

The sperm fraction is ejaculated early in mating and the tie is not essential to conception. Some individuals and some breeds (especially chows) rarely tie, without any deterioration in their fertility.[17]

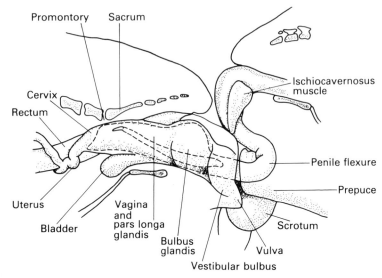

Figure 2.4 The relationships of the male and female genitalia during second-stage coitus. (From Grandage[16])

Aberrations

Aberrations from normal mating behaviour occur quite often. There may be differences in the response from the male as well as from the bitch. These are most evident on the first occasion a dog mates.[18] If dogs are deprived of their normal behaviour, adult mating behaviour may be adversely affected.[3] This may be seen in dogs raised in isolation who then refuse to mate, having inadequate experience of the mating behaviour in other dogs.

Even if the bitch is in full oestrus, the male may have to be stimulated to erect the penis by massage through the prepuce. Introduction of another bitch may stimulate his interest in the first one.[17] Males previously in contact with an aggressive bitch are often nervous and have to be calmed. The opposite may occur when an eager male has to be calmed to avoid frightening the bitch. The male may be sexually inhibited by being away from his home territory, and the presence of an unknown human or the absence of a known one can interfere with sexual behaviour. Some males will only mate in the absence of human beings,[17] but in other cases (such as shy dogs) their presence may be needed.

There are also considerable differences between breeds. For instance, the saluki is difficult to mate with other breeds, perhaps because of a particular sensitivity to the characteristics of its own

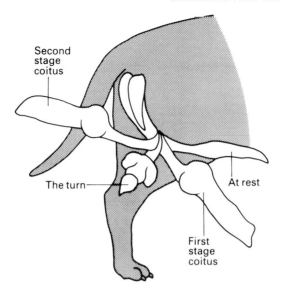

(a)

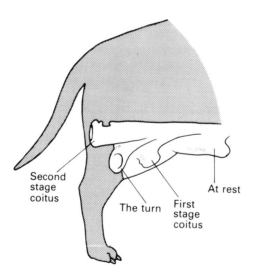

(b)

Figure 2.5 (*a*) Shows the form and curvature of the dog's penis during successive stages of coitus. Notice how penile flexion occurs about a vertical axis so that the dorsal surface remains dorsal. (*b*) Shows the positions of the prepuce during successive stages of coitus. To reach the position characteristic of second stage coitus the prepuce rotates about a transverse axis causing the ventral surface of the prepuce to lie dorsally. (From Grandage[16])

breed during the familiarization period, and thus a failure to recognize other breeds as members of its own species.[7] Bitches may show preferences for individual dogs, even when not in oestrus. The behavioural preference shown by an anoestrous bitch may be obliterated during oestrus and a new pattern of heterosexual affinities and aversions emerge. Bitches may copulate readily with certain males and consistently reject the mating attempts of others.[4,6] A dominant bitch will not permit a less dominant male to mate with her. On the other hand a bitch may lie down in extreme submission when a male dog approaches, even if she is in oestrus.

In some cases there may be premature termination of a lock or intromission with complete erection but without locking. The male dog may then try to copulate again even though this cannot succeed until the first erection ceases. Alternatively the dog may sit licking his penis until it becomes flaccid before another attempt at intromission is made. Some males may stand and continue to make ejaculatory contractions while the bitch shows extreme interest, jumping over him, circling round him and repeatedly licking his penis until it is detumescent. In some cases a lock is broken within a few minutes after intromission. This is described as coitus interruptus.[5] The male dog stands motionless with the back arched, the hindlegs spread and the tail hanging between them, just as during the normal lock. The erection is maintained through the pressure exerted by the prepuce being retracted behind the swollen bulbus glandis. During this simulated lock there are rhythmic ejaculatory contractions without any external stimuli. The duration of such locks varies from 4 to 8 min depending upon the presence or absence of the bitch.

If a dog is not stimulated for a long period by a bitch in oestrus he may mount other males, other animals, or even people or inanimate objects. This abnormal behaviour may be so embarrassing that it has to be eliminated by castration or administration of long-acting progestogens.

Masturbation occurs more frequently in the male dog than in the female. The male may mount other animals, people or such objects as toys and make pelvic thrusts, thus stimulating the prepuce and the penis so that erection and expulsion of prostatic fluid occur. This again my be cured by either castration or administration of progestogens.

The bitch may masturbate by rubbing the vulval region against the floor or other objects. No adequate treatment is available.

REFERENCES

1. Andersen, K. (1977) *Nordic symposium on artificial insemination in dogs and deepfreezing of dog semen.* Uppsala.

2. Andersen, K. (1980) Personal communication.
3. Beach, F.A. (1968) Coital behaviour in dogs. III. Effects of early isolation on mating in males. *Behaviour*, **30**, 218–238.
4. Beach, F.A. (1970) Coital behaviour in dogs. VIII. Social affinity, dominance and sexual preference in the bitch. *Behaviour*, **36**, 131–148.
5. Beach, F.A. (1970) Coital behaviour in dogs. IX. Sequelae to 'coitus interruptus' in males and females. *Physiol. Behav.*, **5**, 263–268.
6. Beach, F.A. and LeBoeuf, B.J. (1967) Coital behaviour in dogs. I. Preferential mating in the bitch. *Anim. Behav.*, **15**, 546–558.
7. Coffey, D.J. (1971) Ethology and canine practice. *J. small Anim. Pract.*, **12**, 123–131.
8. Donovan, C.A. (1967) Some clinical observations on sexual attraction and deterrence in dogs and cattle. *Vet. Med./small Anim. Clin.*, **62**, 1047–1051.
9. Donovan, C.A. (1969) Canine anal glands and chemical signals (pheromones). *J. Amer. vet. med. Assoc.*, **155**, 1995–1996.
10. Dreier, H.K. (1978) Bestimmung des Deckzeitpunktes bei der Hündin. *Kleintier.-Prax.*, **23**, 263–268.
11. Evans, J.M. and Savage, T.J. (1970) The collection of vaginal smears from bitches. *Vet. Rec.*, **87**, 598–599.
12. Fox, M.W. (1972) *Canine Behaviour*, 2nd ed. Charles C. Thomas, Springfield, Illinois.
13. Fuller, J.L. and Fox, M.W. (1969) The behaviour of dogs. In: *The Behaviour of Domestic Animals*, 2nd ed., ed. Hafez, E.S.E., Chapter 14. Baillière, Tindall and Cassel, London.
14. Gill, H.P., Kaufman, C.F., Foote, R.H. and Kirk, R.W. (1970) Artificial insemination of beagle bitches with freshly collected, liquid-stored, and frozen-stored semen. *Amer. J. vet. Res.*, **31**, 1807–1813.
15. Goodwin, M., Gooding, K.M. and Regnier, F. (1979) Sex pheromone in the dog. *Science*, **203**, 559–561.
16. Grandage, J. (1972) The erect dog penis: a paradox of flexible rigidity. *Vet. Rec.*, **91**, 141–147.
17. Harrop, A.E. (1960) *Reproduction in the Dog*. Baillière, Tindall and Cox, London.
18. Hart, B.L. (1967) Sexual reflexes and mating behavior in the male dog. *J. comp. physiol. Psychol.*, **64**, 388–399.
19. van der Holst, W. and Best, A.P. (1976) Een beschouwing over het meest geschikte tijdstip voor de dekking van de teef. *T. Diergeneesk.*, **101**, 658–663.
20. Klötzer, I. (1974) Untersuchungen über den elektrischen Widerstand des Vaginalschleimes der Hündin. *Kleintier.-Prax.*, **19**, 125–133.
21. Laiblin, Ch. and Rohloff, D. (1981) Die künstliche Besamung des Hundes unter besonderer Berücksichtigung der Läufigkeitsdiagnose. *Tierärztl. Prax.*, **9**, 237–244.
22. Schutte, A.P. (1967) Canine vaginal cytology. III. Compilation and evaluation of cellular indices. *J. small Anim. Pract.*, **8**, 313–317.
23. Shille, V.M. (1977) *Nordic symposium on artificial insemination in dogs and deep-freezing of dog semen*. Uppsala.

ADDITIONAL READING

Beach, F.A. and Kuehn, R.E. (1970) Coital behavior in dogs. X. Effects of androgenic stimulation during development on feminine mating responses in females and males. *Hormones and Behavior*, **1**, 347–367.

Beach, F.A., Kuehn, R.E. Sprague, R.H. and Anisko, J.J. (1972) Coital behavior in dogs. XI. Effects of androgenic stimulation during development on masculine mating responses in females. *Hormones and Behavior*, **3**, 143–168.

Beach, F.A., Johnson, A.I., Anisko, J.J. and Dunbar, I.F. (1977) Hormonal control of sexual attraction in pseudohermaphroditic female dogs. *J. comp. physiol. Psychol.*, **91**, 711–715.

Colleen, S., Holmquist, B. and Olin, T. (1981). An angiographic study of erection in the dog. *Urol. Res.*, **9**, 297–302.

Hart, B.L. (1974) Gonadal androgen and sociosexual behaviour of male mammals: a comparative analysis. *Psychol. Bull.*, **81**, 383–400.

Hart, B.L. and Ladewig, J. (1979) Effects of medial preoptic-anterior hypothalamic lesions on development of sociosexual behavior in dogs. *J. comp. physiol. Psychol.*, **93**, 566–573.

King, J.A. (1954) Close social groups among domestic dogs. *Proc. Amer. Philosoph. Soc.*, **98**, 327–336.

Ninomiya, H. (1980) The penile cavernous system and its morphological changes in the erected state in the dog. *Jap. J. vet. Sci.*, **42**, 187–195.

Taradach, C. (1980) Frottis vaginaux chez la chienne beagle. Détermination de la période d'ovulation à l'aide de l'indice éosinophile. *Rev. Méd. vét.*, **131**, 775–782.

Chapter 3
Infertility and Hormone Treatment in the Female

INFERTILITY

Infertility may be caused by anatomical abnormalities, management problems, functional abnormalities, infection or tumours.

First the owner should provide all relevant information by using the gynaecological questionnaire shown in Table 1.8 (page 18). By questioning the owner it should be possible to exclude those cases where the bitch is fertile but the owner is ignorant about the oestrous cycle and the optimum time for breeding. A careful clinical and gynaecological examination should be carried out with all the relevant laboratory tests (see Table 1.7 page 16). It should be kept in mind that the problem may be related to the male, and also that, if it is related to the bitch, therapy should be avoided if the condition is hereditary.

Anatomical abnormalities

Anatomical abnormalities, congenital or acquired, are of only minor importance.

Intersexuality
True hermaphroditism is the presence of both ovaries and testes and abnormal external genitalia. Pseudohermaphroditism occurs either in the female form with ovaries and Müllerian derivatives, but with abnormal development of the Wolffian ducts and urogenital sinus, or in the male form with normal testes but abnormal external genitalia. Intersexuality occurs sporadically and has been described in various breeds.[3,22,34,41,47,54,56,57,59,94,103,111,113]

Administration of sex hormones to a pregnant bitch may cause abnormal development of the genitalia in the fetuses. Exposure of female fetuses between the fourth and seventh weeks of pregnancy to exogenous androgens may result in female hermaphroditism.

Ovarian agenesis, hypoplasia and aplasia of parts of the Müllerian ducts and tubulus
When these occur they can be the cause of infertility. Diagnosis can be made by laparotomy and dye injection or by X-ray examination.

Pregnancy can be achieved if surgical correction of the occlusion is possible and if ovaries are present.

Oestrogens administered to a pregnant bitch during the period of sexual differentiation in the fetuses may result in incomplete or absent connection between the Müllerian ducts and the urogenital sinus, or be the cause of incomplete formation of the vagina, which may also be caused by progestogen treatment of a pregnant bitch.

Hymenal and/or vestibular strictures
These can be the cause of reproductive failure in spite of a normal oestrous cycle. Vestibulo-vaginal stenosis is most often accompanied by clinical signs of vulval pruritus.[61] In some bitches surgical correction is possible, and subsequently mating or artificial insemination can result in pregnancy.

Hypoplasia of the vulva
This can occasionally cause difficulties in intromission and indicates the necessity for artificial insemination. Perivulval dermatitis may accompany an underdeveloped vulva.

Displacement of the vulva in a ventrocranial direction may develop in fetal life on administering progestogen to the pregnant bitch.

Management problems

Mating problems during adolescence (see Chapter 2) may be an indication for the use of artificial insemination.

Bitches kept in isolation from other dogs may show only weak oestrous signs when presented to a male dog for breeding purposes. Some bitches show an objection to a particular male, and in some instances mating can be performed only if the bitch is restrained either physically or medically. Artificial insemination or the use of another male may be necessary.

Bitches first bred at an advanced age can exhibit special problems, as, for example, in the greyhound bitch which is used as a breeding dog after a period of racing.

Functional abnormalities

Delayed puberty and prolonged anoestrum
Some bitches do not have their first oestrus at the anticipated age, presumably because of a hypothalamic–pituitary–gonadal dysfunction caused by poor nutrition, especially a low-protein diet, exposure to cold, or lack of sufficient light. Kennelled dogs often come into puberty at a later age. Bitches of small breeds have a tendency to come into puberty when younger, and bitches in oestrus can stimulate

others to come into oestrus. Heredity may be a reason for prolonged anoestrum in some families as well as in inbred or linebred dogs, because of inadequate pituitary or ovarian function.

Mature dogs having had one or more oestrous cycles can exhibit a period of prolonged anoestrum, either after normal cycles or after pregnancy and lactation. There are variations in the duration of the oestrous cycle, and periods may be so extended that complete cessation of cyclical activity is suspected. As the cause of the prolonged anoestrum may be nutritional, cases of obesity or cachexia should first be corrected.

If nutritional and systemic diseases can be excluded, treatment should not be instituted without bearing in mind the possibility that the condition is hereditary. Therefore, a detailed history should be provided, in conjunction with a careful examination, to exclude both this possibility and anatomical failure. There are no specific symptoms: the vulval region, normally dry and clean in the anoestrous period, might show some dried secretion around the vulva itself, longitudinal folds may be seen in the vagina, and the mucosa has a rosy colour. The cytological picture is dominated by cells from the deeper layers – basal, parabasal, and intermediate – in a mixture of mucus which is basophilic. In some dogs superficial cells may be present as a result of oestrogenic activity caused by weak and probably temporary activity in the ovaries.

Therapy
The best results are obtained in bitches which have been in oestrus before. There seems to be less of a problem in inducing behavioural response than ovulation in mature bitches. Even if ovulation is induced, the number that conceive is lower than normal.

Early experiments were made with injections of urine extracts from both pregnant[71] and menopausal bitches;[70] the result was, respectively, manifestation of some and of all external oestrous symptoms. When the treatment with menopausal urine was combined with injection of LH on the first day of oestrus, it resulted in pregnancy.[70] In another early attempt to mimic the normal hormonal concentrations, a combination of pregnant mare's serum gonadotropin (PMSG) and human chorionic gonadotropin (HCG) was used, resulting in oestrus and pregnancy in some of the mated bitches.[95]

Attempts to induce oestrus have been described using oestrogens, FSH, PMSG, LH, HCG and gonadotropin releasing hormone (GnRH) singly or in different combinations.

Table 3.1. gives the dosages and methods of administration of various hormonal and non-hormonal compounds for bitches with infertility problems caused by delayed puberty, prolonged anoestrum and weak ovarian activity with insufficient pro-oestrum and oestrous

Table 3.1 Treatment with hormones or non-hormonal compounds in bitches with problems of infertility due to delayed puberty, prolonged anoestrum or weak ovarian activity with insufficient pro-oestrous and oestrous symptoms

Condition	Drug	Administration	Results and comments	References
Delayed puberty or prolonged anoestrum	PMSG	250 IU PMSG s.c. daily for 4 days	No encouraging results	1
	PMSG, (HCG)	110 IU/kg i.m. at weekly intervals, maximum 3 times, possibly followed by 500 IU HCG i.v. on the first day of oestrus	Oestrous symptoms and ovulations in all bitches regardless of HCG administration. 4–71 follicles, haemorrhagic discharge not present in all bitches	117
	PMSG, HCG	(a) 500 IU PMSG s.c. for 8–9 days until the onset of oestrus and 500 IU HCG s.c. on day 10	Bodyweight of bitches not stated	21
		(b) 500 IU PMSG s.c. daily for 9 days and 500 IU HCG s.c. on day 10	Of treatments (b), (c) and (d), the best results were obtained after (c) with 9.8 ± 1.5 follicles ovulating (average). (b) and (c): body-weight 7–16 kg. Pro-oestrum shorter than normal in spontaneous pro-oestrum	107
		(c) 250 IU PMSG s.c. daily for 9 days and 500 IU HCG s.c. on day 10		
		(d) 20 IU PMSG/kg s.c. daily for 9 days and 500 IU HCG s.c. on day 10		
		(e) 20–50 IU PMSG/kg i.m. daily for up to 9 days and 500–1000 IU HCG i.m. on first and second days of oestrus	Encouraging results. (f): bodyweight of bitches not stated	31
		(f) 500–1000 IU PMSG i.m. repeated 6 days later and 500–1000 IU HCG i.m. on first and second days of oestrus		
		(g) 44 IU PMSG/kg i.m. for 9 days followed by 500 IU HCG i.m. on day 10	Encouraging results. 60% of the treated bitches ovulated	13
		(h) 44 IU PMSG/kg s.c. for 9 days and 500 IU HCG i.m. on second day of oestrus		
		(i) 250 IU PMSG s.c. daily until onset of oestrus, maximum 20 days, followed by 500 IU HCG s.c. first day of oestrus or on day 21	More or less pronounced oestrous symptoms – often without vulval bleeding –	118

	FSH, LH	(a) FSH s.c. on day 1: 10 mg, day 2: 7.5 mg, days 3 and 4: 5 mg, day 5: 7.5 mg; on day 7: 15 mg LH i.v. (b) 5 mg FSH s.c. for 10 days and then two days later 15 mg LH i.v.	accompanied by development of up to 200 follicles and an ovulation rate of 0–100, bodyweight of bitches 8–13.5 kg Oestrous symptoms, vaginal cornification, follicular growth but no ovulations; bodyweight of bitches 10–15 kg	88
	PMSG, GnRH	(a) 20–50 IU PMSG/kg i.m. daily for up to 9 days and 50 μg GnRH i.m. twice separated 6 h on the first day of oestrus (b) 500–1000 IU PMSG i.m. repeated 6 days later and 50 μg GnRH i.m. twice with an interval of 6 h on the first day of oestrus	Encouraging results. (b): bodyweight not stated	31
	FSH, ECP, LH	20 mg FSH i.m. weekly up to three times and 7 days later 0.5 mg ECP i.m. and 2 days thereafter 5–10 mg LH i.v.	No efficacy data available, bodyweight of bitches not stated	16
	Oestradiol, PMSG	Oestradiol 0.1–0.5 mg s.c./i.m. two to four times at 2–3 days' intervals followed by 25–50 IU PMSG s.c./i.m. every second day 4–8 days after the first signs of pro-oestrum and until oestrus is achieved	Encouraging results. 84% of the treated bitches became pregnant, bodyweight of bitches not stated	9
	Oestrone, PMSG, HCG	300–30 000 μg oestrone in total given during a shorter or longer period and 1000 MU HCG plus 200–400 IU PMSG first day after the pro-oestrous bleeding has stopped	Encouraging results. 86% of the treated bitches became pregnant, bodyweight of bitches not stated	105
	Sexovid	4 mg/kg orally daily for 3–22 days until signs of oestrus	Encouraging results. 71% of the treated bitches became pregnant	64
Weak ovarian activity with insufficient pro-oestrous and oestrous symptoms	PMSG PMSG, oestradiol	50–100 IU PMSG s.c./i.m. with intervals of 3–4 days 50–100 IU PMSG s.c./i.m. with intervals of 3–4 days and then 0.1–0.5 mg oestradiol s.c./i.m., if FSH is without success	Encouraging results. 100% of the treated bitches became pregnant, but the litter size was not optimal in all pregnancies, bodyweight of bitches not stated	9
	PMSG, LH	50–100 IU PMSG s.c./i.m. with intervals of 3–4 days and then 100 IU LH s.c./i.m. 4 days later		

symptoms. The various suggestions comprise administration of FSH/PMSG for a shorter or longer period or until signs of pro-oestrum are obvious, then followed in some cases by administration of LH, HCG or GnRH. In most instances the results have been unsatisfactory as the bitches have shown oestrus but few and sporadic ovulations. Furthermore, injection of HCG may induce formation of antibodies without appreciable changes in the reproductive function.[4]

Other investigators imply that the treatment should be related to the activity of the ovaries.[9,45,46] Thus oestradiol benzoate should be injected in bitches with no ovarian activity, and if the treatment succeeds, according to the characteristic changes in the vaginal smear, PMSG should be administered until there are signs of oestrus.

Administration of GnRH to anoestrous bitches results in a linear response in serum LH.[30] Injections of 5 and 25 μg of GnRH resulted in an increase in the serum levels of LH with a peak of 8.5 ± 3.2 and 18.7 ± 3.2 ng/ml, respectively, within 15 min and a return to pretreatment levels by 2 h after treatment. Injection of 50 μg resulted in a mean peak of 39.7 ± 7.8 ng/ml at 30 min after treatment, and in pretreatment levels again 4 h after treatment. There is no information on ovulation in connection with GnRH treatment. Injection of 2.5 μg/kg releasing hormone causes a significant increase in LH concentration in anoestrous bitches, whereas this is not observed in bitches with a high concentration of progesterone at the time of stimulation.[35]

It is still difficult to judge the value of the various treatments, since sufficient data are sometimes not available. Often information on the symptoms only is reported, whereas other publications include data on the follicular growth or ovulations; only in some very few reports are the number of pregnancies obtained described.

It can be concluded that the best results from treating anoestrous bitches with gonadotropic hormones can be expected after repeated administration of FSH or PMSG followed by LH, HCG or GnRH on the first day of oestrus. Perhaps a better result may be achieved by HCG administration daily for the first 2 or 3 days of oestrus.[118] This suggestion is based on the fact that in many investigations follicle growth was achieved, but only a few follicles followed a single injection of HCG.

Sexovid (*bis*(*p*-acetoxyphenyl)cyclohexylidenemethane) – a synthetic preparation of non-steroidal nature – has been effective in inducing signs of oestrus within 3–22 days of the start of treatment in anoestrous bitches.

Weak ovarian activity
This condition, which is accompanied by insufficient symptoms of pro-oestrum and oestrus, has been treated successfully with FSH, if necessary followed by injection of LH or oestrogens in bitches not responding to the FSH administration.[9]

Prolonged pro-oestrum and oestrus

In bitches with aberrations of the length of pro-oestrum or oestrus, the clinical examination should include a cytological examination of a vaginal smear and perhaps also a hormonal analysis to provide information on the progress of the ovarian events.

In some bitches 'split oestrus' occurs, without interference of fertility, resulting in interruption of the acceptance of the male dog for 1–2 days. A very few other bitches may exhibit polyoestrous activity for a long period, during which they are infertile.

Prolonged pro-oestrum is a condition occurring particularly at the onset of puberty in which the bitch continues to attract and accept males for up to 28 days showing a watery and bloody vaginal discharge. The vulva is not always enlarged, but the cervix is open.

Prolonged oestrus with slight vulva oedema, mucoid vaginal discharge, often accompanied by pruritus and alopecia, can occur in mature bitches for up to 6 months.

Abnormally prolonged oestrous activity for up to 9 months has occurred in cases of ovarian granulosa cell tumours producing oestrogens and progesterone.[76]

Therapy

Table 3.2 reviews various hormones (LH, progestogens, norethisterone) and non-hormonal compounds used for the treatment of these conditions, including the results obtained.

A non-hormonal compound, a cytostatic, 1,4-*bis*-(methan-sulphonyl-oxy)-butan, which causes arrest of spermiogenesis and changes in the ovaries, and thus interruption of reproduction for up to 6 months when administered in high doses to pigeons, has also been investigated in bitches.[8]

Acupuncture has been modified and brought into use in small animals,[14,67,77,91] and successful treatment has been claimed in a small number of bitches with prolonged oestrus after a few applications of hormones in acupuncture points instead of ordinary intramuscular injections.[25]

Prolonged or permanent heat and ovarian cysts

These conditions may occur sporadically, presumably because of the presence of multiple oestrogen-secreting follicular cysts, and are accompanied by sterility. Follicular, luteal and rete cysts can occur in bitches and may be responsible for prolongation of oestrus. Follicular cysts, 1–5 cm in diameter, have been found in 10% of dogs examined post mortem, and luteal cysts have been found in over 20% of bitches.[44] They can be seen in dogs of all ages from 2 to 15 years, but seem to be common in old beagle bitches.[5] Almost filled with fluid, they may double the size of the organ. The treatment is ovariectomy if the cysts are located in one ovary only, and ovariohysterectomy if

Table 3.2 Treatment with hormones or non-hormonal compounds in bitches with problems of infertility due to prolonged pro-oestrum and oestrus, prolonged or permanent heat, ovarian cysts, implantation failure or spontaneous abortion.

Condition	Drug	Administration	Results and comments	References
Prolonged pro-oestrum and oestrus	**Luteinizing hormones** human chorionic gonadotropin (HCG)	(a) 10–20 IU HCG/kg i.v. possibly repeated 2–3 days later		74
		(b) 1000 IU HCG as total dose	Luteinizing hormones are effective	51
	luteinizing hormones (LH)	100–500 IU LH s.c./i.m. daily until arrest of the haemorrhagic discharge		10, 45
	Progestogens progesterone (P)	25–100 mg i.m. single dose	Symptoms disappeared, no matter which progestogen was used, but pyometra developed after treatment in up to 10% of the treated bitches	108
	hydroxyprogesterone acetate (HP)	4.4–13.2 mg/kg orally daily		20
	17α-acetoxyprogesterone	250 mg i.m. single dose		108
	medroxyprogesterone acetate (MAP)	(a) 25–50 mg solution s.c./i.m.		66
		(b) 25–75 mg in suspension		66, 108
		(c) 5 mg/kg orally daily		92
	megestrol acetate (MA)	10–20 mg orally daily for 1–4 days and then 5–10 mg orally daily until entrance to the stage of metoestrum		7
	chlormadinone acetate (CAP)	(a) 8–30 mg i.m. single dose		108, 112
		(b) 2–10 mg orally for up to 20 days		84
	proligestone	33 mg/kg s.c. single dose in small bitches and 10 mg/kg s.c. single dose in large bitches, if no effect treatment should be repeated 2 weeks later with the same dose	Encouraging results	12, 81
	Norethisterone (NET)	0.2–1.0 mg NET/kg orally daily	Symptoms disappeared	93

	Compound	Dosage	Results	Ref.
Non-hormonal compounds				
	1,4-*bis*-(methan-sulphonyl-oxy)-butan	3 mg/kg orally, maximum 50 mg, possibly repeated with half the dose 4–6 days later	Preliminary investigations have given encouraging results	8
	Hormone acupuncture	Administration of hormonal compounds in different acupuncture points	Same results as after normal administration of hormones	25
Prolonged or permanent heat and ovarian cysts	Megestrol acetate (MA)	10–20 mg orally daily for 1–4 days, and thereafter 5–10 mg orally daily in large bitches and 2.5 mg orally daily in small bitches until the stage of metoestrum	Treatment effective and normal cyclic activity obtained	9
	Chlormadinore acetate (CAP)	2–10 mg orally daily until effectual	Encouraging results, pyometra developed in bitches with prolonged oestrus following oestrogen administration for prevention of implantation	84
	Delmadinone acetate (DMA)	0.15 mg DMA/kg	Treatment effective and normal cyclic activity obtained	49
Implantation failure and spontaneous abortion	Progestogens	Cases of implantation failure should be treated from 3–4 days after mating and cases of spontaneous abortion from the beginning of the second week until 1 week before calculated parturition	Treatment effective, but risk of fetal abnormalities and prolonged gestation period if treatment is not stopped in time	6, 45
	progesterone	25–50 mg s.c. one to three times a week		6, 36, 46
	medroxyprogesterone acetate (MAP)	2.5–10 mg orally daily		

the condition is bilateral. Surgical removal of a 1 cm-wide wedge of the greater curvature of polycystic ovaries is reported to have been followed by heat, pregnancy and delivery of living puppies.[110] Injections of luteinizing hormone have not been reported to be effective.

Prolonged or permanent heat may also occur after application of diethylstilboestrol (DES) or oestrogens for prevention of implantation. This type of prolonged heat does not require any kind of treatment, but breeding should be avoided at this time due to an increased risk of developing pyometra.

Dosages for treatment, routes of administration and results are given in Table 3.2.

Implantation failure and spontaneous abortion

Failure of implantation
There are no symptoms other than failure of pregnancy despite normal mating. It must be borne in mind that the reason for the lack of pregnancy may be mating at a non-optimal time. Some bitches may have haemorrhagical vaginal discharge in oestrus proper, others may show no pro-oestrous bleeding, and the optimal time of mating may vary individually by days or even weeks and within the same bitch as judged by the signs of oestrus.

Spontaneous abortion
This may occur at any time during the pregnancy with symptoms of abdominal contractions, expulsion of fetuses (live or dead), and vaginal discharge. In some bitches habitual abortion may occur with the only symptoms being a transient discharge from the vagina, or that whelping fails in spite of a previous positive diagnosis of pregnancy, since the fetuses are often eaten by the bitch.

The cause of abortion may be of fetal, maternal, managerial or infectious origin. The problem has not yet been examined in detail in the bitch, but from knowledge of other species it may be related to fetal abnormalities, hypothyroidism or hypoluteinism in the bitch, calorific or vitamin deficiencies, or specific or non-specific infections.

Specific infections   Brucella canis can cause early embryonic death, abortion at 41st–55th days of pregnancy followed by vaginal discharge containing the organisms for up to 6 weeks, stillbirth, and occasionally delivery of weak puppies.[19,27,28,78,79,106] The infection spreads rapidly among dogs gathered together at shows, at shoots and in kennels. The organisms can be found in the vaginal discharge, milk, semen and urine, and the disease can be spread by contact with aborted fetuses or vaginal discharge or by venereal transmission. Even if tetracyclines or streptomycin can be used for treatment, this will not ensure that a dog does not become a carrier of the infection.

The infection does not appear to be a great hazard to humans although it has occurred in those in close contact with infected dogs. A titre of 1:200 or more found by agglutination test confirms the diagnosis.

Toxoplasma gondii may also cause abortion and stillbirth in bitches.[38,97]

Non-specific infections   In vaginal swabs from normally fertile as well as infertile bitches, a wide range of micro-organisms such as streptococci, staphylococci, *Proteus*, *Pasteurella* and *Escherichia coli* can be isolated.[60,72,85–87,89,102,115] The number of bacteria is found to be greater in bitches with vaginal discharge.[60,86] Organisms such as beta-haemolytic streptococci, *Proteus* and *Pseudomonas aeruginosa* may be of some importance in bitches if they are present in great numbers, and haemolytic streptococci can cause abortion and sterility.[75] Infections with herpesvirus and adenovirus can result in delivery of dead or weak puppies.[39,90,100]

Diagnosis
This is based on physical examination of the bitch together with hormone assay of the plasma thyroid hormone and progesterone content, serological determination of the brucella and toxoplasma titres, histopathological examination and culture of the fetal stomach contents, and, in the future, perhaps also chromosome analysis. In cases of non-specific infections diagnosis can be made by bacteriological culture and swabs from the anterior vagina.

Therapy
This includes administration of antibiotics and, if indicated, oxytocin for emptying the uterus of retained fetuses or fetal membranes, together with individual supportive therapy. If the cause is a non-specific infection, antibiotic therapy can be established after a sensitivity test and should be accompanied by cleaning and disinfection – and in kennels also by suspension of matings for at least 6 months. Future breeding depends on the cause of the abortion. Bitches infected with *B. canis* must not be used for breeding again, but bitches aborted through *T. gondii* infection probably can be used. If the cause of the abortion is found to be a non-specific infection, anterior vaginal swabs should be examined before resumption of breeding activity, and if these are still infected, it must be decided whether the bitch should be excluded from mating or treated systemically with antibiotics during oestrus, insemination and pregnancy. The condition may also be due to low plasma P, and treatment with progestogens can be of value in protecting a future pregnancy (see Table 3.2).

Tumours

Tumours occur in most parts of the reproductive system of the bitch.[18,23,24,96] Ovarian tumours are rather infrequent, the most common being the granulosa–thecal cell tumour and cystadenocarcinoma. Ovarian tumours are often accompanied by combinations of cystic endometrial hyperplasia, vaginal discharge, bilateral symmetrical alopecia or ascites. Most are benign, and early diagnosis and removal may be curative. Diagnosis is based on abdominal palpation, cytological examination of ascitic fluid and radiographic examination. Postsurgical histological examination is necessary to give a prognosis.

Leiomyomas and fibromas are most common in the vagina and vulva, and may be associated with ovarian follicular cysts, oestrogen-secreting tumours and cystic endometrial hyperplasia. Uterine tumours, particularly leiomyomas, are often accompanied by vaginal discharge, ascites, vomiting, anorexia and weight loss.

Transmissible venereal tumours commonly involve the vulva, vagina, prepuce and penis, and are transmitted at coitus by cell transplantation. Symptoms include persistent dripping of blood or serosanguinous fluid from the vagina and sometimes signs of dysuria. Spontaneous regression often occurs without treatment.

HORMONE THERAPY

Treatment with sex hormones has been used in various conditions related to reproduction and the reproductive organs to abolish or suppress symptoms and restore normal conditions.

Pseudopregnancy (false pregnancy, pseudocyesis)

This condition is a normal and non-pathological event occurring in the bitch during metoestrum, intensifying and sometimes prolonging it.[114]

The symptoms begin a few weeks after oestrus and may continue for several weeks beyond the normal pregnancy period; they include 'nest-making', self-nursing, nursing of inanimate objects, vomiting, diarrhoea, polyphagia, anorexia, mammary enlargement, lactation and even tenesmus as though in labour. Pseudopregnancy seems to encourage tumour formation in the mammae, while pregnancy inhibits it.[109]

The cause of this condition has been thought to be a high plasma P, although the intensity and the symptoms are independent of the P concentration, which is comparable with that found in normal bitches at the same time after oestrus.[55] In normal pregnancies the concentration of P decreases during the later part of pregnancy and at parturition, while the concentration of PRL increases. Ovariohysterectomy in bitches in the luteal phase results in a similar decrease in P concentration, and in maintenance of and in some cases even an increase in PRL concentration, which explains why pseudopregnancy may develop in bitches ovariohysterectomized in metoestrum.

Diagnosis
The diagnosis is based upon breeding history and clinical examination.

Differential diagnoses to be considered are normal pregnancy and closed pyometra, which can be excluded by abdominal palpation and radiographical examination.

Therapy
Therapy should include care of the mammary glands and perhaps also administration of a light sedative until anoestrum supervenes. In most cases this treatment is sufficient but steroids may be used to suppress the symptoms (Table 3.3). The mode of action of the steroids is by negative feedback upon the hypothalamic–pituitary glands, and therefore any of the sex hormones may be used, although male steroids are preferable. Because of the possible role of PRL in this syndrome, administration of anti-PRL in combination with symptomatic treatment may be the chosen therapy in future.

Spaying is the only preventive treatment, but the syndrome may recur, if it is performed after hormonally induced regression of the symptoms and it is therefore inadvisable to spay a bitch during a period in which false pregnancy has been treated with steroid preparations.[2]

The cystic hyperplasia–pyometra complex

This condition – characterized by cystic glandular hyperplasia of the endometrium, possibly with accumulation of pus in the uterine cavity – occurs most commonly during the progestational stage of the oestrous cycle in middle-aged and aged bitches either as closed or open pyometra, the latter with vaginal discharge.[26,32,43,48,58,73,92]

The cause of this condition is still obscure, but it has been suggested to be due to excessive hormonal stimulation of the uterus either by progestational or oestrogenic compounds. Although the hyperplasia–pyometra complex can be induced by administration of exogenous P, possibly in combination with oestrogens, the plasma P and oestrogen are normal in cases of pyometra, [15,33,37,40,55] and the primary abnormality may be defective metabolism of these hormones by the target tissue.

The condition occurs in every breed and, although it may occur in year-old bitches, it is found most often in animals over 6 years of age

Table 3.3 Hormones and non-hormonal compounds for treatment of false pregnancy and pyometra in the bitch.

Condition	Drug	Administration	Results and comments	References
False pregnancy	Diethylstilboestrol (DES)	(a) Decreasing doses orally during a 5-day period	Diethylstilboestrol, oestrogens, oestrogen + testosterone, luteinizing hormone and progestogens are quite effective in controlling the symptoms, including milk flow. Oestrogens and pro-gestogens may induce side-effects in uterus, such as hypertrophia and pyometra	74
		(b) 1.2–2.5 mg/kg i.m. single dose		99
	Oestrogens			
	oestrone	1.0–5.0 mg i.m. single dose		99
	oestradiol benzoate	1.0–4.0 mg i.m., repeated 7 days later		6
	oestradiol benzoate + testosterone enanthate	2 mg oestradiol benzoate + 20–45 mg testosterone enanthate		6
	methyloestrenolene (MOE)	(a) 1–3 mg/kg orally daily for 3–5 days		52
		(b) 4–7 mg/kg i.m. daily for 1–2 days		52
	Luteinizing hormone (LH)	5 mg LH i.v.		17
	Progestogens			
	progesterone (P)	10–50 mg i.m. daily until symptoms disappear		99
	hydroxyprogesterone acetate (HP)	3.4–5.5 or 4.4–13.2 mg/kg daily for 4–5 days		20
	delmadinone acetate (DMA)	(a) 0.5–1.1 mg/kg orally daily for 6 days	Return to normal behaviour and termination of galactorrhoea, nesting and nursing behaviour	53
		(b) 2.5–5.0 mg/kg s.c. twice 24 h apart		53
		(c) 2.2 mg/kg s.c. twice 4 days apart		50

	Drug	Dose	Results	Ref
	medroxyprogesterone acetate (MAP)	25–75 mg suspension s.c.		66, 99, 116
	megestrol acetate (MA)	2.5 mg/kg orally for 8 days		99
	chlormadinone acetate (CAP)	2–10 mg orally daily		84
	proligestone	3.3 mg/kg s.c. in small bitches and 10 mg/kg s.c. in large bitches, same dose repeated 2 weeks later if necessary	Encouraging results	81
	Antiprolactin (2-Br-α-ergocryptin)	(a) 20 ng/kg for 4 days (b) 10–20 mg i.m. three to five times 2–3 days apart (c) 10–20 ng/kg orally every second day three to four times	Accompanied by vomiting, so supplementary administration of an antiemetic necessary; encouraging results	11, 69
	Androgens			
	testosterone	(a) 0.5–1 mg/kg i.m. every second or third day (b) 1 mg/kg orally daily	No constant results	74, 80 74
	mibolerone	8–16 µg/kg for 5 days	52% were psychologically normal within 24–36 h, and mammary glands essentially normal 7 days after first treatment	99
Pyometra	Diethylstilboestrol	1–4 mg orally daily for 7 days		49
	Oestradiol	1 mg/kg i.u. every second day for 10–14 days	Only a few bitches treated	62
	Testosterone propionate	25 mg i.m. twice weekly until disappearance of symptoms		101
	Prostaglandin ($PGF_{2\alpha}$-THAM)	250 µg/kg	Encouraging results, but severe side-effects	98

and more commonly in unmated bitches. The symptoms are depression, anorexia, vaginal discharge, polydipsia, polyuria, vomiting, diarrhoea and, less often, bone pain in one or more legs. The vaginal discharge, which appears only if the cervix is open, occurs immediately after oestrus or 1–2 weeks later; it is yellow-green or red-brown in colour with a fetid smell and may (although uncommonly) contain blood from the vagina.

There may be a moderate to marked hyperproteinaemia, and frequently there is hypoalbuminaemia and an elevated blood urea nitrogen. In cases of open pyometra there is a large number of bacteria and neutrophils in the vaginal smear, the amount of the latter indicating the degree of toxicity.

Differential diagnoses are pregnancy, renal failure, diabetes mellitus, hepatic failure and hypoadrenocorticism, vaginitis and vaginal neoplasia and, in those rare cases with symptoms related only to the skeleton, disc prolapse and polyarthritis. A radiographic examination may confirm the diagnosis by showing the enlarged uterus.

Therapy

Surgical as well as pharmacological treatment is used, but the only effective treatment is ovariohysterectomy.

In valuable breeding bitches medical treatment may be tried, but only if the animal is in a very good physical condition. Treatment should then include drainage of the uterus, possibly by the use of a self-retaining Foley catheter (no. 20), and drugs to stimulate uterine motility. Uterine infection should be eliminated by long-term administration of antibiotics chosen after sensitivity tests.

Various hormones and non-hormonal compounds have been used for the treatment of pyometra in bitches (see Table 3.3).

Prostaglandins have been used in a small number of bitches with pyometra.[42,63,68,82,104] The dosage of 0.1–0.23 mg/kg $PGF_{2\alpha}$-THAM is followed by restlessness, signs of severe pain, vomiting and profuse vaginal discharge for a period of 10–30 min. This treatment is unreliable and may either cure the condition or cause deterioration or even rupture of the uterus.

Subcutaneous administration of 250 μg/kg $PGF_{2\alpha}$-THAM in combination with antibiotics has been reported as successful in 7 of 7 bitches with open pyometra. The same treatment gave a similar result in 5 of 6 bitches with endometritis and in 2 of 3 bitches with metritis. Two of the 16 treated bitches died, and of 9 bitches that were followed up 9 showed oestrous symptoms, and 4 of 5 bitches mated became pregnant.[98] In spite of results such as these, and even though the side-effects are temporary, it is too early to decide whether this treatment will be accepted in the future as routine practice.

In another investigation,[83] 13 out of 14 bitches with pyometra recovered after administration of $PGF_{2\alpha}$ and showed oestrus, and 9 of 11 mated bitches became pregnant; pyometra redeveloped in 2 other bitches.

Ovariohysterectomy

Before operating, the body fluid balance should be restored and the circulation stabilized by administration of hypotonic electrolyte or saline solutions, dextrose, or blood or plasma expanders depending on the severity of the case. In bitches with renal disease diuretics may be indicated. Antibiotics and corticosteroids are also indicated together with antishock therapy during and after the operation.

Light inhalation anaesthesia in combination with muscle relaxants and nitrous oxide is used after premedication with atropine and ultrashort-acting barbiturates.

The surgical technique is – with the following modifications – the same as described for spaying (Chapter 7). The midline abdominal incision should be long enough to admit careful movement of the enlarged uterus without damaging it. After ligating and suturing the ovarian attachments and the broad ligaments, the uterus is incised between two Crile forceps, clamped across the uterine body to prevent escape of pus. Any protruding mucosa is excised, the cut surface treated with iodine and the stump closed with an inverted Cushing suture using resorbable material, returned to the abdominal cavity and the abdominal incision sutured.

Vaginitis

Vaginitis is characterized by vaginal discharge, frequent licking of the vulva, perivulval dermatitis and sometimes attraction of male dogs. The vaginal mucosa is cherry-red in the acute state and dark to bluish-red in the chronic state.

Primary vaginitis occurs quite frequently before the first oestrus, and is characterized by a sticky grey to yellow-green discharge on the hair round the vulva, a condition that in most puppies does not require treatment as it disappears after the oestrous period.

Diagnosis is based on the history, vaginal inspection, vaginal smears, bacteriological and haematological examinations, abdominal palpation and radiological examination to exclude vaginal tumours and normal conditions such as pro-oestrous bleeding and lochial discharge, as well as pathological uterine conditions like metritis and pyometra as the cause.

Therapy comprises administration of antibiotics, chosen according to sensitivity tests, locally or systemically, and, bearing in mind the possible side-effects, combined with diethylstilboestrol (DES) in daily doses of 1–5 mg for 5 days.[29,49]

Mammary tumours

These are the most frequent tumours in the bitch and occur especially in older animals. The anterior glands are least commonly affected while the fourth and fifth are often involved. The tumours may be either benign (adenomas, fibromas or mixed tumours) or malignant (carcinomas, sarcomas or mixed tumours), possibly complicated by metastasis in lungs, liver, skin or bone. Early ovariohysterectomy before maturity reduces the frequency of mammary tumours.

Therapy

Tumours are treated by surgical extirpation of the affected glands. Where all glands are to be removed an ovariohysterectomy should be performed at the same time to prevent pregnancy and consequent nursing problems.

Radiation is little used in tumour therapy, immunotherapy is being investigated, and hormones have been tried.

Drostanolone propionate has been used in cases of unspecified inoperable tumours or when the bitch was a bad operative risk. Administration of 100 mg either weekly or every second week has in some cases kept the tumour in abeyance for up to several months, and in a few cases resulted in a decrease in size to operable dimensions.[65]

REFERENCES

1. Allen, W.E. (1982) Attempted oestrus induction in four bitches using pregnant mare serum gonadotrophin. *J. small Anim. Pract.*, **23**, 223–231.
2. Allen, W.E. and Stockman, V. (1979) Pseudopregnancy in the bitch. *Vet. Rec.*, **104**, 220.
3. Allen, W.E. Daker, M.G. and Hancock, J.L. (1982) Three intersexual dogs. *Vet. Rec.*, **109**, 468–471.
4. Al-Kafawi, A.A., Hopwood, M.L., Pineda, M.H. and Faulkner, L.C. (1974) Immunization of dogs against human chorionic gonadotropin. *Amer. J. vet. Res.*, **35**, 261–264.
5. Andersen, A.C. and Simpson, M.E. (1959) In: *The Ovary and Reproductive Cycle of the Dog (Beagle)*. Geron X Inc., Los Altos, California.
6. Arbeiter, K. (1968) Hormone zur Behandlung der Hündin. *Wien. tierärztl. Monatsschr.*, **55**, 587–591.
7. Arbeiter, K. (1968) Die therapeutische Anwendung syntetischer Progestagene bei der Hündin. *Kleintier-Prax.*, **13**, 1–4.
8. Arbeiter, K. (1976) Über ein neues Prinzip zur Behandlung von Gynäkopathien der Hündin. *Prakt. Tierarzt*, **57**, 295–300.
9. Arbeiter, K. and Dreier, H.K. (1972) Pathognostik und Behandlungsmöglichkeiten der Sub-Anöstrie und Anaphrodisie bei Zuchthündinnen. *Berl. Münch. tierärztl. Wochenschr.*, **85**, 341–344.
10. Arbeiter, K. and Geigenmüller, H. (1969) Genitalerkrankungen der Hündin – Diagnose und Therapie. *Wien. tierärztl. Monatsschr.*, **56**, 232–236.
11. Arbeiter, K. and Winding, W. (1977) Zur Behandlung der Lactatio sine graviditate und vom Milchstauungen im Anschluss an die Geburt mit dem Antiprolaktin 2-Br-α-Ergocryptin. *Kleintier-Prax.*, **22**, 271–278.
12. Arbeiter, K., Wollinger, F. and Coreth, H. (1981) Über die Behandlung der

Verlängerten Läufigkeit mit Proligeston (Delvosteron®). *Kleintier-Prax.*, **26**, 3–6.

13. Archbald, L.F., Baker, B.A., Clooney, L.L. and Godke, R.A. (1980) A surgical method for collecting canine embryos after induction of estrus and ovulation with exogenous gonadotropins. *Vet. Med./small Anim. Clin.*, **75**, 228–238.

14. Artmeier, P. and König, H.E. (1978) Zur Ohrakupunktur beim Hund. *Kleintier-Prax.*, **23**, 299–306.

15. Austad, R., Blom, A.K. and Børresen, B. (1979) Pyometra in the dog. III. A pathophysiological investigation. III. Plasma progesterone levels and ovarian morphology. *Nord. Vet.-Med.*, **31**, 258–262.

16. Bardens, J.W. (1968) Sterility in the bitch. In: *Current Veterinary Therapy. III. Small Anim. Pract.*, pp. 681–682. W.B. Saunders, Philadelphia, London, Toronto.

17. Bardens, J.W. (1971) Hormonal therapy for ovarian and testicular dysfunction in the dog. *J. Amer. vet. med. Assoc.*, **159**, 1405.

18. Barrett, R.E. and Theilen, G.H. (1977) Neoplasms of the canine and feline reproductive tracts. In: *Current Veterinary Therapy. VI. Small Anim. Pract.*, 1263–1267. W.B. Saunders Co., Philadelphia, London, Toronto.

19. Barton, C.L. (1977) Canine brucellosis. *Vet. Clin. N. Amer.*, **7**, 705–710.

20. Beard, D.C. (1961) Hydroxyprogesterone acetate: use in estrogenic and progesterogenic states in the bitch. *Small Anim. Clin.*, **1**, 215–218.

21. Bell, E.T. and Christie, D.W. (1971) Gonadotrophin induced ovulation in the bitch. *VII Wld Congr. Fert. Steril.*, Tokyo, p. 26.

22. Bosu, W.T.K., Chick, B.F. and Basrur, P.K. (1978) Clinical, pathologic and cytogenetic observations on two intersex dogs. *Cornell Vet.*, **68**, 375–390.

23. Broadhurst, J.J. (1974) Neoplasms of the reproductive system. In: *Current Veterinary Therapy. V. Small Anim. Pract.*, pp. 928–937. W.B. Saunders, Philadelphia, London, Toronto.

24. Brodey, R.S. and Roszel, J.F. (1967) Neoplasms of the canine uterus, vagina, and vulva: a clinicopathologic survey of 90 cases. *J. Amer. vet. med. Assoc.*, **151**, 1294–1307.

25. Brunner, F. (1976) Akupunktur in der Hundeklinik. *Kleintier-Prax.*, **21**, 182–189.

26. Børresen, B. (1979) Pyometra in the dog. II. A pathophysiological investigation. II. Anamnestic, clinical and reproductive aspects. *Nord. Vet.-Med.*, **31**, 251–257.

27. Carmichael, L.E. (1966) Abortion in 200 beagles (news report). *J. Amer. vet. med. Assoc.*, **149**, 1126.

28. Carmichael, L.E. and Kenney, R.M. (1968) Canine abortion caused by *Brucella canis*. *J. Amer. vet. med. Assoc.*, **152**, 605–616.

29. Carpenter, J.L. (1971) Vaginitis. In: *Current Veterinary Therapy. IV. Small Anim Pract.*, pp. 759–760, W.B. Saunders, Philadelphia, London, Toronto.

30. Chakraborty, P.K. and Fletcher, W.S. (1977) Responsiveness of anoestrous labrador bitches to GnRH (39618). *Proc. Soc. exp. Biol.*, **154**, 125–126.

31. Chakraborty, P.K., Wildt, D.E. and Seager, S.W.J. (1982) Induction of estrus and ovulation in the cat and dog. *Vet. Clin. N. Amer.*, **12**, 85–92.

32. Chaffaux, S. and Thibier, M. (1976) Hormonologie sexuelle de la chienne. Application au pyomètre. *Rec. Méd. vét.*, **152**, 471–481.

33. Chaffaux, S. and Thibier, M. (1978) Peripheral plasma concentrations of progesterone in the bitch with pyometra. *Ann. Rech. vét.*, **9**, 587–592.

34. Chaffaux, S., Mailhac, J.M., Cribiu, E.P., Popescu, C.P. and Cotard, J.P. (1980) L'intersexualité chez le chien (*Canis familiaris*). *Rec. Méd. vét.*, **156**, 179–192.

35. Chaffaux, S., Chassagnite, F. and Thibier, M. (1981) Concentration de LH après stimulation par la gonadolibérine (LRH) chez la chienne et effet de l'ovariectomie. *Rec. Méd. vét.*, **157**, 725–733.

36. Christiansen, Ib J. (1980) Unpublished observations.

37. Christie, D.W., Bell, E.T., Parkes, M.F., Pearson, H., Frankland, A.L. and Renton, J.P. (1972) Plasma progesterone levels in canine uterine diseases. *Vet. Rec.*, **90**, 704–705.
38. Cole, C.R., Sanger, V.L., Farrell, R.L. and Kornder, J.D. (1954) The present status of toxoplasmosis in veterinary medicine. *N. Amer. Vet.*, **35**, 265–270.
39. Cornwell, H.J.C. and Wright, N.G. (1969) The pathology of experimental infectious canine hepatitis in neonatal puppies. *Res. vet. Sci.*, **10**, 156–160.
40. De Coster, R., D'ieteren, G., Josse, M., Jacovljévic, S., Ectors, F. and Derivaux, J. (1979) Aspects clinique, histologique, bactériologique et hormonal de la métrite chronique chez la chienne. *Ann. Méd. vét.*, **123**, 233–247.
41. Dain, A.R. (1974) Intersexuality in a cocker spaniel dog. *J. Reprod. Fertil.*, **39**, 365–371.
42. Davies, G. Ll. (1979) Dinoprost in pyometritis in the bitch. *Vet. Rec.*, **105**, 109.
43. Dow, C. (1958) The cystic hyperplasia–pyometra complex in the bitch. *Vet. Rec.*, **70**, 1102–1108.
44. Dow, C. (1960) Ovarian abnormalities in the bitch. *J. comp. Pathol.*, **70**, 59–69.
45. Dreier, H.K. (1974) Richtige und falsche Hormonanwendung bei der Hündin. *Berl. Münch. tierärztl. Wochenschr.*, **87**, 68–70.
46. Dreier, H.K. (1978) Fruchtbarkeitsstörungen bei der Hündin. *Kleintier-Prax.*, **23**, 315–318.
47. Edols, J.H. and Allan, G.S. (1968) A case of male pseudohermaphroditism in a cocker spaniel. *Aust. vet. J.*, **44**, 287–290.
48. Ewald, B.H. (1961) A survey of the cystic hyperplasia–pyometra complex in the bitch. *Small Anim. Clin.*, **1**, 383–386.
49. Ewing, G. (1973) Treatment of selected reproductive problems in the dog. *Proc. Ann. AVSSBS Conf.*, **1**, 90–94 (cited in Jöchle, W.[66]).
50. Fry, P.D. (1980) Use of delmadinone acetate in pseudopregnancy in the bitch. *Vet. Rec.*, **106**, 297.
51. Gannon, J. (1976) *The Racing Greyhound*, **1**, 12 (cited in Noakes, D.E. (1980) Pathology of reproduction in the dog. *9th Int. Congr. Anim. Reprod.*, Madrid, **vol. I**, 245–260).
52. Gerber, H.A. and Sulman, F.G. (1964) The effect of methyloestrenolone on oestrus, pseudopregnancy, vagrancy, satyriasis and squirting in dogs and cats. *Vet. Rec.*, **76**, 1089–1093.
53. Gerber, H.A., Jöchle, W. and Sulman, F.G. (1973) Control of reproduction and of undesirable social and sexual behaviour in dogs and cats. *J. small Anim. Pract.*, **14**, 151–158.
54. Gerneke, W.H., de Boom, H.P.A. and Heitchen, I.G. (1968) Two canine intersexes. *J. S. Afr. vet med. Assoc.*, **39**, 55–59.
55. Hadley, J.C. (1975) Unconjugated oestrogen and progesterone concentrations in the blood of bitches with false pregnancy and pyometra. *Vet. Rec.*, **96**, 545–547.
56. Hare, W.C.D. (1976) Intersexuality in the dog. *Can. vet. J.*, **17**, 7–15.
57. Hare, W.C.D., McFeely, R.A. and Kelly, D.F. (1974) Familial 78XX male pseudohermaphoditism in three dogs. *J. Reprod. Fertil.*, **36**, 207–210.
58. Hardy, R.M. and Osborne, C.A. (1977) Canine pyometra–a polysystemic disorder. In: *Current Veterinary Therapy. VI. Small Anim. Pract.*, pp. 1229–1234. W.B. Saunders, Philadelphia, London, Toronto.
59. Hinsch, G.W. (1979) Intersexes in the dog. *Teratology*, **20**, 463–467.
60. Hirsch, D.C. and Wiger, N. (1977) The bacterial flora of the normal canine vagina compared with that of vaginal exudates. *J. small Anim. Pract.*, **18**, 25–30.
61. Holt, P.E. and Sayle, B. (1981) Congenital vestibulo-vaginal stenosis in the bitch. *J. small Anim. Pract.*, **22**, 67–75.
62. Huff, R.W. (1969) Some practice tips (medical treatment of pyometra). *Amer. Anim. Hosp. Assoc.*, 36th Ann. Meet., p. 272.
63. Jackson, P.G.G. (1979) Treatment of canine pyometra with dinoprost. *Vet. Rec.*, **105**, 131.

64. Jensen, N.E. (1967) Induction of ovulation by *bis* (*p*-acetoxyphenyl)cyclohexyl-idenemethane (F 6066) in the dog. *Nord. Vet.-Med.*, **19**, 236–239.
65. St. John, S. (1970) Treatment of canine mammary tumors. *Vet. Rec.*, **86**, 479.
66. Jöchle, W. (1975) Hormones in canine gynecology. A review. *Theriogenology*, **3**, 152–165.
67. Klide, A.M. and Kung, S.H. (1977) *Veterinary Acupuncture*. University of Pennsylvania Press.
68. Kongsengen, K. (1979) Treatment of pyometra with prostaglandins to dogs. *Norsk Vet.-T.*, **91**, 524–525.
69. Lavaud, J. (1979) Un nouveau traitement des 'grossesses nerveuses'. *L'Animal de Compagnie*, **14**, 323–327.
70. Leathem, J.H. (1938) Experimental induction of estrus in the dog. *Endocrinology*, **22**, 559–567.
71. Leathem, J.H. and Morrell, J.A. (1939) Induction of mating in the dog with pregnancy urine extract. *Endocrinology*, **24**, 149–156.
72. Ling, G.V. and Ruby, A.L. (1978) Aerobic bacterial flora of the prepuce, urethra, and vagina of normal adult dogs. *Amer. J. vet. Res.*, **39**, 695–698.
73. Low, D.G. (1954) Pyometra in the bitch. *Vet. Med./small Anim. Clin.*, **49**, 527–530.
74. Mann, C.J. (1971) Some clinical aspects of problems associated with oestrus and with its control in the bitch. *J. small Anim. Pract.*, **12**, 391–397.
75. Mantovani, A., Restani, R., Sciarra, D. and Simonella, P. (1961) Streptococcus L infections in the dog. *J. small Anim. Pract.*, **2**, 185–194.
76. McCandlish, I.A.P., Munro, C.D., Breeze, R.G. and Nash, A.S. (1979) Hormone producing ovarian tumors in the dog. *Vet. Rec.*, **105**, 9–11.
77. Milin, J. (1973) L'acupuncture en gynécologie chez la chienne. *L'Animal de Compagnie*, **33**, 293–305.
78. Moore, J.A. and Bennett, M. (1967) A previously undescribed organism associated with canine abortion. *Vet. Rec.*, **80**, 604–605.
79. Moore, J.A. and Gupta, B.N. (1970) Epizootiology, diagnosis, and control of *Brucella canis*. *J. Amer. vet. med. Assoc.*, **156**, 1737–1740.
80. Mulnix, J.A. (1974) False pregnancy. In: *Current Veterinary Therapy. V. Small Anim. Pract.*, p. 926. W.B. Saunders, Philadelphia, London, Toronto.
81. Mundt, S. (1981) Einsatzmöglichkeiten von Proligeston (Delvosteron®) bei Hunden und Katzen. *Prakt. Tierarzt*, **62**, 1058–1061.
82. Nelson, M. (1979) Dinoprost in small animals. *Vet. Rec.*, **105**, 261–262.
83. Nelson, R.W., Feldman, E.C. and Stabenfeldt, G.H. (1982) Treatment of canine pyometra and endometritis with prostaglandin $F_{2\alpha}$. *J. Amer. vet. med. Assoc.*, **181**, 899–903.
84. Oettel, M., Arnold, P. and Arnold, H.-I. (1969) Der Einsatz von Chlormadinonazetat bei Hund und Katze. *Fortpfl. Haust.*, **5**, 358–364.
85. Olson, P.N.S. and Mather, E.C. (1978) Canine vaginal and uterine bacterial flora. *J. Amer. vet. med. Assoc.*, **172**, 708–710.
86. Osbaldiston, G.W. (1978) Bacteriological studies of reproductive disorders of bitches. *J. Amer Anim. Hosp. Assoc.*, **14**, 363–367.
87. Osbaldiston, G.W., Nuru, S. and Mosier, J.E. (1972) Vaginal cytology and microflora of infertile bitches. *J. Amer. Anim. Hosp. Assoc.*, **8**, 93–101.
88. Paisly, L.G. and Fahning, M.L. (1977) Effects of exogenous follicle-stimulating hormone and luteinizing hormone in bitches. *J. Amer. vet. med. Assoc.*, **171**, 181–185.
89. Platt, A.M. and Simpson, R.B. (1974) Bacterial flora of the canine vagina. *Southwest. Vet.*, **27**, 76–77.
90. Poste, G. and King, N. (1971) Isolation of a herpesvirus from the canine genital tract: association with infertility, abortion and stillbirths. *Vet. Rec.*, **88**, 229–233.

91. Rogers, P.A.M., White, S.S. and Ottaway, C.W. (1977) Stimulation of the acupuncture points in relation to therapy of analgesia and clinical disorders in animals. *Vet. Ann.*, **17**, 258–279.
92. Rüsse, M. (1965) Über die Ätiologie der Hyperplasia glandularis cystica endometrii der Fleischfresser, ihre klinischen Symptome und Behandlung. *Die Blauen Hefte*, **30**, 15–18.
93. Rüsse, M. and Jöchle, W. (1963) Über die sexuelle Ruhigstellung weiblicher Hunde and Katzen bei normalem und gestörtem Zyklusgeschehen mit einem peroral wirksamen Gestagen. *Kleintier-Prax.*, **8**, 87–89.
94. Sbernardori, U. (1967) Un caso di intersessualitá nel cane. *Clin. vet. (Milano)*, **90**, 85–89.
95. Scorgie, N.J. (1939) The treatment of sterility in the bitch by the use of gonadotrophic hormones. *Vet. Rec.*, **51**, 265–268.
96. Siegel, E.T. (1977) In: *Endocrine Diseases of the Dog*. Lea and Febiger, Philadelphia.
97. Siim, J.C., Biering-Sørensen, U. and Møller, T. (1963) Toxoplasmosis in domestic animals. In: *Advances in Veterinary Science*, ed. Brandy, C.A. and Jungherr, E.L. **8**, 335–429. Academic Press, New York and London.
98. Sokolowski, J.H. (1980) Prostaglandin F_2alpha-THAM for medical treatment and endometritis, metritis, and pyometritis in the bitch. *J. Amer. Anim. Hosp. Assoc.*, **16**, 119–122.
99. Sokolowski, J.H. (1982) Mibolerone for treatment of canine pseudopregnancy and galactorrhea. *Canine Pract.*, **9**, 6–11.
100. Spalding, V.T., Rudd, H.K., Langman, B.A. and Rogers, S.E. (1964) Isolation of C.V.H. from puppies showing the 'fading puppy' syndrome. *Vet. Rec.*, **76**, 1402–1403.
101. Spy, G. (1966) The results of testosterone propionate treatment in ten cases of pyometra in the bitch. *Vet. Rec.*, **79**, 281–282.
102. Stafseth, H.J., Thompson, W.W. and Neu, L. (1937) Streptococcic infections in dogs. I. 'Acid milk', arthritis and post vaccination abscesses. *J. Amer. vet. med. Assoc.*, **90**, 769–781.
103. Stewart, R.W., Menges, R.W., Selby, L.A., Rhoades, J.D. and Crenshaw, D.B. (1972) Canine intersexuality in a pug breeding kennel. *Cornell Vet.*, **62**, 464–473.
104. Swift, G.A., Brown, R.H. and Nuttall, J.E. (1979) Dinoprost in pyometritis in the bitch. *Vet. Rec.*, **105**, 64–65.
105. Takeishi, M., Kodama, Y., Mikami, T., Tunekane, T. and Iwaki, T. (1976) Studies on reproduction in the dog. XI. Induction of estrus by hormonal treatment and results of the following insemination. *Jap. J. Anim. Reprod.*, **22**, 71–75.
106. Taul, L.K., Powell, H.S. and Baker, O.E. (1967) Canine abortion due to an unclassified Gram-negative bacterium. *Vet. Med./small Anim. Clin.*, **62**, 543–544.
107. Thun, R., Watson, P. and Jackson, G.L. (1977) Induction of estrus and ovulation in the bitch, using exogenous gonadotropins. *Amer. J. vet. Res.*, **38**, 483–486.
108. Tröger, C.P. (1966) Die hormonelle Behandlung von Zyklusstörungen bei der Hündin. *Berl. Münch. tierärztl. Wochenschr.*, **79**, 456–459.
109. Überreiter, O. (1966) Der Einfluss von Trächtigkeit und Scheinträgigkeit auf die Entstehung von Mammatumores bei der Hündin. *Berl. Münch. tierärztl. Wochenschr.*, **79**, 451–456.
110. Vaden, P. (1978) Surgical treatment of polycystic ovaries in the dog (a case report). *Vet. Med./small Anim. Clin.*, **73**, 1160.
111. Vandevelde, J.E. (1965) True hermaphroditism in a dog. *Canad. vet. J.*, **6**, 241–242.
112. Wagner, R. (1968) Präventive und inhibitive Zyklusbeeinflussung bei der Hündin. *Kleintier-Prax.*, **13**, 133–135.

113. Weaver, A.D., Harvey, M.J., Monroe, C.D., Rogerson, P. and McDonald, M. (1979) Phenotypic intersex (female pseudohermaphroditism) in a dachshund dog. *Vet. Rec.*, **105**, 230–232.
114. Whitney, J.C. (1967) The pathology of the canine genital tract in false pregnancy. *J. small Anim. Pract.*, **8**, 247–263.
115. Wilkinson, G.T. (1974) O-groups of *E. coli* in the vagina and alimentary tract of the dog. *Vet. Rec.*, **94**, 105.
116. Withers, A.R. and Whitney, J.C. (1967) The response of the bitch to treatment with medroxyprogesterone acetate. *J. small Anim. Pract.*, **8**, 265–271.
117. Wright, P.J. (1980) The induction of oestrus and ovulation in the bitch using pregnant mare serum gonadotrophin and human chorionic gonadotrophin. *Aust. vet. J.*, **56**, 137–140.
118. Wright, P.J. (1982) The induction of oestrus in the bitch using daily injections of pregnant mare serum gonadotrophin. *Aust. vet. J.*, **59**, 123–128.

ADDITIONAL READING

Allen, W.E. and Dagnall, G.J.R. (1982) Some observations on the aerobic bacterial flora of the genital tract of the dog and bitch. *J. small Anim. Pract.*, **23**, 325–335.
Allen, W.E. and Renton, J.P. (1982) Infertility in the dog and bitch. *Brit. vet. J.*, **138**, 185–198.
Alvarez, R., Mailhac, J.-M. and Chaffaux, S. (1980) Emploi de la bupivacaïne par voie épidurale, pour la chirurgie gynécologique de la chienne. *Rec. Méd. vét.*, **156**, 291–296.
Anon. (1982) Therapeutic use of prostaglandin $F_{2\alpha}$. *J. Amer. vet. med. Assoc.*, **181**, 932–934.
Arbeiter, K. (1966) Klinische Diagnostik der Pyometra–Erkrankung der Hündin. *Wien. tierärztl. Wochenschr.*, **5**, 346–350.
Baker, B.A., Archbald, L.F., Clooney, L.L., Lotz, K. and Godke, R.A. (1980) Luteal function in the hysterectomized bitch following treatment with prostaglandin $F_{2\alpha}$ ($PGF_{2\alpha}$). *Theriogenology*, **14**, 195–205.
Barrett, R.P. (1979) Cytology of intrauterine fluid in a case of pyometra in a dog. *Vet. Med./small Anim. Clin.*, **74**, 63–67.
Barta, M., Archbald, L.F. and Godke, R.A. (1982) Luteal function of induced corpora lutea in the bitch. *Theriogenology*, **18**, 541–549.
Barton, C.L. (1977) Canine vaginitis. *Vet. Clin. N. Amer.*, **7**, 711–714.
Blake, S. Jr. and Lapinski, A. (1980) Hypothyroidism in different breeds. *Canine Pract.*, **7**(2), 48–51.
von Bomhard, D. and Dreiack, J. (1977) Statistische Erhebungen über Mammatumores der Hündinnen. *Kleintier-Prax.*, **22**, 205–209.
Brodey, R.S., Goldschmidt, M.H. and Roszel, J.R. (1983) Canine mammary gland neoplasms. *J. Amer. Anim. Hosp. Assoc.*, **19**, 61–90.
Burke, T.J. (1982) Prostaglandin $F_{2\alpha}$ in the treatment of pyometra-metritis. *Vet. Clin. N. Amer.*, **12**, 107–109.
Christiansen, Ib J. (1982) Bitches, queens and prostaglandins. *Nord. Vet.-Med.*, **34**, 33–38.
Currier, R.W., Raithel, W.F., Martin, R.J. and Potter, M.E. (1982) Canine brucellosis. *J. Amer. vet. med. Assoc.*, **1**, 132–133.
Czernicki, B. and Weiss, R. (1981) Genitale Zytologie und Bakteriologie der Hündin. *Zuchthygiene*, **16**, 79.
Daykin, P.W. (1971) Use and misuse of steroids in canine and feline dermatology. *J. small Anim. Pract.*, **12**, 425–430.
English, P.B. (1983) Antimicrobial chemotherapy in the dog. III. Possible adverse reactions. *J. small Anim. Pract.*, **24**, 423–436.

English, P.B. and Prescott, C.W. (1983) Antimicrobial chemotherapy in the dog. I. Related to body system or organ infected. *J. small Anim. Pract.*, **24**, 277–292.

English, P.B. and Prescott, C.W. (1983) Antimicrobial chemotherapy in the dog: II. Some practical considerations. *J. small Anim. Pract.*, **24**, 371–382.

Faulkner, R.T. and Johnson, S.E. (1980) An ovarian cyst in a West Highland white terrier. *Vet. Med./small Anim. Clin.*, **75**, 1375–1377.

Fidler, I.J., Brodey, R.S., Howson, A.E. and Cohen, D. (1966) Relationship of estrous irregularity, pseudopregnancy, and pregnancy to canine pyometra. *J. Amer. vet. med. Assoc.*, **149**, 1043–1046.

Fischer, K. (1983) Pseudohermaphroditism in a dog. *Vet. Med./small Anim. Clin.*, **78**, 683–685.

Freudiger, U. (1961) Komplikationen nach Endometritisoperationen. *Kleintier-Prax.*, **6**, 101–105.

Furneaux, R.W. (1979) Surgical disorders of the canine vagina and vulva. *Vet. Ann.*, **19**, 245–254.

Hadley, J.C. (1975) The development of cystic endometrial hyperplasia in the bitch following serial uterine biopsies. *J. small Anim. Pract.*, **16**, 249–257.

Hamilton, J.M., Knight, P.J., Beevers, J. and Else, R.W. (1978) Serum prolactin concentrations in canine mammary cancer. *Vet. Rec.*, **102**, 127–128.

Hill, H. and Maré, C.J. (1974) Genital diseases in dogs caused by canine herpesvirus. *Amer. J. vet. Res.*, **35**, 669–672.

Holt, P.E., Long, S.E. and Gibbs, C. (1983) Disorders of urination associated with canine intersexuality. *J. small Anim. Pract.*, **24**, 475–487.

Holzmann, A., Laber, G. and Walzl, H. (1979) Experimentally induced mycoplasmal infection in the genital tract of the female dog. *Theriogenology*, **12**, 355–370.

Homer, B.L., Altman, N.H. and Tenzer, N.B. (1980) Left horn uterine torsion in a non-gravid nulliparous bitch. *J. Amer. vet. med. Assoc.*, **176**, 633–634.

Jackson, P.S., Furr, B.J.A. and Hutchinson, F.G. (1982) A preliminary study of pregnancy termination in the bitch with slow-release formulations of prostaglandin analogues. *J. small Anim. Pract.*, **23**, 287–294.

Janssens, L.A.A. (1981) De behandeling van schijndracht bij de teef met bromocriptine (Parlodel®). *T. Diergeneesk.*, **106**, 767–770.

Kuwabara, S., Yoshida, H., Tanaka, S. and Murasugi, E. (1973) Studies on the artificial of the bitch with gonadotropic hormone. *J. Tokyo Soc. vet. zootech. Sci.*, **19/20**, 128–136.

Lagneau, F. (1971) L'infertilité des carnivores domestiques. *XIX Wld Vet. Congr.*, Mexico, **vol. 1**, 280–283.

Lamatsch, O. (1966) Behandlung der Gebärmutter-Erkrankung beim Hund. *Wien, tierärztl. Monatsschr.*, **5**, 351–353.

Linde, C., Edqvist, L.-E., Ekman, L. and Shille, V. (1978) Endocrine causes for irregularities of the estrous cycle in the bitch and cat. *XIII Nord. Vet. Congr.*, Åbo, 124–127.

MacVean, D.W., Monlux, A.W., Anderson, P.S. Jr., Silberg, S.L. and Roszel, J.F. (1978) Frequency of canine and feline tumors in a defined population. *Vet. Pathol.*, **15**, 700–715.

Medleau, L., Johnson, C.A., Perry, R.L. and Dulisch, M.L. (1983) Female pseudohermaphroditism associated with mibolerone administration in a dog. *J. Amer. Anim. Hosp. Assoc.*, **19**, 213–215.

Mialot, J.P., Lagneau, F., Chaffaux, St. and Badinand, F. (1981) Inhibition de la lactation de pseudogestation chez la chienne par la bromocryptine. *Rec. Méd, vét.*, **157**, 351–355.

Misdorp, W. and Hart, A.A.M. (1979) Canine mammary cancer. I. Prognosis. *J. small Anim. Pract.*, **20**, 385–394.

Misdorp, W. and Hart, A.A.M. (1979) Canine mammary cancer. II. Therapy and causes of death. *J. small Anim. Pract.*, **20**, 395–404.

Monty, D.E., Jr., Wilson, O. and Stone, J.M. (1979) Thyroid studies in pregnant and newborn beagles, using [125]I. *Amer. J. vet. Res.*, **40**, 1249–1256.

Nesbit, G.H., Izzo, J., Peterson, L. and Wilkins, R.J. (1980) Canine hypothyroidism: a retrospective study of 108 cases. *J. Amer. vet. med. Assoc.*, **177**, 1117–1122.

Newman, R.H. (1979) Pyometra and a Sertoli cell tumor in a hermaphroditic dog. *Vet. Med./small Anim. Clin.*, **74**, 1757.

Nielsen, S.W. (1983) Classification of tumors in dogs and cats. *J. Amer. Anim. Hosp. Assoc.*, **19**, 13–60.

Oettel, M. (1971) Die Anwendung von Sexualsteroiden in der Kleintierpraxis. *Mh. Vet.-Med.*, **26**, 225–235.

Oettel, M. (1979) Möglichkeiten zur hormonellen Beeinflussung der Ovulation bei der Hündin. *Mh. Vet.-Med.*, **34**, 942–945.

Oettel, M., Schimke, E. and Zacharias, J. (1969) Klinische Befunde und zellkernmorphologische Untersuchungen beim Mammakarzinom der Hündin. *Arch. exp. Vet.-Med.*, **23**, 549–553.

Olson, P.N.S. (1977) Canine vaginitis. In: *Current Veterinary Therapy. VI. Small Anim. Pract.*, pp. 1235–1236. W.B. Saunders, Philadelphia, London, Toronto.

Priester, W.A. (1979) Occurrence of mammary neoplasms in bitches in relation to breed, age, tumour type, and geographical region from which reported. *J. small Anim. Pract.*, **20**, 1–11.

Prole, J.H.B. and Allen, W.E. (1979) Unusual abortion in a bitch. *Vet. Rec.*, **104**, 558–559.

Renton, J.P., Munro, C.D., Heathcote, R.H. and Carmichael, S. (1981) Some aspects of the aetiology, diagnosis and treatment of infertility in the bitch. *J. Reprod. Fertil.*, **61**, 289–294.

Roehl, J.F. and Simons, K.E. (1982) Surgical correction of a urogenital anomaly in a dog. *Vet. Med./small Anim. Clin.*, **77**, 1375–1377.

Rogers, P.A.M. (1974) Success claimed for acupuncture in domestic animals. A veterinary news item. *Irish vet. J.*, **28**, 182–191.

Rowley, J. (1980) Cystic ovary in a dog: a case report. *Vet. Med./small Anim. Clin.*, **75**, 1888.

Ruckstuhl, B. (1978) Die Incontinentia urinae bei der Hündin als Spätfolge der Kastration. *Schweiz. Arch. Tierheilk.*, **120**, 143–148.

Rüsse, M. (1980) Die Endometritis der Hündin, Ätiologie, Diagnose, Therapie. *Tierärztl. Umsch.*, **35**, 482–483.

Scheel-Thomsen, A. (1959) Ovarian transplantation in the bitch. *Nord. Vet.-Med.*, **11**, 162–182.

Scott, J.R., Anderson, W.R., Kling, T.G. and Yannone, M.E. (1970) Uterine transplantation in dogs. *Gynec. Invest.*, **1**, 140–148.

Stockman, M.J.R. (1963) An unusual breeding history in the bitch. *Vet. Rec.*, **75**, 903.

Stott, G.G. (1974) Granulosa cell islands in the canine ovary. Histogenesis, histomorphologic features, and fate. *Amer. J. vet. Res.*, **35**, 1351–1355.

Sundberg, J.P. (1979) A case of true bilateral hermaphroditism in a dog. *Vet. Med./small Anim. Pract.*, **74**, 477–486.

Varga, B., Horváth and Stark, E. (1977) Effect of PGE_2 and $PGF_{2\alpha}$ on local blood flow and progesterone ovarian secretion. *Acta physiol. Acad. Sci. hung.*, **49**, 369.

Verma, O.P. & Chibuzo, G.A. (1974) Hormonal influences on motility of canine uterine horns. *Amer. J. vet. Res.*, **35**, 23–26.

Voith, V.L. (1980) Functional significance of pseudocyesis. *Mod. vet. Pract.*, **61** (1), 75–77.

Whitney, J.C. (1969) The pathology of unilateral pyometra in the bitch. *J. small Anim. Pract.*, **10**, 223–230.

Whitney, L.F. (1947) Ovarian transplantation in dogs. *Vet. Med.*, **42**, 30–32.

Wingate, M.B., Karasewich, E., Wingate, L., Lauchian, S. and Ray, M. (1970) Experimental uterotubovarian homotransplantation in the dog. *Amer. J. Obstet. Gynec.*, **106**, 1171–1176.

Chapter 4
Andrology of the Normal Male

ANATOMY AND PHYSIOLOGY

The testes and scrotum

The testes, round to oval in shape, are situated in the scrotum with the long oblique axis directed dorsocaudally, and the relatively large epididymis firmly attached dorsolaterally (Figure 4.1). There is great variation in the size of testes between large and small breeds, the average being $3 \times 2 \times 1.5$ cm. In the testes the spermatozoa and the male sex hormone testosterone are produced under influence of the pituitary production of luteinizing hormone (LH) and follicle-stimulating hormone (FSH).

The skin of the scrotum is thin, pigmented, supplied with sweat glands and covered with hair. The absence of subcutaneous fat and the cooling of the arterial blood by passage near to the cooler venous blood in the pampiniform plexus are mechanisms that prevent increases in scrotal temperature, which would lead to degenerative changes in the seminiferous tubules. The distance of the testes from the body can be regulated by the cremaster muscle, which also serves as a thermoregulatory mechanism.

The penis and prepuce

The penis consists of two distinct corpora cavernosa. The posterior part is separated by the median septum penis; anteriorly a bone, the os penis, reaches from the bulbus glandis almost to the tip of the glans, into which it is prolonged by a conical projection of fibrocartilage approximately 0.5 cm long. In large dogs the os penis can be 10 cm or more in length. The caudal three-quarters are grooved for the urethra and its corpus spongiosum, and the dorsolateral concave surface is attached to the erectile tissue of the pars longa penis. The glans penis comprises two parts: the bulbus glandis, consisting of erectile tissue completely surrounding the os penis and urethra, and the pars longa penis, composed of erectile tissue dorsally and laterally only.

The prepuce forms a complete sheath around the anterior part of the penis.

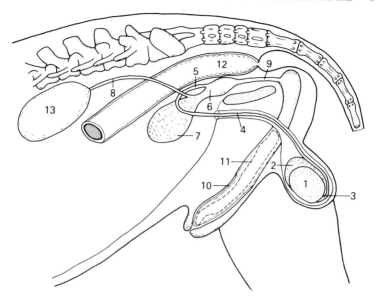

Figure 4.1 Reproductive structure, *in situ*, of the dog. (1) Testis. (2) Caput epididymis. (3) Cauda epididymis. (4) Ductus (vas) deferens. (5) Ampullary gland. (6) Prostate gland. (7) Bladder. (8) Ureter. (9) Urethra. (10) Penis. (11) Baculum. (12) Rectum. (13) Kidney. (From McKeever[59])

The prostate gland

The prostate gland is the only accessory gland in the dog. It is normally situated on the middle of the pelvis resting on the cranial part of the symphysis pelvis, about 1 cm posterior to the neck of the bladder and surrounding the cranial part of the urethra. The prostate gland is symmetrical, globular in shape with a median dorsal furrow dividing it into two halves. It has a tensile elastic consistency and in adult dogs the gland varies from a minimum length, width and thickness of 1.4–1.9 cm to a maximum of 2.5–2.8 cm, with a volume of 6–15 ml, an absolute weight of 1.7–14.5 g, and a weight in relation to the bodyweight of 0.21–0.57 per kg.[12]

The testicles

The descent of the testicles occurs rather early, often in fetal life, and most male puppies are born with the testes in the scrotum.[5] Sometimes the testes may descend later; normally they are present by the age of 6 months, but descent at 8 months has also been reported.[78]

Puberty is reached at an average age of 9 months, but some dogs may be fertile at 6 months. Males normally reach puberty a little later

Table 4.1 Volume of ejaculate, sperm content, motility, percentage of abnormal spermatozoa, size of scrotum and serum concentration of testosterone in beagle dogs at 235 days of age (maturity) and at 375 days of age: average values. (After Takeishi et al.[81])

	Age	
	235 days	375 days
Ejaculate (ml)	0.55	0.62
Spermatozoa per ml	2.48×10^8	7.22×10^8
Motility (%)	83	80
Abnormal spermatozoa (%)	25.8	14.6
Length of left side of scrotum (cm)	3.24	3.46
Length of right side of scrotum (cm)	2.96	3.32
Width of left side of scrotum (cm)	2.06	2.32
Width of right side of scrotum (cm)	2.04	2.29
Serum testosterone (μg/ml)	1.15 ± 0.14	4.09 ± 0.51

than females of the same breed, except in certain breeds (chow chows and salukis), where the male may reach puberty several years later than the bitch.[11]

Testicular biopsies have been used in attempts to determine the approximate age at the time of the first appearance of mature spermatozoa. Spermatogonial activity is found in collies at 5 and 6 months of age and the first spermatozoa at 9–10 months.[32]

Spermatogenesis comprises an 8-stage cycle of the seminiferous epithelium with a total average duration of 13.6 days and a lifespan for primary spermatocytes of 20.9 days, secondary spermatocytes 0.5 days, spermatids with round nuclei 10.5 days, and spermatids with elongated nuclei 10.6 days[31] up to the time they are released into the lumen.

The first ejaculation in beagles has been recorded at 235 days, with volume and spermatozoa concentration increasing up to 375 days of age. In the same period scrotal size and serum testosterone concentration increase, while the percentage of abnormal spermatozoa decreases (Table 4.1).

Sexual maturity and environment

In an attempt to evaluate the influence of the environment, male puppies were reared under different conditions:[4] with their dams and sibs, with their dams, with their sibs but without their dams, with adult male dogs, with cats, and completely isolated without visual contact with other dogs. It was found that even though sexual maturity was not significantly affected, successful copulation was not achieved by males reared without visual contact with other dogs. Furthermore, contact with dams was less important than contact with littermates for the development of successful mating behaviour. It is well known that dogs reared somewhat in isolation from others may exhibit sexual inexperience, including unwillingness or inability to copulate when required later for breeding at 2–3 years of age, although otherwise mature. After preputial massage and collection of semen a few times, normal spontaneous copulation may result.[77]

ANDROLOGICAL EXAMINATION

Table 4.2 shows the various andrological examinations that can be made, the parameters that can be determined, and the pathological conditions that may be found. Where a dog is presented for the evaluation of its soundness for breeding a detailed history and careful clinical examination must be made without overlooking apparently harmless conditions such as, for example, balanoposthitis, since this can be the reason for loss or reduction of the ejaculatory reflex.

History

History must include all available information concerning feeding, management, vaccinations, previous health problems and medical treatments including hormonal treatment, time, preparations and dosages used, previous use for breeding and the breeding activity during the preceding year with the number of bitches served, pregnancies obtained and the litter size.

In the case of a high percentage of unsuccessful matings or pregnancies resulting in small litters, it is also necessary to obtain information concerning breeding and management of the bitch, since these may be the cause of the problem.

Information cannot be accepted without careful consideration; for example, the information that a dog, having sired litters previously, now fails in one or a limited number of cases to impregnate bitches could be explained by the fact that the dog had previously been held responsible for paternity which was actually the result of an unwanted accidental mating with another dog.

Clinical examination

The clinical examination includes both a general and a specific andrological examination. The general examination comprises the current state of health, the constitution and the physical appearance. It can be expanded further if necessary, for example with possible symptoms of infectious disease or locomotor defects caused by lame-

Table 4.2 Andrological examination.

History/region/sample	Method	Parameter	Motive
History	Interview Questionnaire	Previous breeding	Insemination Sterility
Scrotum	Inspection Palpation	Size Symmetry Content Thickness of the skin Injuries Temperature	Presence of testes Cryptorchidism Hypogonadism Hernia Infections Eczema Tumours
Testicles	Palpation Biopsy	Anatomy Dimension Shape Consistency Pain	Sterility Hypogonadism Spermiogenic tissue Infections Tumours
Epididymides	Palpation	Anatomy	Aplasia Infections
Penis Prepuce	Inspection Palpation Mating behaviour Microbiological examination	Anatomy Size Appearance Pathogenic organisms	Hypospadism Phimosis Paraphimosis Trauma Adhesions Persistence of penile frenulum Infections Tumours

Prostate Prostate secretion	Palpation Biopsy Collection of ejaculate Biochemical examination Radiographic examination	Anatomy Size Pathogenic organisms Anions and cations	Hypertrophy Atrophy Cysts Infections Calculi Tumours
Semen	Collection of ejaculate Macroscopical examination Microscopical examination Microbiological examination Biochemical examination	Appearance Volume Concentration Motility Morphology Anion and cations Pathogenic micro-organisms	Insemination Sterility Infections
Mating behaviour	Mating performance	Behaviour	Sterility
Blood/plasma/serum	Hormone assay Haematological examination Chromosome analysis Agglutination test	Hormone content Erythrocytes Leukocytes Blood plates Cell volume Haemoglobin percentage Haematocrit Sedimentation rate Chromosome picture *Brucella* titre	Sterility Pathological conditions Hereditary diseases

ness, swollen joints, painful areas and deformities that can also influence reproductive performance. The specific examination can include inspection and palpation of the reproductive organs, collection and evaluation of semen, and inspection of mating behaviour, and can be expanded into hormonal assays and chromosomal analyses.

Intersexuality has been mentioned in Chapter 3 (page 55).

Scrotum

The scrotum should be inspected and palpated carefully for its size, for lesions, and for the presence of testicles. If the testicles are not descended or are retracted, the scrotum will be smaller than normal, while increased size may occur in pathological conditions of the testicles such as infection and neoplasm in the scrotal tissue. In cases of thick scrotal tissue and traumatic lesions, thermoregulation will be disturbed leading to higher scrotal temperatures which may result in degenerative changes in the testis and arrest of spermiogenesis for up to 3 months. Asymmetry of the scrotum can be caused by unilateral hypogonadism, cryptorchidism or hernia; the latter condition will cause sterility because of a temperature rise caused by the presence of intestines in the scrotum.

Testicles

The testicles are examined for dimension and consistency and for the presence of infections or tumours.

If one or both testicles are not palpable, this may be because the examination takes place before the normal time for descent, because of retraction of the testicles or cryptorchidism.

Retraction of the testicles

This is caused by a highly effective cremaster reflex. The testicles can often be manipulated down in the scrotum for palpation in normal dogs; this is not possible in cryptorchids. Retraction of the testes does not cause sterility, since the reflex becomes less sensitive with increasing age, and the testicles will normally descend without treatment.[78]

Cryptorchidism

This is inherited in most cases. It may be unilateral or bilateral, completely abdominal with testes, epididymides, etc. situated within the abdominal cavity, partly abdominal with only part of the testes or the appendages intra-abdominal or in the inguinal canal, or inguinal with the testes in the inguinal canal.[5,15,65,68,79,85] The failure of normal testicular descent occurs twice as frequently on the right side as the left, with a right:left ratio of 2.3:1 for inguinal and 2.0:1 for abdominal retention.[67]

Even though the testicles in most dogs are palpable at birth or within the first 6–8 weeks of age, a definite diagnosis of cryptorchidism should not be made until the dog is 6 months old. Due to the raised temperature in a retained testicle, spermatogenesis does not occur, leading to sterility in a bilaterally cryptorchid dog. The spermatozoa are normal in the descended testicle in a unilaterally cryptorchid dog.

Cryptorchidism results in lowered fertility and a variably lowered ejaculation reflex. Badinand et al.[6] examined 47 cryptorchid dogs and found that a bilaterally cryptorchid dog may ejaculate but without any spermatozoa in the ejaculate, whereas less than half of the unilaterally cryptorchid dogs ejaculate and 27% of these have no sperm in the ejaculate. Altogether 69% of all the unilaterally cryptorchid dogs investigated did not give any spermatozoa. A normal volume was found in only 8%, and there was a normal concentration of spermatozoa in 3 of 17 unilaterally cryptorchid dogs; normal volume and numbers of spermatozoa were found in the ejaculate from 1 of the 17 dogs.

Cryptorchid dogs should not be used for breeding. Castration is to be advised, even though the use of chorionic gonadotropin has been shown to be successful in making the testicles descend. This is because of, firstly, the hereditary nature of cryptorchidism and, secondly, the fact that a high frequency of Sertoli cell tumours (SCT) and seminomas (SEM) occur in the retained testicles of cryptorchid dogs. The incidence of inguinal testicular tumours is found to be twice that of abdominal testicular tumours.[67] SCT and SEM occur with equal frequency in cryptorchid dogs,[42] and develop at an earlier age than in dogs with testes situated normally in the scrotum.[66] Cryptorchidism with SCT and SEM is followed by the same symptoms as tumours in descended testes.

In unilaterally cryptorchid dogs therapy may comprise removal of the undescended testicle and vasectomy on the other side, thus preventing the spread of the condition without interfering with the hormone production in the scrotal testicle.

Hypogonadism

Hypogonadism can occur either primarily, with the testes remaining small and immature after puberty, or secondarily as in dogs with Cushing's syndrome.

Orchitis

This is clinically obvious with one or both testicles increased in size, and in the acute state it also presents with increased heat and pain and a changed consistency of the affected testicle.[55] In the chronic state the texture is firm and the shape nodular.

In cases of orchitis serological tests for *Brucella canis* infection should be made.

Tumours
Interstitial cell tumours (ICT), SCT and SEM occur fairly commonly in the testes of dogs more than 7 years of age.[45] The lowest mean age is found in dogs with SCT whereas SEM and ICT occur in progressively older dogs.[57] In the adult the incidence of testicular tumours is approximately 16%.[16] Cryptorchids appear to have a 13.6 times higher risk of testicular tumours than normal dogs, and male dogs with inguinal hernia have a 4.7 times increased risk of testicular tumours.[42] The most common form, mainly seen in the mature dog, is the interstitial cell tumour consisting of enlarged Leydig's cells obviously without functional significance. SCT are often associated with feminization and attraction of other male dogs, loss of libido, gynaecomastia and atrophy of the unaffected testis. Feminization due to SCT occurs more often in cryptorchid dogs than in dogs with testes in the scrotum. Thus, feminization is found in 16.7% of dogs with scrotal SCT, 50.0% of dogs with inguinal SCT, and 70.4% of dogs with abdominal SCT.[66] In cases without metastases it can be expected that feminization signs disappear 2–6 weeks after castration.[91] SEM are also common; they are usually benign, derived from the seminiferous tubules and they cause no functional disturbance in the dog.

Epididymides
The epididymides are examined for anatomical defects such as unilateral or bilateral aplasia and for evidence of infection.
  Epididymitis is not uncommon in dogs and is often accompanied by orchitis. It may be caused by a non-specific infection or be the result of distemper.

Penis and prepuce
The penis and prepuce are examined for anatomical abnormalities, injuries,[50,51] infectious disease and tumours.

Trauma of the penis
This may be caused by contusion or laceration, the latter often being accompanied by profuse bleeding. Contusions are treated by cleansing and administration of antibiotics locally and parenterally, and lacerations are sutured to control haemorrhage. Tranquillizers and antibiotics are given as necessary. If the penile urethra is damaged a Foley catheter is inserted to allow drainage of subcutaneous tissue contaminated by urine.
  Surgical intervention[49] usually results in complete recovery from traumatic fracture of the os penis.

Phimosis, preputial adhesions and persistent frenulum of the penis
This can make erection impossible or so painful that libido is reduced.
These abnormalities can be corrected surgically.[51]

Paraphimosis
Paraphimosis is a condition where the penis cannot be retracted. The
cause may be phimosis, strangulation (for example by a rubber band),
balanoposthitis, trauma of the penis, masturbation or fracture of the
os penis. Treatment depends on the cause.

Balanoposthitis
This occurs commonly with symptoms varying from minimal to copious purulent discharge from the preputial orifice, often followed by
excessive licking of the genitalia.[56]
    Administration of an antibiotic solution into the prepuce after sen-
sitivity testing is usually effective, but in refractory cases 2% copper
sulphate can be used, and similarly copper sulphate crystals may be
used for cauterization of the lymphoid follicles.

Venereal tumours
If transmitted at mating these are usually benign. The symptoms are
single or multiple small nodules on the mucosa of the penis and
prepuce, irritation of the region and a bloodstained discharge. Spontaneous regression may occur but because of the pain and bleeding
the tumours may have to be removed surgically or irradiation therapy
instituted.[7,16]

Prostate
The prostate is examined by abdominal and rectal palpation for the
presence of infection, hypertrophy, cysts and tumours. Marked
enlargement of the gland, whatever the reason, may lead to such
symptoms as weakness in the hindlegs, walking with the back arched
and a stiff, stilted gait, and pain when urinating and defecating which
often results in constipation.[17]

Prostatic hypertrophy
This occurs frequently in older dogs. The gland is symmetrically
increased in size with a smooth surface. Administration of 5–25 mg
diethylstilboestrol (DES) or 0.25–2.0 mg oestradiol cypionate may
lead to cessation of symptoms for a period from a few months up to
several years.[13] Subcutaneous administration of 2–3.6 ml medroxyprogesterone acetate (MAP) has resulted in disappearance of symptoms within 3–6 days and with an effect lasting at least 3 months.[13]
Castration is an effective treatment leading to permanent atrophy of

the gland. Experimentally induced hypertrophy can be avoided by simultaneous injection of cyproterone acetate.[86,87]

Prostatic atrophy
Atrophy results from castration or reduced function of the interstitial testicular cells.

Prostatic cysts
These often cause more enlargement of the gland than does hypertrophy and a weight of up to 250 g has been reported.[72] Treatment is by aspiration of the cysts or surgical extirpation.

Prostatitis
Prostatitis can occur at any age and is accompanied by a swelling of the gland which makes palpation very painful. In the acute state the gland is warm, and pus may appear in the urethra in cases of suppurative prostatitis. When abscesses are present, the gland may be huge and irregularly shaped, often with adhesions to other organs.

For diagnosis of prostatitis examination of an ejaculate and transperineal or transabdominal biopsies from the gland may be of use.[90] Micro-organisms in the ejaculate may originate from the prostate or from urethral or preputial contamination during collection. A careful sampling technique, as well as a cytological examination and quantitative bacterial cultures, are therefore necessary to decide whether an ejaculate is normal or indicative of prostatic inflammation. Gram-positive organisms and squamous cells with a small number of neutrophils may be of urethral or preputial origin in dogs without prostatic lesions, whereas large numbers of organisms, particularly if Gram-negative, or organisms isolated in pure culture, are suggestive of prostatic infection.[8]

Antibiotic therapy, administered in large doses for a long period of time, perhaps combined with oestrogen treatment, is possible, but surgical intervention may be the best solution, either as castration or marsupialization.[35,44]

Prostatic calculi
These may occur as exogenous stones from the urinary tract or as endogenous stones, and thus cause the same symptoms as in other prostatic diseases. The diagnosis is made by rectal palpation and radiographic examination, and treatment is surgical removal.

Tumours
Tumours may cause irregular enlargement of the gland often accompanied by adhesions to the surrounding tissue. The most frequently occurring tumours are adenocarcinomas and sarcomas.[7,16,61,84,92]

Testicular biopsy

In efforts to solve the problem of sterility in male dogs a biopsy of the testicles can be of value in diagnosis. Specimens can be obtained either by incision or needle biopsy. The material should be obtained without applying pressure to the testes or causing any trauma likely to produce artifacts, fixed immediately after removal, stained and examined.[69]

Biopsy sampling is not advisable in fertile dogs that are soon to be used for breeding as it may lead to a decrease in semen quality and possibly also to subfertility for a period because of inflammation and increased scrotal temperature.[54] It is common practice to biopsy only one testis, although this carries the risk of missing a localized lesion.[53]

Incision

A biopsy specimen can be taken by incision, under general anaesthesia. The testicle is moved forward under the skin to a position slightly lateral to the point where an incision for castration is normally made, and a 2 cm long incision made in the non-scrotal skin followed by a 1.5 cm incision in the tunica vaginalis. An ultrafine blade is used to incise the tunica albuginea and for removing the clump of seminiferous tubules which normally then bulge from the incision.

Needle biopsy

A needle or punch biopsy can be performed under deep sedation or anaesthesia by the use of a needle directed posteriorly and diagonally within the ventral hemisphere of the testis, but this technique carries the risk of permanent damage to the testis. An aspiration biopsy taken through a fine needle is a much simpler method for obtaining sufficient testicular tissue.[43] Atrophy of the testicle has been seen after needle biopsy, but usually there is initial haemorrhage and subsequent necrosis and atrophy of the site while the surrounding tissue remains normal.[46] There is no adverse effect on the size of the testis or on androgen production as judged from the concentration of circulating plasma testosterone.

Hormone assay

Testosterone and LH

The overall testosterone concentration in the first 2 days of life is 0.15 ng/ml.[41] Maximal testosterone concentration is correlated with the completion of testicular growth and the appearance of mature spermatozoa.[48] The average concentration of testosterone in plasma is 560.866 ± 395.488 ng/100 ml in dogs from 1 to 5 years old, with a decrease after the third year of age.[70] In intact male dogs the testosterone concentration ranges from 0.4 to 6.0 ng/ml plasma, and the LH concentration from 0.2 to 12.0 ng/ml over a 24 h period. A diurnal rhythm has not been found. Release or administration of LH stimulates secretion of testosterone from the testes,[27] and an approximate 50 min interval is found between corresponding LH and testosterone peaks.[21] In castrated dogs the LH concentration is higher (9.8 ± 2.7 ng/ml) than in the intact dog in spite of undetectable plasma testosterone levels.[21] The rates of testosterone secretion vary considerably in adult mongrel dogs, and there seems to be a trend towards a higher secretion of testosterone in early spring compared with late winter.[28]

Semen evaluation

Evaluation of the semen is carried out after clinical examination. It must be emphasized that evaluation of a single sample may give a false impression, especially if the dog has recently been used extensively for breeding. In such cases a rest period of 4–5 days should be allowed before collection of a new sample. In dogs not used for breeding for a long time, several samples must be taken before the ejaculate can be considered representative.

Semen collection

Individual differences exist in the way that dogs react to collection. Some will ejaculate in any environment while others do so only in familiar surroundings – at home, in the presence of their owner, etc. Some dogs will have an erection and ejaculate without any stimuli from an oestrous bitch, while others will do so only in the presence of an oestrous bitch. This requirement is more pronounced at the first collection. Some dogs will ejaculate only while mounting a bitch in season, others will do so standing beside or behind the bitch. Even in the presence of a bitch in oestrus, dogs will sometimes not mount either because of weakness or stiffness in their hindlegs, or because of aggressiveness in the bitch, or the memory of previous attempts at mounting an aggressive bitch. Some dogs will not ejaculate at the first attempt, and collection may also be difficult from an affectionate dog. Loud conversation and repeated orders to the dog are inadvisable, and those present must act calmly and completely ignore the dog's behaviour. Sometimes several attempts are needed before collection is successful.

The best circumstances for obtaining a sample are familiar surroundings with the dog on the floor, or on a table with a non-slip surface, in the presence of an oestrous bitch and with as few people present as possible. All manipulations must be calm and gentle, noise and disturbance must be avoided and the procedure must not be hurried. The dog should be allowed to investigate the room. Sometimes young, shy dogs must be placed behind the bitch to encourage them to mount. The bitch should be prevented from biting or gripping the dog with her teeth. Inexperienced young dogs especially may be frightened and therefore refuse to ejaculate. The bitch should be prevented from lying down during collection and it may be necessary to support her under the chest, either by the left hand or by an assistant's knee, who kneels on the right side of the bitch while holding her head with the right hand. The bitch may also be tethered to a ring about 0.6 m up on the wall.

Several methods have been advocated for collection of semen: for example, digital massage, use of an artificial vagina or electrical vibrator, and electroejaculation; any of these can be used. Some are claimed to be better than others, although only few comprehensive investigations have been made. Boucher et al.[14] found that manipulation by hand is superior to an artificial vagina, and digital manipulation has proven especially reliable for untrained dogs; some dogs may even show a very definite preference for a gloved hand holding the penis.[58] The equipment used must be clean and prewarmed. Young and timid dogs often need to be encouraged before collection by manual stimulation a number of times, thereby increasing their interest in the sexual process.[77] Some dogs may exhibit very vigorous and continuous coital movements when ejaculating the first fraction of semen, thus making collection very difficult. In these cases the use of an artificial vagina is to be preferred for facilitating the collection and reducing the risk of the collection funnel damaging the penis.

Semen can be collected from most dogs and often, at least after some training, without a teaser bitch for stimulating activity, although the latter is essential for some dogs, particularly young, timid or inexperienced ones. The use of a teaser bitch also increases the concentration of spermatozoa in the ejaculate. Three methods of collecting semen – an artificial vagina without a teaser bitch and manipulation without and with a teaser bitch – gave a number of spermatozoa per ejaculate of $10^6 \times 289$, $\times 314$ and $\times 528$ and a number of motile spermatozoa per ejaculate of $10^6 \times 65$, $\times 260$ and $\times 396$, respectively.[14]

Digital manipulation   For the collection of semen by this method various kinds of equipment have been used (Figure 4.2): for example, glass or plastic tubes and funnels,[14,29] a 15 cm long glass cylinder continuing as a narrower 6–9 cm long graduated lower part,[62] and a graduated glass tube.[2,20]

The dog is allowed to mount the bitch, and, even if she is not in oestrus, ejaculation sometimes follows massage of the prepuce. The semen is collected from the left side of the dog with the equipment in the left hand and the right hand fitted with a glove grasping the penis

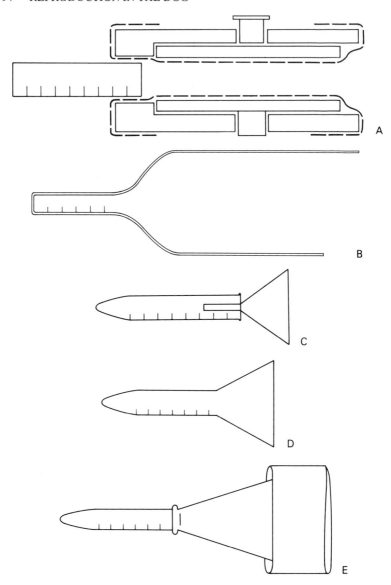

Figure 4.2 Examples of equipment for semen collection: (A) Artificial vagina, ad modum Harrop.[38] (B) Glass cylinder, 15 cm long, continuing as a narrower 6–9 cm long graduated lower part, ad modum Peters.[62] (C) Funnel and graduated test tube (glass or plastic) which can be held in the left hand for semen collection, ad modum Foote.[29] (D) Graduated glass tube, ad modum Andersen.[2] (E) A 15 ml plastic centrifuge tube connected with an apex rubber cone, ad modum Seager.[76] (From M. Schmidt, 1983)

gently. The prepuce is gently drawn backwards behind the bulbus glandis, and with the thumb and forefinger encircling the penis this is fixed with a firm pressure beyond the bulbus resulting in full erection. Usually erection follows spontaneously, but massage of the prepuce is required in some dogs. The prepuce cannot be slipped over the bulbus if there is full erection. Separation of the dogs may then be necessary, and another attempt at collection made and after subsidence of the erection.

Within 20 s ejaculation starts and the ejaculatory reflex is maintained for 1–22 min.[22] The penis must not be allowed to touch the equipment as a hard contact may inhibit later collections. Because of the fractionated ejaculation, separate collection of each fraction is possible by changing the collection funnels. After the second, sperm-rich fraction the dog will try to lift one hindleg over the arm of the person who is collecting the semen, as always happens under natural conditions when the male dog lifts the leg and turns around. The collection should then be interrupted to avoid unwanted admixture of secretion from the prostate gland. Collection of semen is often facilitated if the penis is grasped with a hand from behind between the hindlegs of the dog. When the coital movements have ceased, the penis is deflected downwards and backwards, enabling the collection to be easily monitored and stopped when the ejaculate becomes clearer and more transparent.

Artificial vagina   The most widely used type is Harrop's artificial vagina.[38,40] Another type was described by Hancock and Rowlands.[37] A more simple model consisting of a plastic syringe case and a surgical glove has been used.[33] Other simple and very effective equipment is described by Seager and Fletcher.[75]

Harrop's artificial vagina is composed of an outer rubber casing, 15 cm long and with a 5 cm lumen. A latex liner can be fitted and hot water introduced between the outer covering and the lining. Between the two layers a rubber bladder is fitted, connected to the outside by a valve through which air may be pumped, so that the lumen of the vagina can be varied and a pulsating movement communicated to the inner layer. The artificial vagina may be used without lubrication,[9] or the liner may be lubricated with Vaseline.[40] The bulbus glandis is grasped and, as soon as erection occurs after a few massaging movements, the penis is directed into the artificial vagina where further stimulation is obtained by the pulsation of the liner. Fractionated collection can be obtained by changing collection tubes. When thrusting movements cease the artificial vagina is moved backwards between the dog's hindlegs to simulate the position of the penis during natural mating, and is maintained there until ejaculation has ended.[9] Some dogs need to get accustomed to the artificial vagina before they can ejaculate, but it is a very useful instrument for long-term collections.

Seager's equipment for collection of semen from 10–75 kg dogs consists of a 15 ml graduated plastic centrifuge tube attached to the narrow end of a conical rubber sheath similar to that used as a collection funnel in bovine artificial vaginas.[74-76] Only a small amount of lubricant is needed inside the sheath. After preparation of the teaser bitch as described above, the collection apparatus is held in the left hand to prevent a drop in the temperature of the equipment. As in digital collection the prepuce is drawn gently backwards behind the bulbus glandis before full erection occurs. The collection sheath is placed gently over the penis about three-quarters of the way to where penis is encircled by the fingers, and full erection is obtained by gentle digital pressure. The penis is then directed downwards and backwards between the hindlegs. When the clear third fraction of the ejaculate appears in the collection tube, the sheath is gently removed from the penis.

Vibrator   Ejaculation has been obtained by the use of an electric vibrator fitted with a special collecting cup attached to the head of the vibrator. By placing the vibrating cup on the glans penis, ejaculation has been induced in nearly 65% of dogs within 15 s to 7 min.[71] The volume of the ejaculates ranged from 1.5 to 28.0 ml with a sperm concentration of 60 000–400 000/mm³.

Electroejaculation   This has been tried in dogs,[19,58] but it is not recommended, partly because of the accompanying pain, and thus the need for general anaesthesia, and partly because the ejaculate contains only few spermatozoa. To avoid urine in the ejaculate the bladder should be drained prior to employing electrical stimulation.

Ejaculation
The ejaculate comprises three distinct separate fractions. The first fraction is a clear fluid containing few, if any, spermatozoa; it is small in volume, probably made up of secretion from the urethral gland[36,40,60] and/or from the prostate gland.[3] The second fraction is milky in colour, its opacity depending on the sperm concentration, and contains the greatest part of the spermatozoa. The third and largest fraction, originating from the prostate gland, is a clear fluid containing only few spermatozoa.

The ejaculation of each of the three fractions is separated by 10–20 s, thus making fractionated collection possible.

Collection
Collection of semen every second day is found to be near the maximum frequency; and collections made more often tend to deplete the

reserve of spermatozoa.[13] By increasing the frequency of collection, the concentration of spermatozoa decreases, whereas neither the volume of the ejaculate nor the percentage of motile spermatozoa is altered. Even though libido remains high when the dog ejaculates once or twice a day, the number of spermatozoa decreases after four successive collections to less than half the original number and remains low until the dog is rested.[14] Collection of more than one or two successive ejaculates from a dog with only a 5–10 min rest period between is impossible, more because of loss of libido than through exhaustion of the genital tract secretions.[9] Accordingly it seems reasonable to use a dog to stud two or three times per week, because more frequent use over a longer period may lead to a decrease in semen quality. More frequent collections should therefore be made only over a short period.

The quality of the semen as measured by sperm concentration and output is found to be highest in the spring.[80] In some investigations a similar result has been found for volume, motility and percentage of normal spermatozoa,[52,82] but there is no seasonal change in libido or plasma testosterone concentration.[80]

The volume of the total ejaculate, as well as of the single fractions, may vary. In dogs of larger breeds the total ejaculate comprises 30 ml or more, whereas in smaller dogs it consists of only a few millilitres. When collecting semen by massage it has been found that the ejaculatory reflex ranges from 60 s to 20 min and 25 s; the total volume averages 5.38 ml (1–22.5 ml) for dogs weighing ≤ 20 kg and 12.75 ml (2–45 ml) for dogs weighing > 20 kg.[22,26] When using an artificial vagina for collection an ejaculation time of 11 ± 0.2 min and a total ejaculate of 8.1 ± 0.14 ml/10 kg bodyweight were found.

The duration of ejaculation and the volume of the single fractions collected by use of an artificial vagina and by digital manipulation are shown in Table 4.3.

A significant correlation of −0.68 appears to exist between the ejaculate volume and the sperm concentration, whereas collection time and volume of the ejaculate are not correlated.[63]

Normally the ejaculate is grey to white in colour depending on the content of the spermatozoa and the admixture of leukocytes or other cells. In cases of aspermia the ejaculate is watery and clear, sometimes with a light yellow colour, which may also be seen if the ejaculate is mixed with urine or pus. A red or pink colour occurs in ejaculate from dogs with prostatitis or balanoposthitis.

The volume of ejaculate increases linearly with the prostatic weight in dogs with normal or glandular hyperplastic prostates, whereas this correlation is not observed in those with cystic hyperplasia, in which there is a reduction of 50% in the volume of the semen produced for each gram of prostate.[93]

Table 4.3 Duration of ejaculation and volume of each of the 3 fractions of ejaculates collected by use of an artificial vagina and by manipulation. (After Bartlett[9] and Dubiel[22])

| | Artificial vagina | | Manipulation | | |
| | Duration | Volume (ml) | Duration | | Volume (ml) |
			Average	Range	
First fraction	2.7 s ± 0.3 s	0.9 ± 0.0	13.5 s	5–90	0.1–3.0
Second fraction	5.2 s ± 0.4 s	2.6 ± 0.1	54.4 s	5–200	0.5–4.0
Third fraction	8 min ± 0.1 min	9.2 ± 0.3	6 min 55 s	1–20 min	1–30

Concentration

The number of spermatozoa is determined by a haemocytometer, a spectrophotometer or an automatic cell counter. When using a haemocytometer, the semen is gently mixed and drawn up into the red blood cell-diluting pipette to the 0.5 mark or the 1.0 mark, depending upon the sperm concentration. After wiping the tip of the pipette, formol saline or eosin-diluting fluid, which makes the sperm more visible, is drawn up to the 101.0 mark. The pipette is then shaken, and the first three to four drops are discarded. The haemocytometer is loaded on both sides, and the spermatozoa counted in the four corner squares each consisting of 16 small squares, the count including the spermatozoa lying on the left-side line and the top line in each small square. In order to find the sperm count per millilitre, the total number of spermatozoa in the four corners is multiplied by 500 000 if the dilution rate is 1:200 (semen drawn up to the 0.5 mark), or multiplied by 250 000 if the dilution rate is 1:100.

For estimation of sperm concentration by means of a spectro-photometer, a 2.9% sodium citrate diluent is used at a dilution rate depending upon the opacity of the ejaculate as judged by eye.[30]

In the unfractionated ejaculate the number of spermatozoa is about 500×10^6, but a total number of 5940×10^6 has been observed.[18] A concentration of $247 \pm 9.9 \times 10^6$/ml has been found in the second fraction corresponding to a total number of spermatozoa of $637 \pm 31.7 \times 10^6$, and a concentration of $10 \pm 0.8 \times 10^6$/ml in the third fraction.[9] The presence of spermatozoa in the third fraction has also been reported in other investigations, and may be due to the mixing of fractions in the urethra or the artificial vagina.[63]

Motility

A drop of semen placed on a microscope slide and covered with a coverslip, both at a temperature of 38 °C, should be examined at a

$100\times$ magnification. The number of motile spermatozoa should be evaluated immediately after collection. It normally ranges from 80 to 90%, and motility below 60% must be regarded as abnormal. The motility status (the progressive motility) can be expressed according to a scale:[76]

0 all dead
1 slight side-to-side movements, no forward progression
2 rapid side-to-side movements, no forward progression
3 rapid side-to-side movements, occasional forward progression
4 slow steady forward progression
5 rapid steady forward progression

Another scale expresses the motility based on the percentage of motile spermatozoa:[89]

n all dead
0 only oscillatory movements
1 20% of the spermatozoa motile
2 40% of the spermatozoa motile
3 60% of the spermatozoa motile
4 80% of the spermatozoa motile
5 all spermatozoa motile

In dog semen the wave motion characteristic of bovine semen is not seen. A progressive movement of the spermatozoa is normal, whereas a side-to-side movement is not.

Morphology
The average dimensions of live spermatozoa with no detectable abnormalities are: total length 68 ± 0.3 μm, head length 7 ± 0.0 μm, head width 5 ± 0.1 μm, midpiece length 11 ± 0.2 μm and tail length 50 ± 0.3 μm.[9] Harrop[39] found that the total length varies from 55 to 65 μm.

The percentage of 'dead' (eosin-stained) spermatozoa per ejaculate at room temperature varies from 15% in the second fraction to 20% in the total ejaculate. Storage at 37 °C for up to 2 h increases this percentage.[9]

By counting at least 100 spermatozoa the number of spermatozoa with morphologically atypical features can be recorded. Abnormalities are most often located in the head and the tail and are divided into primary and secondary types (Table 4.4 and Fig. 4.3). Primary abnormalities originate from spermatogenesis, while secondary abnormalities develop during the passage through the epididymis or during ejaculation. The number of secondary abnor-

Table 4.4 Survey of the principal primary and secondary sperm abnormalities.

	Abnormality	
Location	Primary	Secondary
Head	Macrocephalus Microcephalus Double heads Pointed heads Indented heads Opaque heads	Free heads Bent heads Detached heads Swollen acrosomes Detaching acrosomes
Neck	Thickened neck Eccentric insertion	Disintegration of the neck region Cytoplasmic droplets
Midpiece	Irregularly thickened midpiece Irregularly thinned midpiece Coiled midpiece Kinked midpiece Double midpiece	Bent midpiece Extraneous material surrounding midpiece Proximal cytoplasmic droplets Mid-cytoplasmic droplets Distal cytoplasmic droplets
Tail	Thin tail Double tail Triple tail	Coiled tail Looped tail Kinked tail Folded tail Detached tail Cytoplasmic droplets

malities increases with increasing sexual abstinence. These have been found to have an average frequency of 13.29% with a range of 2.4–88.0%.[23] In the same investigation the percentage of primary abnormalities ranged from 0.8 to 8.0. Bartlett[9] found – based on 30 ejaculates, 10 from each of 3 dogs – that the abnormalities comprise 15 ± 1.6% distributed as follows: head 11 ± 1.4%, neck 5 ± 1.5%, midpiece including cytoplasmic droplets 21 ± 5.0% and excluding cytoplasmic droplets 2 ± 0.5% and tail abnormalities 7 ± 0.8%. No significant difference was found between dogs except for abnormalities of the neck and midpiece including cytoplasmic droplets.

In cases of brucellosis, sperm abnormalities occur about 5 weeks after infection and reach maximal intensity 3 weeks later. The first abnormalities seen are immature spermatozoa with retained perinuclear sheaths, deformed acrosomes, swollen midpieces and retained protoplasmic droplets. Abnormalities found 15 weeks after infection are bent tails, detached heads, and head-to-head agglutination occurring together with inflammatory cells.[34]

Biochemistry
The composition of dog semen is shown in Table 4.5. Sodium is the

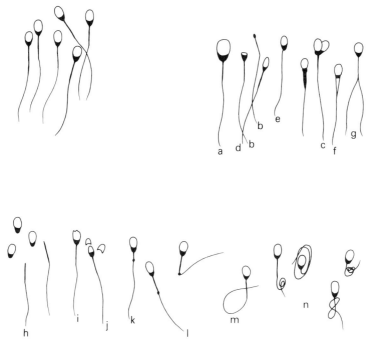

Figure 4.3 Examples of primary and secondary sperm abnormalities. (a) Macrocephalus. (b) Microcephalus. (c) Double heads. (d) Pointed heads. (e) Eccentric insertion. (f) Double midpiece. (g) Double tail. (h) Free heads. (i) Swollen acrosomes. (j) Detaching acrosomes. (k) Mid-cytoplasmic droplets. (l) Distal cytoplasmic droplets. (m) Coiled tail. (n) Looped tail. (From M. Schmidt, 1983)

main cation in both the unfractionated ejaculate and the single fractions. Potassium, magnesium and calcium occur in much lower concentrations. The main anion is chloride, and in all fractions there is a low concentration of total reducing sugar, fructose, and lactic and citric acids.[1,10,18,64,88]

The enzymes alanine and aspartate aminotransferases and alkaline and acid phosphatases occur in seminal plasma in mean concentrations of 12, 71, 9705 and 2012 MU/ml, whereas the concentrations in the sperm fraction are 180, 190, 129 and 26 MU/ml, respectively.[47] In dogs 6–16 months of age investigated at monthly intervals the acid phosphatase activities were found to increase from 7.3 to 70.1 MU/ml,[83] and the serum testosterone from 0.36 to 3.64 ng/ml. The concentrations of acid and alkaline phosphatases are likely to depend on the activity of the prostate gland and may be a useful index of prostatic function. There seems to be some connection between these

Table 4.5 The composition of dog semen. (After Wales and White[88]). Mean values = the standard deviation are given with the number of replications in parentheses. Except for reducing sugar, fructose, lactic acid, glycerylphosphorylcholine and total protein, the values were obtained by analysis of all 3 fractions from the same ejaculate on each occasion. All the phosphorus estimations on centrifuged fluid were made on all fractions of the same 4 ejaculates.

Constituent or property	Pre-sperm fraction	Sperm-bearing fraction	Prostatic fluid
Sperm count (10^6 ml)	—	505 ± 206 (16)	—
pH		6.3 ± 0.2 (17)	6.8 ± 0.3 (8)
Sodium (m Eq/l), centrifuged fluid	145 ± 6 (5)	122 ± 8 (5)	156 ± 11 (5)
Potassium (m Eq/l), centrifuged fluid	14.6 ± 2.1 (5)	12.9 ± 1.8 (5)	7.0 ± 3.2 (5)
Magnesium (m Eq/l), centrifuged fluid	0.62 ± 0.20 (5)	0.46 ± 0.26 (5)	0.17 ± 0.09 (5)
Calcium (m Eq/l), centrifuged fluid	0.63 ± 0.37 (5)	0.37 ± 0.07 (5)	0.24 ± 0.07 (5)
Chloride (m Eq/l), centrifuged fluid	141 ± 2 (4)	125 ± 3 (4)	148 ± 4 (4)
Total orcinol reactive carbohydrate (mg%)	97 ± 52 (20)	341 ± 109 (20)	189 ± 67 (20)
Total reducing sugar (mg%)	14 ± 7 (5)	9 ± 11 (10)	4 ± 2 (6)
Fructose (mg%)	2 (5)	1 (14)	1 (10)
Lactic acid (mg%)	7 ± 3 (4)	5 ± 1 (4)	8 ± 6 (5)
Citric acid (mg%)	3 ± 4 (5)	4 ± 2 (5)	1 ± 0.02 (5)
Total protein (mg%), ethanol precipitate	1.3 ± 0.7 (6)	3.7 ± 1.4 (16)	2.8 ± 1.1 (21)
Urea (mg%), centrifuged fluid	68.8 ± 24.2 (3)	69.4 ± 25.1 (3)	72.0 ± 17.3 (3)
Ammonia (mg%), centrifuged fluid	9.8 ± 2.1 (3)	14.7 ± 1.0 (3)	18.6 ± 5.1 (3)
Total phosphorus (mg%), whole fluid	17.9 ± 16.9 (5)	154.6 ± 34.2 (5)	7.8 ± 2.8 (5)
Total phosphorus (mg%), centrifuged fluid	18.8 ± 3.2 (4)	92.7 ± 14.8 (4)	7.8 ± 20 (4)
Acid insoluble phosphorus (mg%), whole fluid	9.4 ± 9.8 (4)	53.6 ± 21.2 (4)	5.5 ± 1.6 (4)
Acid insoluble phosphorus (mg%), centrifuged fluid	1.7 ± 0.8 (4)	7.8 ± 0.8 (4)	4.8 ± 1.3 (4)
Lipid phosphorus (mg%), whole fluid	0.5 ± 0.4 (4)	5.4 ± 1.2 (4)	0.2 ± 0.2 (4)
Lipid phosphorus (mg%), centrifuged fluid	0.4 ± 0.2 (4)	0.5 ± 0.2 (4)	0.1 ± 0.2 (4)
Inorganic phosphorus (mg%)	3.9 ± 2.4 (4)	6.5 ± 1.5 (4)	1.6 ± 0.5 (4)
Acid soluble phosphorus (mg%), centrifuged fluid	16.2 ± 3.0 (4)	82.3 ± 22.0 (4)	5.4 ± 3.9 (4)
Glycerylphosphorylcholine (mg%)	5 ± 6 (5)	176 ± 31 (8)	20 ± 9 (8)

phosphatase concentrations and the incidence of abnormal or damaged spermatozoa, and there also seems to be a correlation between the enzymes and the sperm motility. In dogs with fertility disorders, large differences are found in the amount of phosphatase present in the semen.[24]

In cryptorchid dogs the concentrations of fructose and citric acid are raised, whereas the concentration of lactic acid is lowered, and in addition the level of acid phosphatase is higher than in normal dogs.[6]

The biochemical characteristics after vasectomy alter, with a significant fall in the calcium content because of the absence of spermatozoa and testicular fluid, which normally contribute a considerable amount of calcium. The lactic acid content also rises significantly, probably due to the non-utilization of lactic acid because there are no spermatozoa.[18]

Electrophoresis

By means of electrophoresis the seminal plasma separates into 3 fractions, both in fertile dogs and in those with fertility disorders.[25] In old dogs only a slight electrophoretic band, diffuse towards the cathode, is visible.

REFERENCES

1. De Abreu, J.J., Bruschi, J.H., Vieira, R.M., Bruschi, M.C.M. and Megale, F. (1980) Physical morphological and biochemical aspects of the 'German shepherd' dog semen. *9th Int. Congr. Anim. Reprod.*, Madrid, Proc. **vol. III**. Symposia, 302
2. Andersen, Kj. (1969) Artificial insemination in dog. *Medlemsbl. norske Vet.-Foren.*, **21**, 482–484.
3. Andersen, Kj. (1978) Personal communication.
4. Antonov, V.V. and Khananashvili, M.M. (1973) Significance of early individual experience in the establishment of sexual behaviour in male dogs. *Zh. vyssh. nerv. Deyat. Pavlova*, **23**, 68–73.
5. Ashdown, R.R. (1963) The diagnosis of cryptorchidism in young dogs: a review of the problem. *J. small Anim. Pract.*, **4**, 261–263.
6. Badinand, F., Szumowski, P. and Breton, A. (1972) Étude morphobiologique et biochimique de sperme de chien cryptorchide. *Rec. Méd. vét.*, **148**, 655–689.
7. Barrett, R.E. and Theilen, G.H. (1977) Neoplasms of the canine and feline reproductive tracts. In: *Current Veterinary Therapy. VI. Small Anim. Pract.*, pp. 1263–1267. W.B. Saunders, Philadelphia, London, Toronto.
8. Barsanti, J.A., Shotts, E.B., Prasse, K. and Crowell, W (1980) Evaluation of diagnostic techniques for canine prostatic diseases. *J. Amer. vet. med. Assoc.*, **177**, 160–163.
9. Bartlett, D.J. (1962) Studies on dog semen. I. Morphological characteristics. *J. Reprod. Fertil.*, **3**, 173–189.
10. Bartlett, D.J. (1962) Studies on dog semen. II. Biochemical characteristics. *J. Reprod. Fertil.*, **3**, 190–205.
11. Beaver, B.V. (1977) Mating behavior in the dog. *Vet. Clin. N. Amer.*, **7**, 723–728.
12. Berg, O.A. (1958) The normal prostate gland of the dog. *Acta Endocrinol. (Kbh.)*, **27**, 129–139.

104   REPRODUCTION IN THE DOG

13. Betts, C.W. and Finco, D.R. (1974) Diseases of the canine prostate gland. In: *Current Veterinary Therapy. V. Small Anim. Pract.*, pp. 938–941. W.B. Saunders, Philadelphia, London, Toronto.
14. Boucher, J.H., Foote, R.H. and Kirk, R.W. (1958) The evaluation of semen quality in the dog and the effects of frequency of ejaculation upon semen quality, libido, and depletion of sperm reserves. *Cornell Vet.*, 48, 67–86.
15. British Veterinary Association (1955) Cryptorchidism in the dog. *Vet. Rec.*, 67, 472–474.
16. Broadhurst, J.J. (1974) Neoplasms of the reproductive system. In: *Current Veterinary Therapy. V. Small Anim. Pract.*, pp. 928–937. W.B. Saunders, Philadelphia, London, Toronto.
17. Chaffaux, S. (1979) Pathologie de la prostate de chien. *Rec. Méd. vét.*, 155, 421–427.
18. Chatterjee, S.N., Meenakshi, Sharma, R.N. and Kar, A.B. (1976) Semen characteristics of normal and vasectomized dogs. *Ind. J. exp. Biol.*, 14, 411–414.
19. Christensen, G.C. and Dougherty, R.W. (1955) A simplified apparatus for obtaining semen from dogs by electrical stimulation. *J. Amer. vet. med. Assoc.*, 127, 50–53.
20. Christiansen, Ib J. and Schmidt, M. (1980) Deepfreezing of dog semen. *Kgl. Vet.-og Landbohøjsk., Inst. Sterilitetsforskn. Årsberetn.*, 23, 69–75.
21. DePalatis, L., Moore, J. and Falvo, R.E. (1978) Plasma concentrations of testosterone and LH in the male dog. *J. Reprod. Fertil.*, 52, 201–207.
22. Dubiel, A. (1972) Studies on semen collection by masturbation method and on ejaculation reflex in dogs. *Weterynaria Wroclau*, 29, 225–234.
23. Dubiel, A. (1973) Evaluation of semen properties in dogs with reference to fertility. *Weterynaria Wroclau*, 30, 181–191.
24. Dubiel, A. (1973) The activity of gamma-glutamyl-transpeptidase, acid and alkaline phosphatase in the semen of fertile dogs and with some disturbances in fertility. *Med. weteryn.*, 29, 679–681.
25. Dubiel, A. (1975) Electrophoretic investigations on the semen plasma of dogs, both fertile and with fecundity disorders. *Pol. Arch. weteryn.*, 17, 699–706.
26. Dubiel, A. (1976) Evaluation of semen properties and ejaculation reflex in dogs with reference to fertility. *VIII Int. Congr. Anim. Reprod.*, Krakow, Proc. vol. I. *Communication Abstracts*, 75.
27. Eik-Nes, K.B. (1962) Secretion of testosterone in anesthetized dogs. *Endocrinology*, 71, 101–106.
28. Eik-Nes, K.B. (1971) Production and secretion of testicular steroids. *Rec. Progr. Horm. Res.*, 27, 517–535.
29. Foote, R.H. (1968) Artificial insemination of dogs. In: *Current Veterinary Therapy. III. Small Anim. Pract.*, pp. 686–689. W.B. Saunders, Philadelphia, London, Toronto.
30. Foote, R.H. and Boucher, J.H. (1964) A comparison of several photoelectric procedures for estimating sperm concentration in dog semen. *Amer. J. vet. Res.*, 25, 558–561.
31. Foote, R.H., Swierstra, E.E. and Hunt, W.L. (1972) Spermatogenesis in the dog. *Anat. Rec.*, 173, 341–352.
32. Ford, L. (1965) Evidence of spermatogenesis in various breeds and cross-breeds of dogs. *Proc. LA Acad. Sci.*, 28, 117–120 (*Anim. Breed. Abstr.*, 1967, 35, 143).
33. Frye, F.L., Hoeft, D.J., Cucuel, J.-P.E. and Hardy, R.J. (1970) Canine artificial vagina. *J. Amer. vet. med. Assoc.*, 156, 1578–1579.
34. George, L.W., Duncan, J.R. and Charmichael, L.E. (1979) Semen examination in dogs with canine brucellosis. *Amer. J. vet. Res.*, 40, 1589–1595.
35. Gourley, I.M.G. and Osborne, C.A. (1966) Marsupialization – a treatment for prostatic abscess in the dog. *Anim. Hosp.*, 2, 100–105.
36. Hamner, C.E. (1970). The semen. In: *Reproduction and Breeding Techniques for Laboratory Animals*, ed. Hafez, E.S.E. Lea and Febiger, Philadelphia. Chapter 3.

37. Hancock, J.L. & Rowlands, I.W. (1949) The physiology of reproduction in the dog. *Vet. Rec.*, **61**, 771–779.
38. Harrop, A.E. (1954) A new type of canine artificial vagina. *Brit. vet. J.*, **110**, 194–196.
39. Harrop, A.E. (1955) Some observations on canine semen. *Vet. Rec.*, **67**, 494–498.
40. Harrop, A.E. (1960) *Reproduction in the Dog*. Baillière, Tindall and Cox, London.
41. Hart, B.L. and Ladewig, J. (1979) Serum testosterone of neonatal male and female dogs. *Biol. Reprod.*, **21**, 289–292.
42. Hayes, H.M. and Pendergrass, T.W. (1976) Canine testicular tumors: epidemiologic features of 410 dogs. *Int. J. Cancer*, **18**, 482–487.
43. Hendricks, F.B., Lambird, P.A. and Murphy, G.P. (1969) Percutaneous needle biopsy of the testis. *Fertil. and Steril.*, **20**, 478–481.
44. Hoffer, R.E., Dykes, N.L. and Greiner, T.P. (1977) Marsupialization as a treatment for prostatic disease. *J. Amer. Anim. Hosp. Assoc.*, **13**, 98–104.
45. James, R.W. & Heywood, R. (1979) Age-related variations in the testes and prostate of Beagle dogs. *Toxicology*, **12**, 273–279.
46. James, R.W., Heywood, R. and Fowler, D.J. (1979) Serial percutaneous testicular biopsy in the Beagle dog. *J. small Anim. Pract.*, **20**, 219–228.
47. James, R.W., Heywood, R. and Street, A.E. (1979) Biochemical observations on beagle dog semen. *Vet. Rec.*, **104**, 480–482.
48. James, R.W., Crook, D. and Heywood, R. (1979) Canine pituitary–testicular function in relation to toxicity testing. *Toxicology*, **13**, 237–247.
49. Jeffery, K.L. (1974) Fracture of the os penis in a dog. *J. Amer. Anim. Hosp. Assoc.*, **10**, 41–44.
50. Johnston, D.E. (1965) Repairing lesions of the canine penis and prepuce. *Mod. vet. Pract.*, **46**(1), 39–46.
51. Johnston, D.E. and Archibald, J. (1974) Male genital system. In: *Canine Surgery*, 2nd Archibald ed., Amer. Vet. Publ. Inc., Drawer KK, Santa Barbara, California, Chapter 16.
52. Kuroda, H. and Hiroe, K. (1972) Studies on the metabolism of dog spermatozoa. I. Seasonal variation on the semen quality and of the aerobic metabolism of spermatozoa. *Jap. J. Anim. Reprod.*, **17**, 89–98.
53. Larsen, R.E. (1977) Evaluation of fertility problems in the male dog. *Vet. Clin. N. Amer.*, **7**, 735–745.
54. Larsen, R.E. (1977) Testicular biopsy in the dog. *Vet. Clin. N. Amer.*, **7**, 747–755.
55. Lein, D.H. (1977) Canine orchitis. In: *Current Veterinary Therapy. VI. Small Anim. Pract.*, pp. 1255–1259. W.B. Saunders, Philadelphia, London, Toronto.
56. Leiniger, S.R., Chabot, J.F. and Westcott, W. (1971) Balanoposthitis. In: *Current Veterinary Therapy. IV. Small Anim. Pract.*, pp. 741–742. W.B. Saunders, Philadelphia, London, Toronto.
57. Lipowitz, A.J., Schwartz, A., Wilson, G.P. and Ebert, J.W. (1973) Testicular neoplasms and concomitant clinical changes in the dog. *J. Amer. vet. med. Assoc.*, **163**, 1364–1368.
58. Macpherson, J.W. and Penner, P. (1967) Canine reproduction. I. Reaction of animals to collection of semen and insemination procedures. *Canad. J. comp. Med.*, **31**, 62–64.
59. McKeever, S. (1970) Male reproductive organs. In: *Reproduction and Breeding Techniques for Laboratory Animals*, ed. Hafez, E.S.E., Lea and Febiger, Philadelphia. Chapter 2.
60. Nales, N.G. (1957) Dilution and storage of canine semen. *Rev. Patron. Biol. Anim.*, **3**, 189–236. (*Anim. Breed. Abstr.* 1958, **26**, 197, no. 984).
61. O'Shea, J.D. (1963) Studies on the canine prostate gland. II. Prostatic neoplasms. *J. comp. Pathol.*, **73**, 224–252.
62. Peters, H. (1943) Zur Technik der Spermauntersuchung beim Hunde. *Berl. Münch. tierärztl. Wochenschr. and Wien tierärztl. Monatsschr.*, 253–258.

63. Power, J.H. (1962) A study of canine semen: physical factors which may affect the electrolyte concentration in the ejaculate. *Irish vet. J.*, **16**, 226–231.
64. Power, J.H. (1964) A study of canine semen: the relationship between chloride content and ejaculate volume. *Irish vet. J.*, **18**, 41–48.
65. Pullig, T. (1953) Cryptorchidism in cocker spaniels. *J. Hered.*, **44**, 250 and 264.
66. Reif, J.S. and Brodey, R.S. (1969) The relationship between cryptorchidism and canine testicular neoplasia. *J. Amer. vet. med. Assoc.*, **155**, 2005–2010.
67. Reif, J.S., Maguire, T.G., Kenney, R.M. and Brodey, R.S. (1979) A cohort study of canine testicular neoplasia. *J. Amer. vet. med. Assoc.*, **175**, 719–723.
68. Rhoades, J.D. and Foley, C.W. (1977) Cryptorchidism and intersexuality. *Vet. Clin. N. Amer.*, **7**, 789–794.
69. Rowley, M.J. and Heller, C.G. (1966) The testicular biopsy: surgical procedure, fixation, and staining technics. *Fertil. and Steril.*, **17**, 177–186.
70. Salutini, E., Ghilarducci, P. and Biagi, G. (1977) Radioimmunoassay for plasma testosterone in young male dogs. *Ann. Fac. Med. vet., Torino*, **30**, 245–275.
71. Schefels, W. (1969) Die Spermagewinnung beim Rüden mit Hilfe eines Vibrators. *Dtsch. tierärztl. Wochenschr.*, **76**, 289–290.
72. Schlotthauer, C.F. (1937) Diseases of the prostate gland in the dog. *J. Amer. vet. med. Assoc.*, **90**, (new ser. **43**), 176–187.
73. Schuberth, B., Schuberth, G. and Weiger, G. (1978) Zur Therapie der Prostatahypertrophie des Hundes mit Medroxyprogesteron. *Kleintier-Prax.*, **23**, 331–332.
74. Seager, S.W.J. (1977) Semen collection and artificial insemination of dogs. In: *Current Veterinary Therapy. VI. Small Anim. Pract.*, pp. 1245–1251. W.B. Saunders, Philadelphia, London, Toronto.
75. Seager, S.W.J. and Fletcher, W.S. (1973) Progress on the use of frozen semen in the dog. *Vet. Rec.*, **92**, 6–10.
76. Seager, S.W.J. and Platz, C.C. (1977) Collection and evaluation of canine semen. *Vet. Clin. N. Amer.*, **7**, 765–773.
77. Seager, S.W.J. and Platz, C.C. (1978) Semen collection in the inexperienced male dog. *Appl. Anim. Ethol.*, **4**, 91–96.
78. Siegel, E.T. (1977) *Endocrine Diseases of the Dog*. Lea and Febiger, Philadelphia.
79. Sittmann, K. (1980) Cryptorchidism in dogs: Genetic assessment of published data. *9th Int. Congr. Anim. Reprod.*, Madrid, Proc. **vol. III**. Symposia, 247.
80. Taha, M.B., Noakes, D.E. and Allen, W.E. (1981) The effect of season of the year on the characteristics and composition of dog semen. *J. small Anim. Pract.*, **22**, 177–184.
81. Takeishi, M., Toyoshima, T., Ryo, T., Takematsu, S., Miki, H. and Tsunekane, T. (1975) Studies on the reproduction of the dog. VI. Sexual maturity of male beagles. *Bull. Coll. Agric. Vet. Med.*, Nihon Univ., **32**, 213–223.
82. Takeishi, M., Iwaki, T., Ando, Y., Hasegawa, M. and Tsunekane, T. (1975) Studies on the reproduction of the dog. VII. The seasonal characteristics of the dog semen. *Bull. Coll. Agric. Vet. Med.*, Nihon Univ., **32**, 224–231.
83. Takeishi, M., Tanaka, N., Imazeki, S., Kodama, M., Tsumagari, S., Shibata, M. and Tsunekane, T. (1980) Studies on the reproduction of the dog. XII. Changes in serum testosterone level and acid phosphatase activity in the seminal plasma of sexually mature male beagles. *Bull. Coll. Agric. Vet. Med.*, Nihon Univ., **37**, 155–158. (*Anim. Breed Abstr.*, 1980, **48**, 903, no. 7526).
84. Taylor, P.A. (1973) Prostatic adenocarcinoma in a dog and a summary of ten cases. *Canad. vet. J.*, **14**, 162–166.
85. Technical Development Committee (1954) Cryptorchidism with special reference to the condition in the dog. *Vet. Rec.*, **66**, 482–483.
86. Tunn, U., Senge, Th., Schenck, B. and Neumann, F. (1979) Biochemical and histological studies on prostates in castrated dogs after treatment with androstanediol, oestradiol and cyproterone acetate. *Acta Endocrinol.*, **91**, 373–384.

87. Tunn, U., Senge, T., Schenck, B. and Neumann, F. (1980) Effects of cyproterone acetate on experimentally induced canine prostatic hyperplasia. *Urol. int. (Basel)*, **35**, 125–140.
88. Wales, R.G. and White, I.G. (1965) Some observations on the chemistry of dog semen. *J. Reprod. Fertil.*, **9**, 69–77.
89. Walton, A. (1933) *The Technique of Artificial Insemination.* Imperial Bureaux of Animal Genetics. Oliver and Boyd, Edinburgh (cited in Harrop, A.E.[40]).
90. Weaver, A.D. (1977) Transperineal punch biopsy of the canine prostate gland. *J. small Animal. Pract.*, **18**, 573–577.
91. Weaver, A.D. (1983) Survey with follow-up of 67 dogs with testicular Sertoli cell tumours. *Vet. Rec.*, **113**, 105–107.
92. Weiss, E. (1962) Das Prostatakarzinom des Hundes. *Berl. Münch. tierärztl. Wochenschr.*, **75**, 145–150.
93. Wheaton, L.G., de Klerk, D.P., Strandberg, J.D. and Coffey, D.S. (1979) Relationship of seminal volume to size and disease of the prostate in the Beagle. *Amer. J. vet. Res.*, **40**, 1325–1328.

ADDITIONAL READING

Allison, R.M.M., Watson, A.D.J. and Church, D.B. (1983) Pituitary tumour causing neurological and endocrine disturbances in a dog. *J. small Anim. Pract.*, **24**, 229–236.

Balke, J. (1981) Persistent penile frenulum in a cocker spaniel. *Vet. Med./Small Anim. Clin.*, **76**, 988–990.

Baumans, V. and Wensing, C.J.G. (1982) The role of the testis in the regulation of the descensus testis in the dog. (122nd Meet. Anat. Ass.) *Acta morphol. neerl.-scand.*, **20**, 264.

Begg, T.B. (1963) Persistent penile frenulum in the dog. *Vet. Rec.*, **75**, 930–931.

Berigaud, A. and Maynard, Ph. (1979) Torsion testiculaire chez un chien. *Point vét.*, **9**, 25–27.

Boyden, T.W., Pamenter, R.W. and Silvert, M.A. (1980) Testosterone secretion by the isolated canine testis after controlled infusions of hCG. *J. Reprod. Fertil.*, **59**, 25–30.

Carmignani, G., Belgrano, E., Puppo, P. and Bentivoglio, G. (1979) Hodenautotransplantation. Eine experimentelle Untersuchung am Hund. *Akt. Urol.*, **10**, 321–327.

Cochran, R.C., Ewing, L.L. and Niswender, G.D. (1981) Serum levels of follicle stimulating hormone, luteinizing hormone, prolactin, testosterone, 5α-dihydrotestosterone, 5α-androstane-3α, 17β-diol, 5α-androstane-3β, 17β-diol, and 17β-estradiol from male beagles with spontaneous or induced benign prostatic hyperplasia. *Invest. Urol.*, **19**, 142–147.

Copland, M.D. and Maclachlan, N.J. (1976) Aplasia of the epididymis and vas deferens in the dog. *J. small Anim. Pract.*, **17**, 443–449.

Dixit, V.P., Gupta, R.A. and Kumar, S. (1980) Effects of alloxan diabetes on testicular function of male dog, *Canis familiaris. Ind. J. exp. Biol.*, **18**, 459–462.

Folman, Y., Haltmeyer, G.C. and Eik-Nes, K.B. (1972) Production and secretion of 5α-dihydrotestosterone by the dog testis. *Amer. J. Physiol.*, **222**, 653–656.

Froman, D.P. and Amann, R.P. (1983) Inhibition of motility of bovine, canine and equine spermatozoa by artificial vaginal lubricants. *Theriogenology*, **20**, 357–361.

Günzel, A.-R. and Krause, D. (1981) Einfluss verschiedener Entnahmefrequenzen auf Spermamerkmale von Beagle-Rüden. *Zuchthygiene*, **16**, 70–71.

Hart, B.J. and Ladewig, J. (1980) Accelerated and enchanced testosterone secretion in juvenile male dogs following medial preoptic-anterior hypothalamic lesions. *Neuroendocrinology*, **30**, 20–24.

Heywood, R. and Sortwell, R.J. (1971) Semen evaluation in the beagle dog. *J. small Anim. Pract.*, **12**, 343–346.

Hrudka, F. and Post, K. (1982) Leukophages in canine semen. *Andrologia*, **15**, 26–33.

Isaacs, J.T., Isaacs, W.B., Wheaton, L.G. and Coffey, D.S. (1980) Differential effects of estrogen treatment on canine seminal plasma components. *Invest. Urol.*, **17**, 495–498.

Joshua, J.O. (1975) 'Dog pox': some clinical aspects of an eruptive condition of certain mucous surfaces in dogs. *Vet. Rec.*, **96**, 300–302.

Kelly, D.F., Long, S.E. and Strohmenger, G.D. (1976) Testicular neoplasia in an intersex dog. *J. small Anim. Pract.*, **17**, 247–253.

Kibble, R.M. (1978) Guidelines for semen collection and artificial insemination in the dog. *Aust. vet. Pract.*, **8**, 141–144.

Kirby, F.D. (1980) A technique for castrating the cryptorchid dog or cat. *Vet. Med./small Anim. Clin.*, **75**, 632.

Lindberg, R., Jonsson, O.-J. and Kasström, H. (1976) Sertoli cell tumours associated with feminization, prostatitis and squamous metaplasia of the renal tubular epithelium in a dog. *J. small Anim. Pract.*, **17**, 451–458.

McLain, D.L. (1982) Surgical treatment of perineal prostatic abscesses. *J. Amer. Anim. Hosp. Assoc.*, **19**, 794–798.

Mialot, J.-P. (1980) Examen de l'appareil génital du chien. *Point Vét.*, **10**, 65–72.

Miller, T.B., Smith, E.R. and Pebler, R. (1980) Prostatic fluid secretion in the dog. *Fed. Proc.*, **39**, 378, abstract no. 574.

Morris, B.J. (1983) Fetal bone marrow depression as a result of Sertoli cell tumor. *Vet. Med./small Anim. Clin.*, **78**, 1070–1072.

Nitschke, Th. (1966) Zur Frage der Vena profunda glandis des Rüden. *Zbl. Vet.-Med. A*, **13**, 474–476.

Olar, T.T., Amann, R.P. and Pickett, B.W. (1981) Relationship between total scrotal width and daily sperm output in dogs. *J. Anim. Sci.*, **53**, Suppl. 1, p. 354.

Panel Report (1976) Cryptorchidism in dogs. *Mod. vet. Pract.*, **57** (2), 137–138.

Ryer, K.A. (1979) Persistent penile frenulum in a cocker spaniel. *Vet. Med./small Anim. Clin.*, **74**, 688.

Schörner, G. (1978) Zuchttauglichkeitsuntersuchung beim Rüden. *Kleintier-Prax.*, **23**, 329–330.

Stubbs, A.J. and Resnick, M.I. (1978) Protein electrophoretic patterns of canine prostatic fluid. *Invest. Urol.*, **16**, 175–178.

Taha, M.B. and Noakes, D.E. (1982) The effect of age and season of the year on testicular function in the dog, as determined by histological examination of the seminiferous tubules and the estimation of peripheral plasma testosterone concentrations. *J. small Anim. Pract.*, **23**, 351–357.

Taha, M.B., Noakes, D.E. and Allen, W.E. (1981) Some aspects of reproductive function in the male Beagle at puberty. *J. small Anim. Pract.*, **22**, 663–667.

Teunissen, G.H.B. and van der Horst, C.J.G. (1968) Investigations into the occurrence of steroids in abnormal canine testicles. *T. Diergeneesk.*, **93**, 1237–1249.

Treu, H., Reetz, I., Wegner, W. and Krause, D. (1976) Andrologische Befunde in einer Merlezucht. *Zuchthygiene*, **11**, 49–61.

Turk, J.R., Turk, M.A.M. and Gallina, A.M. (1981) A canine testicular tumor resembling gonadoblastoma. *Vet. Pathol.*, **18**, 201–207.

Wales, R.G. and White, I.G. (1958) The interaction of pH, tonicity, and electrolyte concentration on the motility of dog spermatozoa. *J. Physiol. (Lond.)*, **141**, 273–280.

Weiss, E. and el Etreby, M.F. (1966) Zellkernmorphologische Geschlechtsbestimmung an der normalen und veränderten Prostata des Hundes. *Zbl. Vet.-Med. A*, **13**, 283–288.

Weller, R.E. and Palmer, B. (1983) Metastatic seminoma in a dog. *Mod. vet. Pract.*, **64**(4), 275–278.

Wildt, D.E., Baas, E.J., Chakraborty, P.K., Wolfle, T.L. and Stewart, A.P. (1982) Influence of inbreeding on reproductive performance, ejaculate quality and testicular volume in the dog. *Theriogenology*, **17**, 445–452.

Wolff, A. (1981) Castration, cryptorchidism and cryptorchidectomy in dogs and cats. *Vet. Med./small Anim. Clin.*, **76**, 1739–1741.

Young, A.C.B. (1979) Two cases of intrascrotal torsion of a normal testicle. *J. small Anim. Pract.*, **20**, 229–231.

Chapter 5
Infertility and Hormone Treatment in the Male

INFERTILITY

In dogs with a history of reduced fertility or sterility, a clinical and andrological examination should be carried out as described in Chapter 4. Infertility in male dogs with no causal anatomical, infectious or neoplastic conditions may be classified according to the history and the findings of the andrological examination as follows.[1]

Primary infertility with azoospermia

This occurs in young, clinically normal dogs with normal libido. The testes are small and soft. The ejaculate is watery and opalescent, containing no spermatozoa, but only epithelial cells and a few erythrocytes and leukocytes.

Primary infertility with abnormal spermatozoa

This is found in young dogs after a few infertile matings. The dog is apparently in good health and clinically normal but there is an increased number of abnormal spermatozoa. The type of abnormality varies from case to case and the aetiology is probably complex.

Permanently lowered fertility with abnormal spermatozoa

Here the male dog has a lowered fertility. The clinical condition is normal, but the testes are small, the number of spermatozoa and their motility are low, and there is often an increased number of immature spermatozoa and spermatozoa with coiled tails.

Primary fertility with increasing azoospermia

In these cases the dog has been fertile, but recent matings have been infertile. The dog is in good physical condition, the reproductive organs are normal, but one or both testes have atrophied within 1–2 months. Spermatozoa are few or absent. Where there is obvious testicular atrophy the condition is permanent.

Lowered fertility due to a decreased number of ejaculated spermatozoa and sometimes lowered libido may be caused by hypothyroidism. Lowered numbers of spermatozoa and testicular atrophy occur in about 30% of dogs with canine Cushing's syndrome, where the secretion of gonadotropin is suppressed by the high level of serum cortisol.[16] Estimation of protein-bound iodine may be of diagnostic aid in canine hypothyroidism.[3]

Primary fertility with an increasing number of abnormal spermatozoa

The physical condition of the dog appears to be normal as do the reproductive organs. In the ejaculate the concentration of spermatozoa is normal or reduced, motility is reduced, and the number of abnormal spermatozoa, such as immature spermatozoa and those with coiled tails or acrosomal defects, is increased.

Therapy

In cases of primary infertility, either acquired or congenital, the prognosis for a return to fertility is pessimistic.

If the infertility is of hereditary origin the owner should be advised not to use the dog for breeding, and in dogs with cryptorchidism castration should be advised. Treatment of more or less permanent infertility should not be seriously considered, and avoided if possible, since a pregnancy may make the problem more serious.

Infections causing, for example, orchitis and epididymitis are usually accompanied by infertility in spite of antibiotic treatment.

Previously fertile dogs may become azoospermic through degeneration of the majority of the seminiferous tubules and this may be revealed by biopsy,[8] but recovery after a period of several months' rest has been described.[5]

Routine examination of stud dogs may reveal abnormalities of spermatogenesis early enough to allow rest to be given while there is still a chance of complete recovery.

Hormones

Follicle-stimulating hormone (FSH) or luteinizing hormone (LH)

In cases where clinical examination does not reveal any cause for the infertility, gonadotropins, FSH or LH, in a dose of 100–500 IU, have been used for a period of 1 month without improvement,[9] although in 2 cases which were followed by testicular biopsies and histological examination the numbers of spermatocytes and spermatids were increased, but no mature spermatozoa were present.

Human chorionic gonadotropin (HCG) and pregnant mare serum gonadotropin (PMSG)

When administered to puppies as well as adult dogs HCG or PMSG results in increased testosterone values.[15,17] Treatment of normal fertile dogs with HCG increases the volume of the ejaculate, especially the third fraction, but libido is unchanged during the treatment period.[17]

Testosterone

This has been tried in a few dogs in attempts to solve problems of impotence.[5] It has also been used for direct androgenic stimulation of the testes, but without any success because of suppression of the pituitary gonadotropins. The use of testosterone in the dog to produce a rebound effect may be of value, as has been shown in cases of oligospermia in men.[12]

Testosterone phenylpropionate injected into normal fertile dogs increases the volume of the ejaculate by increasing the volume of the third fraction, but it has no effect on libido.[17]

Synthetic androgens

The use of synthetic androgens has not been reported in dogs, but they have been found to be of value in treating oligospermia in men.

Chlomyphen

Although used for the treatment of oligospermia in men, this has not yet been investigated in the dog.

Releasing hormones

When administered to adult dogs they increase plasma testosterone levels,[15] but no information is available on increased fertility following this treatment.

HORMONE TREATMENT

Various abnormalities of reproduction may indicate treatment with sex hormones.

Exaggerated sexual behaviour

In this condition various symptoms such as excessive sexual interest in bitches in oestrus, sexual activity towards other males, frequent licking of the genitalia, mounting humans or other animal species, restlessness, aggressiveness, and even epileptiform seizures may be seen.[6,11,14]

A complete history of the frequency and duration of symptoms, from the beginning of the condition, should be obtained and a careful clinical examination performed.

Treatment of this problem and its rationale are poorly understood, but hereditary disposition, hormonal imbalance, inadequate adaptation to other animals and mental disorders may be aetiologically associated.

Therapy
Therapy consists of castration (see Chapter 7), administration of tranquillizers or hormones.

Progestogens

Medroxyprogesterone acetate (MAP)   This reduces abnormal sexual behaviour in castrated dogs.[10] The dose is 2–4 mg/kg subcutaneously, and may have to be repeated every 6–12 months.[13]

Used in a dose of 4 mg/kg it does not influence testicle size, testicular consistency or any semen character, but reduces the mean testosterone level by 58%.[18]

Megestrol acetate (MA)   If MA is administered orally in a daily dose of 2–4 mg/kg, reduced after an 8-day interval to 0.5–1 mg/kg daily, then this latter dose given once or twice a week is effective for maintenance therapy.[13]

Methyloestrenolene (MOE)   When used orally in a dose of 1–10 mg/kg per day for 5–7 days or 6 mg/kg subcutaneously once this has been shown to be effective with only a slight tendency to relapse.[2,7]

Chlormadinone acetate (CAP) and Δ¹-chlormadinone acetate   These are effective in most animals against hypersexuality, aggressiveness and epileptiform seizures when administered intramuscularly or subcutaneously at intervals of 3–4 weeks, and against hypersexuality and behavioural disturbances when administered at intervals of 3–4 months.[6]

Delmadinone acetate (DMA)   This can also be of use for the treatment of hypersexuality. Libido is reduced within 2–5 days of the beginning of treatment,[17] as is the total volume of the ejaculate; this is due mainly to a reduction in the third fraction and, in combination with a lower sperm concentration, results in a smaller sperm output.

REFERENCES

1. Bane, A. (1970) Sterility in male dogs. *Nord. Vet.-Med.,* **22**, 561–566.
2. Brass, W. and Horzinek, I. (1971) Erfahrungen mit Δ¹-Chlormadinonacetat bei der Epilepsie des Hundes. *Dtsch. tierärztl. Wochenschr.,* **78**, 302–304.
3. Bullock, L. (1970) Protein bound iodine determination as a diagnostic aid for canine hypothyroidism. *J. Amer. vet. med. Assoc.,* **156**, 892–900.
4. Engle, J.B. and Macbrayer, R. (1940) Impotence in the dog: its treatment with male hormone substance. *J. Amer. vet. med. Assoc.,* **97**, 158–161.

5. Evans, J. and Renton, J.P. (1973) A case of azoospermia in a previously fertile dog with subsequent recovery. *Vet. Rec.*, **92**, 198–199.
6. Ficus, H.J. and Jöchle, W. (1970) Erfahrungen mit Antiandrogenen (Δ¹-chlormadinonazetat und Chlormadinonazetat) beim Rüden. *Kleintier-Prax.*, **15**, 32–39.
7. Gerber, H.A. and Sulman, F.G. (1964) The effect of methyloestrenolone on oestrus, pseudopregnancy, vagrancy, satyriasis and squirting in dogs and cats. *Vet. Rec.*, **76**, 1089–1093.
8. Hadley, J.C. (1972) Spermatogenic arrest with azoospermia in two Welsh springer spaniels. *J. small Anim. Pract.*, **13**, 135–138.
9. Harrop, A.E. (1966) The infertile male dog. *J. small Anim. Pract.*, **7**, 723–725.
10. Hart, B.L. (1979) Indications for progestin therapy for problem behavior in dogs. *Canine Pract.*, **6**(5), 10–12.
11. Holzmann, A. (1975) Hypersexualität beim Rüden. *Wien. tierärztl. Monatsschr.*, **62**, 355–356.
12. Larsen, R.E. (1977) Evaluation of fertility problems in the male dog. *Vet. Clin. N. Amer.*, **7**, 735–745.
13. Pemberton, P.L. (1976) Use of progestagens in certain behavioural abnormalities in dogs and cats. *Refresher course for veterinarians*, Proc. No. 30, University of Sydney.
14. Schörner, G. (1978) Hypersexualität beim Rüden. *Kleintier-Prax.*, **23**, 271–272.
15. Schörner, G., Choi, H.S. and Bamberg, E. (1977) Plasmatestosterongehalt beim Rüden nach Behandlung mit PMSG, HCG und LH-RH. *Wien. tierärztl. Monatsschr.*, **64**, 153–156.
16. Siegel, E.T. (1977) *Endocrine Diseases of the Dog*. Lea and Febiger, Philadelphia.
17. Taha, M.B., Noakes, D.E. and Allen, W.E. (1981) The effect of some exogenous hormones on seminal characteristics, libido and peripheral plasma testosterone concentrations in the male beagle. *J. small Anim. Pract.*, **22**, 587–595.
18. Wright, P.J., Stelmasiak, T., Black, D. and Sykes, D. (1979) Medroxyprogesterone acetate and reproductive processes in male dogs. *Aust. vet. J.*, **55**, 437–438.

ADDITIONAL READING

Allen, W.E. and Longstaffe, J.A. (1982) Spermatogenic arrest associated with focal degenerative orchitis in related dogs. *J. small Anim. Pract.*, **23**, 337–343.
Beaver, B.V. (1982) Hormone therapy for animals with behavior problems. *Vet. Med./small Anim. Clin.*, **77**, 337–338.
Campbell, W.E. (1975) Mounting and other sex-related problems. *Mod. vet. Pract.*, **56**, 6, 420–422.
Flashman, A.F. (1967) Behavioural abnormalities of the dog. (AVA conference, Melbourne, 1967). *Aust vet. J.*, **43**, 524–529.
Heger, C., May, J. and Mechow, A. (1979) Sexuelle Ruhigstellung von Rüden durch den Zirbeldrüsenextrakt Epigen®. *Prakt. Tierarzt*, **8**, 666–668,
Ibach, B., Passia, D., Weissbach, L. and Goslar, H.G. (1978) The effect of low-dose HCG on the testis of prepubertal dogs. *Int. J. Androl.*, **1**, 509–522.
Meier, K.U. (1968) Klinisch-chemische Befunde in Blut und Liquor bei Hunden mit epileptiformen Krämpfen. *Kleintier-Prax.*, **13**, 135–138.
Meinecke, B. (1976) Retrograde Ejakulation beim Rüden. *Zuchthygiene*, **11**, 122–123.
Poduschka, W. (1978) Spontanejakulationen bei einem Cockerrüden. *Säugetierkundliche Mitteilungen*, **26**, 173–177.
Sandelien, H. (1981) Behandling av epilepsi hos hund. *Norsk Vet.-T.*, **93**, 6.
Worden, A.N. (1959) Abnormal behaviour in the dog and cat. *Vet. Rec.*, **71**, 966–981.

Chapter 6
Artificial Breeding and Embryo Transfer

ARTIFICIAL BREEDING

Various factors may cause dog owners to request artificial insemination with either fresh or deepfrozen semen, but if there is any risk or possibility of spreading hereditary diseases it should not be used. Intensive utilization of the semen as in cattle breeding is not possible.

In a normal ejaculate there are on average, 500 million spermatozoa, and at least 200 million motile spermatozoa are needed to obtain optimal results after artificial insemination using fresh semen. When semen is frozen, the number of motile spermatozoa is lowered during the process of freezing and thawing. More semen, and thus more inseminations, are therefore needed for normal fertility.

Several surveys of the results of preservation of semen and insemination with fresh and frozen semen are available in the literature.[11,14,19,24,37,42,43]

Indications for insemination

Bitch
1. Various congenital or acquired abnormalities.
2. A narrow vagina in young bitches resulting in pain and even refusal to mate in spite of normal maturity and heat.
3. Scar tissue from a previous difficult whelping, polyps or an unperforated hymen, which may make intromission of the penis impossible or painful.
4. Lowered fertility caused by chronic vaginitis.[3]
5. A nervous bitch or one with behavioural problems that refuses to mate or acts aggressively towards the male although in heat.
6. In cases where a bitch and a male dog are housed together, but kept apart during several oestrous periods, there may be problems when they are to be mated.[26]

Dog
1. Mating difficulties occurring in spite of normal spermatogenesis and libido.
   (a) Stiffness of the hindquarters and weak hindlegs. This may occur especially in young males or smaller breeds with extremely short legs.

(b) Inbreeding and degeneracy.
(c) Too early erection of the penis, making intromission impossible because of the swelling of the bulbus glandis.[25]
2. Import and quarantine regulations may be complied with more easily and cheaply by the use of deepfrozen semen.
3. The spread of infectious diseases may be prevented by using artificial inseminaton; this applies to the spread of genital infection as well as other more generalized diseases which can be spread via semen.
4. The risk of exposing a valuable stud dog to injuries and infectious diseases during breeding can be avoided.
5. Champion sires can be used more extensively and utilized better.
6. Champion sires can be used after their death.

Some of the above indications are of theoretical interest only, and in such cases insemination should not be performed.

Handling of semen

Fresh semen
Fresh semen for insemination can be used undiluted or diluted.

Undiluted fresh semen
After cooling the ejaculate in a water-bath and storage at 5–10 °C the spermatozoa can survive in the undiluted ejaculate for a short time, up to a maximum of 21 h.

Diluted fresh semen
If the semen cannot be used shortly after collection the survival of the spermatozoa can be improved by diluting the ejaculate.[15-17] The dilution rate varies from 1:3 to 1:8 depending on the initial concentration of spermatozoa.

After dilution semen may be stored for several hours without any serious decrease in fertility. The same rates of pregnancy have been found in bitches inseminated with diluted semen used immediately after collection and those inseminated with semen diluted and stored for 24 h at 5 °C.[21] Diluted semen shipped from England to New Zealand and used for insemination 2 days after collection resulted in pregnancy.[22] The maximum time for storage of diluted semen is 4–5 days. The survival of the spermatozoa depends on the diluent or extender used as well as the time of storage.[39]

Various extenders, if necessary with antibiotics added, can be used for dilution of semen:

1. Skimmed milk after heating to 92–94 °C for 10 min.
2. Sterile homogenized milk containing 2% fat.

3. Sodium citrate solution (2.9%) 80% and egg yolk 20%.
4. IVT (Illinois Variable Temperature) extender.
5. TRIS (hydroxymethyl-aminomethane)-egg extender with 8% glycerol and 20% egg yolk.

Insemination technique
The insemination equipment consists of a plastic bovine insemination tube of half the usual length attached to a disposable syringe with a short rubber connector, a vaginoscope, and lubricant (Figure 6.1).

Saline solution is used for washing the vulval region, which is then dried. Disinfectant should not be used since it may interfere with the viability of the spermatozoa.

The bitch is placed on a convenient table with a non-slippery surface and the vulva is exposed by drawing the tail to one side. A large bitch can be allowed to stand on the floor with the hindquarters elevated at an angle of about 60° and the head and the neck fixed between the knees of an assistant who exposes the vulva by spreading the hindlegs slightly and drawing the tail to one side.

Before slow aspiration of the semen 1 ml of air is drawn into the syringe, making expulsion of the total amount of aspirated semen possible. A small amount of the lubricating jelly is applied to the insemination tube, which is inserted into the vagina through a vaginoscope or along a lubricated index finger inserted into the vagina dorsally over the brim of the pelvis. When the rod is clear of the pelvic brim, it is tipped and gently pushed to the cervix where the semen is deposited slowly. If the bitch is standing, this final direction

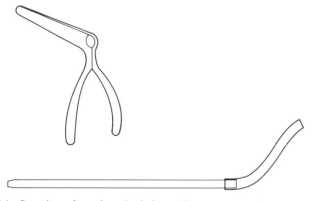

Figure 6.1 Speculum of metal or plastic for vaginoscopy and a plastic bovine insemination tube of half the normal length, i.e. 17 cm, with a diameter of 6 mm for intravaginal insemination. The inseminator is fitted with a rubber tube for connection to the syringe containing the semen. (From M. Schmidt, 1983)

is horizontal, whereas it is nearly vertical in a bitch with elevated hindquarters. The cervical area is about 12–14 cm from the vulva in medium-sized breeds, ± 4 cm in large and small breeds. After gently withdrawing the tube the hindquarters should be elevated for 4–6 min to prevent the semen flowing out; this may also be done by inserting a finger into the vagina. When handling the bitch afterwards, urination should be prevented and elevation of the fore-end avoided for some time.

According to Knaus[27] deposition of semen in the cervix of the uterus is to be preferred because of the presence of micro-organisms in the vagina.

Only the second fraction of the ejaculate, diluted or undiluted, is normally used for insemination. The dose depends on the size of the bitch, and ranges from 1.5 to 10 ml.[13] The third fraction of the ejaculate has been used successfully as a diluent to make up the optimal volume.[33] The number of motile spermatozoa required in fresh semen meant for insemination ranges from 60×10^6 [13] to the optimum of 200×10^6.[21] Thus a pregnancy rate of 20% has been obtained by using 50×10^6 motile spermatozoa in diluted semen, while insemination with 200×10^6 motile spermatozoa resulted in a pregnancy rate of 80%.

Deepfrozen semen
If the semen is frozen, it may be stored for years without loss of fertility; pregnancy has been obtained after insemination with semen deepfrozen for 35 months.[41] The same conception rates may be achieved after insemination with either fresh or deepfrozen semen. Only a slight decrease in motility occurred after storage at -196 °C for 8 years.[45]

It is essential to collect the ejaculate in fractions, because the first harms the spermatozoa during the freezing procedure[18] and the prostatic secretion is unnecessary.[41]

The semen fraction is diluted with either TRIS-egg extender containing 8% glycerol and 20% egg yolk;[7] TRIS-egg extender with 20% egg yolk combined with TRIS-egg extender containing 16% glycerol and 20% egg yolk,[12] egg yolk–lactose–glycerol extender containing 11% lactose, 4% glycerol and 20% egg yolk;[40] or Triladyl concentrate with similar sperm preservation qualities.[36] Glycerol and dimethyl sulphoxide (DMSO) both possess a cryoprotective quality, but that of glycerol is better than DMSO. Glycerol should be added to the semen at 5 °C rather than at 25 °C.[38]

Antibiotics may be added to the extender, for example 0.1 g/100 ml dihydrostreptomycin,[2] or penicillin and streptomycin.[40]

The rate of dilution, 1:3 up to 1:8, depends on the concentration; there should be 100–150 million motile spermatozoa per ml after dilution.

The semen can be frozen in ampoules,[16,31,32] in straws[2,4-9,12,46] or in pellets.[20,23,30,38,40-42,45,48] Deepfrozen semen can be identified most reliably by marking the initials of the dog on the straws.

Deepfreezing in straws (paillettes)
Dilution is performed in two steps.[12] The first is made with TRIS–fructose–citrate with 20% egg yolk and no glycerol, and a dilution rate of 1:1 or 1:2 depending on the concentration of the ejaculate. After equilibration for 60–120 min at 5 °C a second dilution is performed with TRIS–fructose–citrate with 20% egg yolk and 16% glycerol, resulting in a final dilution of 1:3–1:5.

After the semen has been drawn into straws (medium or mini, each containing 0.5 ml and 0.25 ml, respectively), it is further equilibrated for 60 min at 5 °C and the freezing procedure carried out in a box made of 2.5 cm thick polystyrene with the straws placed 4–6 cm above an 8 cm deep layer of liquid nitrogen. After 10–15 min the straws are transferred direct to the liquid nitrogen.

Deepfreezing in pellets
The method described by Seager et al.[45] involves an initial dilution of 1:1 with egg yolk–lactose–glycerol extender at room temperature within 20 min after collection. After equilibration on an aliquot shaker for 60 min at 5 °C a second dilution is performed followed by further equilibration for 30 min. Freezing is carried out by placing the sample, in single droplets, in holes pressed into a block of solid carbon dioxide.

Insemination technique
It is essential that thawing should be rapid. Pellets thawed in 2.5 ml of 37 °C physiological saline have shown the best recovery rate.[45] Similarly straws have been thawed successfully in water at 37 °C for 30 s,[12] but quick thawing at a high temperature is preferred by other investigators.[1] Good results are reported after thawing medium straws for 6.5 s[7] and mini straws for 5 s in 75 °C water.[1]

The dose of semen should be 1.5 ml containing 150 million spermatozoa, of which at least 50–70% must be motile. This is in accordance with Anderson[6] and Seager et al.[45]

It is important to inseminate immediately after thawing the semen because the survival time of spermatozoa in vitro after freezing and thawing is decreased. As the survival time in vivo is not known, daily inseminations for 3–4 days are recommended.[9]

The same equipment can be used as for insemination with fresh semen; the semen is deposited in the vagina with the hindquarters of the bitch elevated at an angle of 60° during insemination and for the first 5 min afterwards.[45] It has also been suggested that the best results are achieved after intrauterine insemination.[8] However this

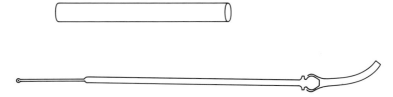

Figure 6.2   Plastic vaginoscope and metal inseminator for intrauterine insemination. The diameter of the vaginoscope is 15 mm and the length depends on the size of the bitch. The inseminator consists of a part 5 cm long, 2 mm diameter for introduction into the cervix and uterus, and a part 10, 15 or 25 cm long, 3 mm diameter, depending on the size of the bitch. The inseminator is fitted with a rubber tube for connection to the syringe containing the semen. (From M. Schmidt, 1983)

procedure is difficult because of the dorsal fold in the anterior part of vagina.[8,34,35] This pseudocervix makes the inspection of the cervix and introduction of the insemination tube difficult, and a special metal inseminator (Figure 6.2) may be used with a plastic speculum.[7,8] The speculum is passed along the dorsal wall of vagina until it reaches the posterior part of cervix, which is fixed by digital pressure through the abdominal wall. The catheter is then passed through the speculum and introduced, via the external orifice and the short cervical canal, into the corpus uteri where the semen dose is deposited. The procedure is performed by digital control through the abdominal wall. In some nervous and excited bitches it may be necessary to use tranquillizers when introducing the catheter.[9] An inseminator has been devised which fixes the cervix when correctly placed by manual manipulation through the abdominal wall.[34]

Results

It is reported that insemination with deepfrozen semen results in the same conception rate as for fresh semen, but at a lower rate than with natural mating.[45] Normal litters are born with puppies having the same viability, birthweight, sex distribution and frequency of abnormalities as with natural mating. When using deepfrozen semen the litter size varies from 1 to 9 puppies.[8] Fifth generation frozen-semen puppies have been obtained, as have pregnancies from semen stored in liquid nitrogen for a period of 12 h to 5 years.[44]

EMBRYO TRANSFER

Embryo transfer has succeeded in dogs with collection and transfer made during natural cycles.[28,29] Synchronization was within ± 4 days based on the first day of standing heat, and 72 embryos were recovered in 26 collections. In 7 operations 37 embryos were transferred resulting in 3 pregnancies and 4 offspring.

Embryo transfer has also been performed in bitches during induced oestrus 9 days after ovulation.[47] The recovery rate of ova was only 5.1%, but the number of puppies born corresponded to the number of eggs transferred.

A surgical method for collecting embryos from anaesthetized bitches after induction of oestrus and ovulations with gonadotropins is described by Archbald.[10] A number of embryos (range 4–23) were recovered in 6 of 10 bitches treated with PMSG and HCG.

REFERENCES

1. Aamdal, J. and Andersen, K. (1968) Fast thawing of semen frozen in straws. *Zuchthygiene*, **3**, 22–24.
2. Andersen, J.B. (1975) Personal communication.
3. Andersen, K. (1969) Artificial insemination in the bitch. *Medlemsbl. norske Vet.-Foren.*, **21**, 482–484.
4. Andersen, K. (1972) Fertility of frozen dog semen. *Acta vet. scand.*, **13**, 128–130.
5. Andersen, K. (1972) Fertility of frozen dog semen. *VII Int. Congr. Anim. Reprod., München*, **2**, 1703–1706.
6. Andersen, K. (1974) Intrauterine insemination with frozen semen in dogs. **12**. *Nord. Vet. Congr., Reykjavik*, 153–154.
7. Andersen, K. (1975) Insemination with frozen dog semen based on a new insemination technique. *Zuchthygiene*, **10**, 1–4.
8. Andersen, K. (1976) Artificial uterine insemination in dogs. *VIII Int. Congr. Anim. Reprod., Krakow*, 960–963.
9. Andersen, K. (1977) *Nordic symposium on artificial insemination in dogs and deepfreezing of dog semen.* Uppsala.
10. Archbald, L.F., Baker, B.A., Clooney, L.L. and Godke, R.A. (1980) A surgical method for collecting canine embryos after induction of estrus and ovulation with exogenous gonadotropins. *Vet. Med./small Anim. Clin.*, **75**, 228–238.
11. Bahlau, E. (1958) *Untersuchungen über die Konservierung von Hundesperma.* Thesis, München.
12. Christiansen, Ib J. and Schmidt, M. (1980) Deep-freezing of dog semen. *Kgl. Vet- og Landbohøjsk., Inst. Sterilitetsforskn., Årsberetn.*, **23**, 69–75.
13. Dubiel. A. (1973) Observations on artificial insemination of dogs. *Med. weteryn.*, **29**, 551–553. (*Anim. Breed. Abstr.*, 1974, **42**, 183–184, no. 1584).
14. Fielden, E.D. (1971) Artificial insemination in the dog. *New Zealand Vet. J.*, **19**, 178–184.
15. Foote, R.H (1964) The effects of electrolytes, sugars, glycerol, and catalase on survival of dog sperm stored in buffered-yolk mediums. *Amer. J. vet. Res*, **104**, 32–36.
16. Foote, R.H. (1964) Extenders for freezing dog semen. *Amer. J. vet. Res.*, **104**, 37–40.
17. Foote, R.H. and Leonard, E.P. (1964) The influence of pH, osmotic pressure, glycine, and glycerol on the survival of dog sperm in buffered-yolk extenders. *Cornell Vet.*, **54**, 78–89.
18. Fougner, J.A. (1977) *Nordic symposium on artificial insemination in dogs and deepfreezing of dog semen.* Uppsala.

19. Fougner, J.A. and Andersen, K. (1979) Artificial insemination with fresh and deepfrozen semen in the bitch – the actual situation in Norway. *Meet. Nord. Kennel Clubs, Oslo.*
20. van Gemert, W. (1970) Diepvries-pups. *T. Diergeneesk.*, **95**, 697–699.
21. Gill, H.P., Kaufman, C.F., Foote, R.H. and Kirk, R.W. (1970) Artificial insemination of beagle bitches with freshly collected, liquid-stored, and frozen-stored semen. *Amer. J. vet. Res.*, **31**, 1807–1813.
22. Harrop, A.E. (1963) Canine artificial insemination between England and New Zealand. *J. small Anim. Pract.*, **4**, 351–353.
23. Heidrich, S. (1977) *Ein Beitrag zur Tiefgefrierung von Rüdensperma.* Thesis, Berlin.
24. Hendrikse, J. (1962) De voortplanting en kunstmatige inseminatie bij honden. *T. Diergeneesk.*, **87**, 1553–1565.
25. Hermansson, K.A. (1935) Artificial insemination of the bitch. *Vet. J.*, **91**, 15–17.
26. Kibble, R.M. (1969) Artificial insemination in dogs. *Aust. vet. J.*, **45**, 194–199.
27. Knaus, E. (1975) Die künstliche Besamung des Hundes bei bestehenden Deckhindernissen. *Wien. tierärztl. Monatsschr.*, **62**, 353–354.
28. Kraemer, D.C., Flow, B.L., Schriver, M.D., Kinney, G.M. and Pennycook, J.W. (1979) Embryo transfer in the nonhuman primate, feline, and canine. *Theriogenology*, **11**, 51–62.
29. Kraemer, D.C., Kinney, G.M. and Schriver, M.D. (1980) Embryo transfer in the domestic canine and feline. *Arch. Androl.*, **5**, 111.
30. Lees, G.E. and Castleberry, M.W. (1977) The use of frozen semen for artificial insemination of German shepherd dogs. *J. Amer. Anim. Hosp. Assoc.*, **13**, 382–386.
31. Martin, I.C.A. (1963) The freezing of dog spermatozoa to −79 °C. *Res. vet. Sci.*, **4**, 304–314.
32. Martin, I.C.A. (1963) The deep-freezing of dog spermatozoa in diluents containing skim-milk. *Res. vet. Sci.*, **4**, 315–325.
33. Miljkovic, V., Pavlovic, D., Mrvos, D., Olujic, M. and Sindelic, V. (1975) Artificial insemination of dogs. *Vet. Glas.*, **29**, 751–755. (*Anim. Breed Abstr.*, 1976, **44**, 254, no. 2351).
34. Obel, N. (1977) *Nordic symposium on artificial insemination in dogs and deepfreezing of dog semen.* Uppsala.
35. Pineda, M.H., Kainer, R.A. and Faulkner, L.C. (1973) Dorsal median postcervical fold in the canine vagina. *Amer. J. vet. Res.*, **34**, 1487–1491.
36. Platz, C.C., Chakraborty, P.K., Wildt, D.E. and Seager, S.W.J. (1978) Comparison of semen extenders for freezing canine semen. *J. Anim. Sci.*, **47**, Suppl. 1., p. 100, Abstr. no. 108.
37. Rohloff, D. (1977) Zum Stand der künstlichen Besamung beim Hund. *Kleintier-Prax.*, **22**, 289–292.
38. Rohloff, D., Laiblin, Ch. and Heidrich, S. (1978) Untersuchungen über die Gefrierschutzwirkung von Glycerin und DMSO bei der Tiefgefrierung von Rüdensperma. *Berl. Münch. tierärztl. Wochenschr.*, **91**, 31–33.
39. RoyChoudhury, P.N. and Dubay M.L. (1974) Observations on dog semen. I. Dilution and conservation in three extenders. *Zootecn. e Vet.*, **29**, 117–120.
40. Seager, S.W.J. (1969) Successful pregnancies utilizing frozen dog semen. *A.I. Digest*, **17**, 6 and 16.
41. Seager, S.W.J. (1976) Freezing and transportation of dog semen. *VIII Int. Congr. Anim. Reprod.*, Krakow, 1251–1252.
42. Seager, S.W.J. and Fletcher, W.S. (1973) Progress on the use of frozen semen in the dog. *Vet. Rec.*, **92**, 6–10.
43. Seager, S.W.J. and Platz, C.C. (1977) Artificial insemination and frozen semen in the dog. *Vet. Clin. N. Amer.*, **7**, 757–764.
44. Seager, S.W.J. and Platz, C.C. (1978) Semen freezing in the dog. *Cryobiology*, **15**, 687, Abstr. no. 23.

45. Seager, S.W.J., Platz, C.C. and Fletcher, W.S. (1975) Conception rates and related data using frozen dog semen. *J. Reprod. Fertil.*, **45**, 189–192.
46. Takeishi, M., Mikami, T., Kodoma, Y., Tsunekane, T. and Iwaki, T. (1976) Studies on the reproduction in the dog. VIII. Artificial insemination using the frozen semen. *Jap. J. Anim. Reprod.*, **22**, 28–33.
47. Takeishi, M., Akai, R., Tsunekane, T., Iwaki, T. and Nakanowatari, K. (1980) Studies on the reproduction in dogs: a trial of ova transplantation in dogs. *Jap. J. Anim. Reprod.*, **26**, 151–153.
48. Vaske, T.R., Moraes, H.F., Romáo, A.R., Blasi, A.C., Perassi, P. and Aun, G.C. (1979) Besamungserfolg mit tiefgefrorenem Samen bei einer Hündin. *Kleintier-Prax.*, **24**, 277–280.

Chapter 7
Limitation of Fertility in the Female and the Male

THE FEMALE

Limitation of fertility might be required for various reasons: by the owner to avoid pregnancy in an individual bitch, because of concern at the growing population of pets and strays or the problems of providing sufficient food, or to avoid pollution and prevent dog bites.[8,42,44,47,60,128,142] Usually one thinks of the female when speaking of limitation of reproductive capacity but restrictions can also be placed on the male.[12,37,69-73,143,148,157]

Fertility can be limited by contraception, i.e. prevention or suppression of oestrus or prevention of implantation, or by termination of pregnancy (Table 7.1).

Contraception

Contraception can be established surgically, physically, chemically, or in other ways and, depending on the method, the result may or may not be reversible.

If the bitch will be required later for breeding, the owner must give serious consideration as to whether contraception should be used at all, by whatever method. Although, at least theoretically, it should be possible to re-establish tubal passage after ligation, and although there is much information about pregnancies in bitches after chemical contraception for either short or long periods, there can be no guarantee that normal fertility can be re-established if requested.

Before contraception is instituted it is of the greatest importance to give all the relevant information to the owner on whether the method proposed is reversible and whether complications may occur. The expenses in connection with surgical and chemical contraception should also be specified. It is unlikely that any single technique will satisfy all dog owners. Some prefer permanent prevention of oestrus, while others require a contraceptive which prevents oestrus for 6–12 months and is reversible, and yet others are willing to administer an oral contraceptive daily. Methods involving incorporation of a contraceptive agent in the food are controversial since dosage cannot be controlled and there is also a risk of human consumption of the food. In many countries there is no free choice between surgical and chemical contraception because of regulations or tradition.

Table 7.1 Methods for limitation of fertility in the bitch.

Contraception	Surgical	Ovariohysterectomy
		Ovariectomy
		Hysterectomy
		Salpingectomy
		Ovarian autograft
	Physical	Intravaginal device
	Chemical	
	prevention of oestrus	Progestogens
		Androgens
		Non-hormonal compounds
	suppression of oestrus	Progestogens
	prevention of implantation	Diethylstilboestrol
		Oestrogens
		Progestogens
	Irradiation	X-ray exposure of ovaries
	Immunological	Immunization against hormones
		Antizona pellucida vaccine
	Asexuality	Androgens/progestogens to puppies
Termination of pregnancy	Surgical	Ovariohysterectomy
		Ovariectomy
	Chemical	Corticosteroids
		Prostaglandins
		Non-prostaglandins

Surgical contraception

This can take the form of ovariectomy, hysterectomy, ovariohysterectomy, tubal ligation, or ovarian autograft possibly combined with
hysterectomy. Removal of the ovaries ends their hormone production
and prevents oestrus, whereas extirpation of the uterus, besides being
contraceptive, excludes the risk of uterine disorders. Ovarian autograft implies a normal production of ovarian hormones and, if used
with hysterectomy for the prevention of uterine disorders, could be a
future method of choice; at present it is still in the experimental stage.

Ovariohysterectomy

This operation is the most effective method for controlling reproduction. It is irreversible and excludes the risk of uterine disorders. However, for emotional or economic reasons it is not often used in

bitches. The operation is fairly safe, but is most hazardous during oestrus and pregnancy and in old obese bitches.

The best time for spaying a mature bitch is 3–4 months after the first or second oestrus, or after weaning when lactation had ceased. Ovariohysterectomy should not be performed until the bitch has had her first oestrous period, or in oestrous bitches, because of the increased vascularity and turgidity of the genital tract. It results in a small but significant gain in weight.[67,76] Total food intake and intake per kg of bodyweight both increase, although not significantly, presumably due to removal of the source of female sex hormones. This is normally the only complication following the operation, but on some occasions haemorrhage, abdominal adhesions and granulomas of the uterine or ovarian stump may occur: these may be limited by proper operative procedure and by the use of absorbable ligatures.[109] Pyometritis and recurrent oestrus may occur because of residual uterine or ovarian tissue; not uncommonly bilateral symmetrical dermatoses in the lumbar region and hindquarters accompanied by scratching and rubbing follow.[92] Temporary urinary incontinence is occasionally seen, but normally resolves spontaneously. Incontinence more often develops later in life becoming gradually more marked and, with it, perivulval dermatitis develops. The condition has been reported to occur more frequently in large than in small breeds, and more often as a complication in bitches being ovariectomized (13%) than in those ovariohysterectomized (10%).[120] Urinary incontinence is reported to be controllable by administration of diethylstilboestrol (DES),[109] oestradiol valerate,[120] or medroxyprogesterone acetate (MAP).[2,120]

First the bitch should be examined physically, so that the correct anaesthetic can be chosen, and then fasted for at least 12 h. Many anaesthetics, for example, ether and halothane, will give suitable abdominal muscle relaxation.

The abdominal wall is prepared in the usual way, the urinary bladder drained, and the bitch placed on her back either horizontally or, more conveniently, with the head slightly lower than the body.

Through a 2–5 cm long midline incision, made caudally from the umbilicus, the left uterine horn (situated near the left flank) may be hooked up and withdrawn by the use of a tenaculum. The left ovary, which is most easily located, is ligated after clamping across the ovarian bursa, and the ovarian blood supply in the broad ligament then tied with a double ligature of chromic catgut, size 0, through an opening made in the ligament caudal to the ovary. The ovarian attachment is sutured distal to the ligature, ensuring that all ovarian tissue is removed and that there is no haemorrhage, before the ligated stump is returned to the abdominal cavity. The right ovary is located by following the uterus, and similar ligatures and incisions made. The uterine body and the uterine vessels are then withdrawn for ligation. After having sutured the uterus cranial to the ligatures and checked carefully for haemorrhage, the stump is returned to the abdominal cavity and the abdominal incision sutured in the usual way.

Salpingectomy
Salpingectomy – tubal ligation – is used in human surgery but has not been much used in bitches. It might, however, be less risky because the operation is almost bloodless, does not change the hormonal balance and thus maintains the normal character of the bitch, and is less expensive and time-consuming than ovariohysterectomy. Ligating and bisecting the Fallopian tubes can be performed through an incision about 5 cm long in the linea alba about 2.5 cm posterior to the umbilicus. A piece of each Fallopian tube can easily be removed between two catgut ligatures.[123] Experiments with laparoscopy in conjuction with electrocautery might lead to a new effective and alternative technique of ovariectomy.

Ovarian autograft
In a primary investigation both ovaries were removed from the abdominal cavity and cut into 1–2 mm thick slices held together by strands of tissue, and then put into subserosal pouches raised by blunt dissection on the greater curvature of the stomach.[117] This resulted in the suppression of oestrus, but the bitches could be trained and used like intact animals whereas ovariectomized bitches were inferior.

Physical contraception

Intravaginal devices
As an alternative to surgical spaying the use of an intravaginal device made of silastic and polyethylene tubing placed in the vestibule and vagina has been tried. Properly inserted, it ought to prevent the insertion of the penis as well as the formation of the coital lock and thus ejaculation. But because up to 50% of bitches fitted with this device form the coital tie, resulting in a pregnancy rate of 25%, a new device, fitted with a copper ring, is being investigated. This ring should act as a spermicide, and a 100% successful pregnancy control has been claimed.[80]

Chemical contraception
Chemical contraception may be brought about by chemicals which prevent or suppress oestrus or prevent implantation. The chemicals used are described below, and the dosages are given in Table 7.2.
  Chemical control of oestrus differs from chemical contraception in humans, where oestrogens are widely used to prevent ovulation. In

Table 7.2 Chemical contraception. Prevention or suppression of oestrus and prevention of implantation.

Condition	Drug	Administration	References
Prevention of oestrus	Progestogens		
	progesterone repositol (P)	2–3 mg/kg i.m. every 2–3 weeks	98
	17α-acetoxyprogesterone	(a) 2.5 mg/kg in dog feed	16
		(b) 4 mg/kg daily orally as a suspension	16
	medroxyprogesterone acetate (MAP)	30–100 mg s.c. every 5–6 months	79
	megestrol acetate (MA)	2.2 mg/kg orally daily for 8 days	10, 19
	chlormadinone acetate (CAP)	1.5–3 mg/kg i.m. every 4–6 months	12, 69, 127
	delmadinone acetate (DMA)	(a) 0.25–0.5 mg/kg orally once a week	45, 54
		(b) 1.5–2.0 mg/kg s.c. twice a year in large thin bitches, and 2.0–2.5 mg/kg s.c. twice a year in small fat bitches	45, 54
	proligesterone	10–13 mg/kg s.c. in large bitches and 20–33 mg/kg s.c. in small bitches, retreated after 3 and 4 months, thereafter every 5 months	102, 103, 104, 105
	melengestrol acetate (MGA)	200–800 μg orally daily	136
	Androgens		
	testosterone	Implants releasing 759 μg/4.5 kg daily	129
	mibolerone	30–180 μg orally daily	132
	Non-hormonal compounds		
	6,6-spiroethylene-19-norspiroxenone	0.9 mg/kg orally daily	95
Suppression of oestrus	Progestogens		
	medroxyprogesterone acetate (MAP)	5 mg orally daily up to 1 month	96, 141
	megestrol acetate (MA)	2.2 mg/kg orally for 8 days	19

	norethisterone acetate (NET)	1.175 mg/kg orally daily for 21–30 days	28, 58, 115, 116, 122
	delmadinone acetate (DMA)	(a) 1.0–2.5 mg/kg orally daily for 6 days	54
		(b) 2.5–5.0 mg/kg s.c. twice within 24 h	54
	proligesterone	10–30 mg/kg s.c. single dose	103, 104
Prevention of implantation	Diethylstilboestrol (DES)	(a) 1–2 mg orally for 5 days	77
		(b) 1–2 mg/kg i.m. + 0.4 mg/kg orally for 5 days	18, 100, 156
	Diethylstilboestrol repositol	1 mg/kg i.m. single dose	77
	Oestrogens		
	oestradiol 17-cyclopentylpropionate (ECP)	<5 days after mating: 0.25–2.0 mg i.m. single dose	77
		>5 days after mating:	
		(a) 0.25–2.0 mg i.m. twice separated 5 days	77
		(b) 0.25–2.0 mg i.m. + 1–2 mg orally for 5 days	77
	oestradiol benzoate	(a) 2.5 mg i.m. single dose within 2–3 days after mating	
		(b) 0.5–3.0 mg i.m. three times with intervals of 48 h, within 4–10 days after mating	7
		(c) 1–3 mg s.c. three times within 5 days	2
		(d) 2–4 mg i.m. + 1–2 mg s.c. twice with intervals of 48 h	119
	oestradiol valerate	(a) 3.0–7.0 mg injected once between days 4 and 10	7
		(b) 1–1.5 mg/10 kg, max 5 mg, s.c./i.m. on days 1, 2 or 3	119
	mestranol	(a) 0.5–4.5 mg orally on day 5	75
		(b) 5 mg/kg orally on days 6 or 21	78
	Progestogens		
	megestrol acetate (MA)	0.8 mg/kg/day orally throughout oestrus	33

bitches a double effect is required, partly for the prevention of ovulation and partly for the reduction of oestrous behaviour, so progestogens especially have been employed, although androgens and other compounds have been tried.

There are other differences between chemical contraception in humans and bitches due to differences in the effect of the chemicals on the two species. Norethisterone acetate (NET) is 5–20 times and progesterone 1–6 times more potent in humans than in dogs when calculated on the basis of mg/kg of bodyweight.[56] In humans chlormadinone acetate (CAP) is about 5 times more effective than NET, but in dogs CAP is 225 times more active on the endometrium than NET.[64] Furthermore, the administration of progestogens in high dosages, especially of agents with a structure similar to testosterone, should be avoided during pregnancy because of the risk of masculinization of the external genitalia of female puppies.[32]

The terms used are defined as follows:[23] *Prevention* of oestrus is the result of administration of a contraceptive agent to the normal anoestrous bitch. *Delay* indicates the increased interval between administration of a preventive dose and the occurrence of the next normal oestrus. *Suppression* is the abolishment of the signs of pro-oestrum or oestrus with a treatment starting when the first signs of pro-oestrum or oestrus appear – in other words, bloody discharge with or without acceptance of the male. *Postponement* indicates the increased interval between the effective administration of a suppressive dose and the recurrence of normal oestrus. Even if suppression results in absence of oestrus for several months, this term is used.

Prevention of oestrus by progestogens
The mechanism of action of progestogens is still unknown, but certainly one or more of the following processes is involved: inhibition of gonadotropic hormones, including FSH, LH and PRL; local prevention of ovarian follicular growth, oestrogen secretion and ovulation; and inhibition of sexual behaviour.

The correct time for treatment (anoestrum) is when the bitch shows no symptoms of pro-oestrum or oestrus caused by oestrogen secretion, such as vulval oedema, hyperaemia and vaginal discharge, and examination of a vaginal smear shows this to be entirely free from erythrocytes. However, it has to be pointed out that erythrocytes are occasionally absent in pro-oestrous smears.[6]

A depot injection of progestogen exerts its effect over a variable period of time, so predicting the time of the next oestrus is difficult, as is the timing of subsequent injections if a build-up of the drug is to be avoided. This problem is of major importance in individuals receiving high doses of progestogens as well as in bitches of small breeds.

Oestrus is said not to be prevented if treatment is begun during the incubation period of some unspecified viral infections because lesions may be produced in the stomach, intestines, kidney, bladder and perhaps even in the bones or on the skin. These reactions inhibit the hormone-blocking action of the progestins.[45] Failure in contraception may also be induced by contact with other bitches in oestrus.[102]

When using progestogens, the risk of inducing pathological changes in the endometrium must be kept in mind. Uterine effects of progestogens, such as megestrol acetate (MA),[33,99] MAP,[1,13,61,110,137–139] 17α-acetoxyprogesterone,[5,16] melengestrol acetate (MGA),[55] and CAP,[64] when administered over a long period or in high dosages are similar to those of P,[20,39] namely: (1) increased weight due to grossly enlarged, corkscrew-like uterine horns; (2) varying amounts of clear, thick, greyish-yellow fluid; (3) endometrial glandular hyperplasia in varying degrees; (4) thickening of the endometrium; (5) inflammatory cell infiltration and necrosis of endometrial structures; and (6) dystrophic calcification. Long-term administration of progestogens stimulates mammary development. In early stages or with moderate dosages the mammary stimulation resembles that seen in the normal cyclic changes of oestrus and pregnancy,[153] but oestrous control established for as little as 4 years has been found to be accompanied by development of mammary tumours.[107] Hyperplastic nodules and malignant changes may follow long-term stimulation of progestogens.[153]

Progesterone   Progesterone in a crystalline suspension, progesterone repositol, can prevent oestrus if injected intramuscularly every 2 or 3 weeks.[98]

17α-Acetoxyprogesterone   This progestogen will prevent oestrus when given orally, either incorporated in the feed or as a suspension. After cessation of treatment the bitch returns to a normal fertile cycle and can produce normal healthy litters, and no side-effects are seen.[16]

Medroxyprogesterone acetate (MAP) (6α-methyl-17α-acetoxyprogesterone)   The first use of MAP was in a suspension, but this had certain unfortunate side-effects such as endometritis and pyometra.[31,81,159] Since then an aqueous form has been used resulting in fewer complications. The metabolic clearance rate of MAP has been found to be one-half that of P (696 ± 51 l/day versus 1332 ± 59 l/day).[121]

Subcutaneous administration of 30–100 mg MAP, depending on the size of the bitch, every 5–6 months can effectively prevent oestrus, and if the first injection is given in anoestrum, the following injections will be made in anoestrum, so that side-effects seldom occur.

The long-term effect of MAP is due to its low solubility in body fluids. Up to 7 months will elapse before the majority of a 50 mg subcutaneous dose is excreted, and oestrus is delayed for approximately 12.5 months (range 6–26.5).[17,94] Bitches given this dosage return to oestrus, conceive and whelp healthy puppies, but it is difficult to predict when they will return to oestrus because after excretion of most of the drug a period of adjustment is required before another oestrus occurs.

If administered subcutaneously, MAP should be injected in an obscure, relatively hairless site (such as the inner fold of the flank or medial aspect of the thigh) because it may cause undesirable thinning of the hair and skin over the injection site. To avoid these complications intramuscular application may be preferred.[93]

In some trials administration of MAP as a depot injection has induced cystic endometrial hyperplasia, endometritis and pyometra but only under conditions of continued dosage at a high level and when potentiated by previous or simultaneous injections of oestrogens,[13,29,52,53,99] whereas no clinical cases of endometritis have been reported after treatment with 50 mg.[79,89] The use of MAP in high, frequently repeated doses over a long period results in a dose-related incidence of mammary tumours,[46] presumably as a result of MAP-induced elevation in growth hormone (GH) production[26] or the involvement of abnormal pituitary secretion.[27]

Megestrol acetate (MA) (6-methyl-Δ6-17-α-acetoxyprogesterone)
MA prevents oestrus effectively. Adverse effects are minimal[10,19,40,84] but pyometra may occur,[24] and the use of this potent progestogen has to be limited to short periods to avoid the possibility of side-effects,[21] i.e. the drug should not be used for more than 32 days and only every 6 months. After cessation of treatment bitches will show fertile oestrus, but the interval is unpredictable.

Chlormadinone acetate (CAP) Administered intramuscularly in a dosage of 1.5–3 mg/kg CAP can prevent oestrus.[12,69,127]

Delmadinone acetate (DMA) (Δ-6-chlor-6-dehydroacetoxy-progesterone) DMA can prevent oestrus either by oral administration once a week or subcutaneously twice a year.[45,54] For the best result it is recommended to inject three-quarters of the dose subcutaneously and a quarter intramuscularly.[54] There is no report of endometrial disturbances in bitches after DMA administration no matter whether in anoestrum, pro-oestrum or metoestrum.

Proligesterone (14α-17α-propylidenedioxy-progesterone) Proligesterone is a highly effective progestin, claimed to be safe at practically any time of the cycle without any deleterious long-term effects on the endometrium.[9,102-105,146]

Proligesterone is very effective if the treatment is initiated in anoestrum, and after cessation of the treatment more than 90% of the bitches come into oestrus within 9–12 months with normal fertility, parturition and litter size.

The hyperplasia–pyometra complex was diagnosed in only very few bitches treated with proligesterone; most often those bitches treated previously with other progestogens for oestrous control exhibited this syndrome. Other side-effects such as abnormal behaviour, weight increase and discoloration of the hair at the injection site may occur. The development of hyperplastic mammary nodules and tumours does not seem to be influenced by the administration of proligesterone.[103,106,107]

Melengestrol acetate (MGA) When given orally MGA has prevented oestrous activity in bitches for a treatment period of 243 days without impairing fertility or interfering with the subsequent ability to return to cyclic activity. The interval from the end of treatment to the recurrence of oestrus ranged from 62.5 to 157 days depending on the dose administered.[136]

MGA has also been used effectively as an implant, but this treatment was accompanied by endometrial hyperplasia.

Prevention of oestrus by androgens

Testosterone implants Capsules containing testosterone, 4 cm long and made of silicone rubber tubing, have been tried as a contraceptive device when surgically inserted in the flank under general anaesthesia. For prevention of oestrus a daily release of 759 µg testosterone per 4.5 kg bodyweight is required.[129] Anoestrum has been maintained for 420–840 days in bitches with these implants. The effect probably takes place via the hypothalamic–pituitary gonadotropin releasing system. After removal of the implants, normal fertile oestrus occurs after a latent period of 34–291 days.

Provided implants are used in adult bitches only, this seems to be an effective, simple and long-lasting contraceptive method. There are side-effects in the form of masculinization with enlargement of the clitoris, possibly with traumatic lesions of the exposed clitoral tissue. In most of the bitches there is periodic vaginal discharge, but urination patterns and bodyweight are not affected and the endometrial hyperplasia–pyometra complex has not been reported in connection with this treatment.

Mibolerone (17β-hydroxy-7α,17dimethylstr-4en-3-one, dimethylnortestosterone) Mibolerone, an androgenic–anabolic steroid, is effective and apparently reliable for oestrus prevention.[131,133,140] The serum concentration in dogs treated with mibolerone seems to depend on the geographical location of the bitch when treated.[82] Mibolerone acts specifically by blocking the ovulatory release of LH. Mibolerone may induce premature epiphyseal closure, enlargement of the clitoris, and vaginitis in prepubertal bitches. Although it has been administered at dose levels of 20 or 60 μg per day during the whole pregnancy period without any adverse effects on conception, pregnancy or parturition, it should not be administered to pregnant or lactating animals, since it may interfere with lactation[135] or cause masculinization of female fetuses.

Prevention of oestrus by non-hormonal compounds

6,6-Spiroethylene-19-norspiroxenone Administered orally for 90 days this resulted in a 100% prevention of oestrus without evidence of cystic endometrial hyperplasia or pyometra. Mammary biopsies were also normal.[95]

Suppression of oestrus by progestogens

Oestrus can be suppressed by short-term administration of progestogens at the beginning of pro-oestrum, and the postponement lasts as long as oral administration of progestogens is maintained. Among indications for this kind of treatment are participation in dog shows and during the hunting and racing seasons. A greyhound bitch with normal cyclic activity can participate in a much smaller number of races in her racing life than a male dog, because she is not allowed to take part in an official race whilst in oestrus (British NGRC rules). Furthermore she is not allowed to race for 10 weeks starting from the first day of pro-oestrum.[116]

The great disadvantage of suppressing oestrus is that continuous administration of the drug is required, but at the same time it has the advantage that the drug is rapidly expelled from the body after the end of treatment.

Drugs used to prevent heat can also be used to suppress it, including already visible signs such as bleeding and attractiveness to male dogs. But the practitioner must be aware that many more difficulties can arise in conjunction with the administration of progestins for suppression of oestrus than for its prevention. Correct timing of the treatment is essential. Treatment given too early, though arresting follicular growth for a while, allows continued growth after withdrawal of the drug. Treatment established too late in pro-oestrum or during oestrus is often inadequate to arrest all signs, and oestrus can then reappear shortly after termination of the treatment, as sometimes occurs in attempts to postpone it.

Medroxyprogesterone acetate (MAP) When administered orally starting 5 days before the effect is required, MAP is effective, and the treatment should continue as long as suppression is desired. Treatment is safe if it does not exceed 1 month.[96,141]

Megestrol acetate (MA) MA is effective for suppression of oestrus when given orally in early pro-oestrum.[19]

Norethisterone acetate (NET) This will suppress oestrus, and is presumed not to affect either racing performance or reproductive functions.[28,58,115,116,122]

Delmadinone acetate (DMA) Administered orally DMA will suppress oestrus within 3 days of initiation of treatment. No side-effects have been reported.[54]

Proligesterone When administered in pro-oestrum this results in the disappearance of signs of heat within 5 days. Retreatment after 3 months maintains oestrus prevention in most bitches.[104]

Prevention of implantation
After unwanted matings, efforts can be made either to prevent implantation or terminate pregnancy. Because of their side-effects many of the available methods should be used only if it is known for certain that mating has taken place – that is, if copulation has been observed by the owner or some person with reliable knowledge of the reproductive biology of the dog. If there is the slightest doubt a vaginal smear must be examined. If the cell picture is characteristic of oestrus or metoestrum, the possibility of a fertile mating exists, even if spermatozoa are not present. Even if this is the case, one must consider whether treatment should be begun with the attendant risk of side-effects, or whether the practitioner should wait for the result of an examination for pregnancy about 30 days later. If this examination is positive, the decision must be made as to whether pregnancy should be allowed to continue or be interrupted surgically or medically.

After ovulation and fertilization the ova remain in the oviduct, for a few days, there developing to the morula stage, and then descend to the uterus where implantation takes place at about the 21st or 22nd day after the beginning of signs of oestrus.

Various kinds of oestrogen have been used to prevent implantation. The way in which they act has not been completely elucidated. Oestrogens provoke oedema and proliferation of the endometrium and might arrest the passage of the eggs to the uterus. The prolonged stay in the oviduct may result in degeneration of the egg and this, together with the endometrial proliferation, may be the reason for implantation not taking place. To achieve the best result with the lowest risk of side-effects, treatment with oestrogen should be begun between 3 and 5 days after mating. Administration too early may have no effect on pregnancy and too late may result in pyometra.

If treatment is to be applied, the owner must be informed of the risks and the fact that oestrus will often be prolonged, or the bitch may appear to be out of oestrus for a few days and then come into oestrus again. Treatment with oestrogens should not be used in bitches meant for later breeding because of the risk of endometritis, pyometra and sterility, all of which may occur regardless of dosage. The possibility of complications increases the later the medication is begun.[119] The risk of pyometra also seems to be greater if the bitch is mated during prolonged or induced heat.

In some dogs the use of oestrogens can lead to damage of the bone marrow in the form of increased granulopoiesis and decreased megakaryocytopoiesis leading to leukocytosis and thrombocytopenia. Total suppression of haematopoiesis and persistent thrombocytopenia may develop, and death due to haemorrhagic diathesis will follow.[30,87,124,147]

Diethylstilboestrol (DES)
Diethylstilboestrol has been used for mismating therapy at an initial dose of 1–2 mg/kg intramuscularly followed by oral administration of 0.4 mg/kg for 5 days.[18,100,156]

Oestrogens

Oestradiol 17-cyclopentylpropionate (ECP)   This can be administered effectively as a single dose within 5 days after mating, but if treatment is instituted later then either a second injection must be given 5 days after the first, or it must be followed by oral administration of diethylstilboestrol for 5 days[18,38,77,83,91,97,101,154] (see Table 7.2).

Oestradiol benzoate or oestradiol valerate   These can be administered as single or multiple injections.

Mestranol (17α-ethinyloestradiol-2 methyl ether)   This can be used orally on day 5 and results in degeneration or retarded development of the fetus and the prevention of implantation.[75] The effect is not related to the dose. According to Kennelly[78] treatment on day 6 results in an embryonic loss of 95.6%, whereas the loss was only 67.3% after treatment on day 21 compared to a loss of 34.5% in a control group.

Progestogens

Megestrol acetate (MA)   When given orally throughout oestrus this will prevent pregnancy by inhibition of ovulation without suppression of oestrous symptoms and libido.[33]

Irradiation

X-ray exposure of ovaries, if performed via laparotomy and direct irradiation of the ovaries, has caused infertility, whereas single-dose irradiation in adult bitches using a non-surgical procedure has not been effective.[85,86] There is at present insufficient knowledge of the value of single-dose irradiation as a method of treatment of prepubertal puppies but this may provide a practical procedure for controlling fertility in the future.

Immunological contraception

If it were possible to immunize against gonadotropin-releasing hormones, reproduction could then be controlled through regulation of the release of FSH and LH. Investigations are being carried out with a view to preventing pregnancy by using an antiprogesterone monoclonal antibody.[160]

The use of an anti-zona pellucida vaccine may also be possible in the future. High concentrations of antisera against canine ovaries can prevent *in vitro* fertilization of the oocytes.[90] Pregnancy can be prevented by immunization with zona antigens even as late as the first day of oestrus, and fertility returns when immunizations stop and the titre decreases in serum.[128]

Induction of asexuality

So far, efforts to produce sterile female puppies by injection of androgens or progestogens within 48 h of birth have not been successful. Testosterone propionate (TP), depo-testosterone cyclopentylpropionate (TCP) as well as MAP have been used, but these have produced only enlargement of the clitoris in all those treated, while the age at puberty and subsequent fertility were identical to those of untreated bitches.[161]

Termination of pregnancy

Surgical termination of pregnancy

If this is preferred, as for example by ovariohysterectomy, it should be performed during the first half of the pregnancy in order to minimize complications.[77]

Bilateral ovariectomy on days 30–56 of pregnancy results in the termination of pregnancy within 8 days.[130]

Chemical termination of pregnancy

Chemical compounds and dosages for induction of abortion are shown in Table 7.3.

Table 7.3   Chemical termination of pregnancy.

Drug	Administration	References
HORMONES:		
Corticosteroids		
dexamethasone	2–5 mg i.m. for 10 days from days 30 or 45	3
NON-HORMONAL COMPOUNDS:		
Prostaglandins		
PGF$_{2\alpha}$	(a) 20 μg/8 h/72 h i.m. between days 33 and 53	25
	(b) 30 μg/12 h/72 h i.m. between days 33 and 53	25
Fluprostenol	(a) 12.5–20 μg/kg s.c. between days 14 and 25	68
	(b) 10 μg/kg by a 24 h-release plastic intravaginal device days 1 to 28	68
Cloprostenol	(a) 10–40 μg/kg s.c. between days 14 and 28	68
	(b) 10 or 25 μg/kg by a 24 h-release plastic intravaginal device days 1 to 26	68
TPT aqueous	(a) 200 μg s.c. between days 20 and 33	151, 152
	(b) 10 μg/h/24–48 h s.c. between days 20 and 30	152
	(c) 20 μg/kg s.c. on day 43	151
TPT PEG 400	200 μg s.c. between days 20 and 30	152
TPT methylester PEG 400	200 μg s.c. between days 20 and 30	152
Non-prostaglandins		
L-10492	12.5 mg s.c. before or 50 mg s.c. after day 30	49
L-10503	12.5 mg s.c. before or 50 mg s.c. after day 30	49
L 12717/DL 717-IT	0.5–1 mg/kg s.c./i.m. on day 20	50
DL 204-IT	6.25 mg/kg s.c./i.m. on day 20	51
Malucidin	20 mg/kg i.v. in early pregnancy or 22 mg/kg i.v. in later pregnancy	155
NDTC	2 mg/kg i.v. between day 40 and day 53	118, 145

TPT: dl-9α,11α,15α-trihydroxy-16-phenoxy-17,18,19,20-tetranorprosta 4,5,13-*trans*-trienoic acid used either in an aqueous solution or dissolved in polyethylene glycol 400 (PEG 400)
L-10492: 2-(3-methoxyphenyl)-5H-s-triazolo(5,1-a)isoindole
L-10503: 2-(3-methoxyphenyl)5,6-dihydro-s-triazolo(5,1-a)isoquinoline
L 12717: 2-(4-chlorophenyl)-s-triazole[5,1-a]isoquinoline
DL 204-IT: 2-(3-ethoxy-phenyl)-5,6-dihydro-s-triazole[5,1-a]isoquinoline
NDTC: N-desacetyl-thiocolchicine

Corticosteroids
Daily intramuscular administration of dexamethasone for 10 days from day 30 of pregnancy resulted in intrauterine death and resorption of the fetuses; an identical treatment from about day 45 resulted in the birth of dead fetuses on days 55 and 59 of pregnancy.[3] Further tests of repeatability and investigations of possible side-effects are needed before this method can be used as a routine.

Prostaglandins
It is hoped that prostaglandins, widely used for induction of abortion and parturition in large animals, will be useful for the same purposes in bitches, and tests have been made of various of them.

$PGF_{2\alpha}$ Dogs are very sensitive to prostaglandins, and the differences between the ineffective luteolytic dose of $PGF_{2\alpha}$-THAM (0.25 mg/kg), the effective luteolytic and abortifacient dose (1 mg/kg) and the median lethal dose (5.13 mg/kg) in the bitch are narrow.[134] Administration of $PGF_{2\alpha}$ as well as its analogues is followed by salivation, vomiting, diarrhoea and urination for a period of up to 3 h. In cases of intoxication these symptoms are accompanied by hyperpnoea, ataxia and pupillary dilatation followed by constriction. Intravenous injection of $PGF_{2\alpha}$ to pregnant bitches, 2 × 2 mg with 7 h intervals, caused a transient drop in the concentration of plasma P with a return to normal concentrations within 24–48 h, and none of the bitches aborted.[74] A transient drop in the concentration of P was seen in the plasma of non-pregnant bitches after 3 intramuscular injections of 5 mg $PGF_{2\alpha}$ with intervals of 1 and 24 h.[66]

Prolonged intramuscular administration of either 20 μg/kg $PGF_{2\alpha}$ with intervals of 8 h or 30 μg/kg every 12 h during a period of 72 h resulted in a decrease in the P concentration and abortion in 4 of 8 bitches. If treatment started on days 33–53 of pregnancy, the concentration of P in the plasma was depressed to 0.6–1.4 ng/ml 56–80 h after the start of the treatment. In bitches not aborting the concentration was only depressed to 2.1 ng/ml, and then it rose to 5–10 ng/ml, a concentration which was maintained until the normal prepartum decline.[25] Total luteolysis was achieved after a treatment extended over 72 h with a total of 180 μg/kg $PGF_{2\alpha}$ being given.

Fluprostenol and cloprostenol   These have been investigated for terminating pregnancy, resulting in termination in 6 of 22 bitches (27%) treated before day 25 of gestation and in 16 of 20 bitches (80%) treated on or after day 25.[68] When the two preparations were administered intravaginally instead of subcutaneously, in the form of a 24 h-release plastic device, the necessary prolonged luteolytic effect was achieved and unacceptable side-effects were avoided.

TPT   This, a synthetic analogue (dl-9α,11α,15α-trihydroxy-16-phenoxy-17,18,19,20-tetranorprosta 4,5,13-*trans*-trienoic acid), has been shown to terminate pregnancies when injected subcutaneously, 20 μg/kg, within the period days 30–43.[151] Abortion was obvious 5.4 ± 1.4 days after treatment. Of dogs treated on days 20–22, some continued to term, but others – though diagnosed pregnant by palpation on day 28 of pregnancy – did not and the products of conception might thus have been resorbed. Dogs treated on day 9 of pregnancy continued to normal parturition at term.

In bitches aborting or resorbing fetuses the concentration of P in plasma falls precipitously during the first 8 h after treatment and continues to decline thereafter. A shortening of the interoestrous interval (average 39.8 days) compared to the pretreatment cycle was observed in bitches in which pregnancy was terminated prematurely, whereas no difference was observed in bitches which continued till term.[152] In bitches not aborting, a similar decrease was seen during the first 4 h after treatment with a return to normal pregnant levels within 24–72 h.

In an effort to increase the efficiency of terminating early pregnancies, attempts have been made to extend the duration of action of TPT by using a minipump, by changing the formulation medium and by using the methyl ester.[152] This has resulted in an extension of the luteal suppression and increased percentages of pregnancy terminations without any change in the duration of side-effects.

It must be stressed that this treatment, just like treatment with the other prostaglandins, has side-effects such as profuse salivation within 5 min after injection and vomiting with or without diarrhoea within 15–30 min after injection. Observable side-effects are present for up to 3 h after treatment.

It seems likely that further investigations will make it possible to use prostaglandins or their analogues for the termination of early pregnancies without any, or with only minimal, side-effects, and provide another tool for effective control of the pet population.

Non-prostaglandins

L-10492 and L-10503   Galliani and Lerner[49] found that two non-hormonal compounds, L-10492 (2-(3-methoxyphenyl)-5H-s-triazolo(5,1-a)isoindole) and L-10503 (2-(3-methoxyphenyl-5,6-dihydro-s-triazolo(5,1-a)isoquinoline), can terminate pregnancy in the bitch. The effect depends on the dose and time of administration. Subcutaneous injection of 12.5 mg during the first half of pregnancy (days 13–30) terminates pregnancies by causing degeneration and resorption of the products of fertilization. In the second half of gestation a dose four times greater is needed for interruption of pregnancy. The bitches treated with L-10503 returned to a normal oestrus after 5–8 months, conceived, had normal pregnancies and delivered normal litters. The treatment was accompanied by temporary dose-related side-effects such as loss of appetite, reduction of bodyweight and diarrhoea. The mode of action is not yet clear but is presumed to be via the prostaglandins in the placenta.

Recently, experiments on the induction of abortion with two other non-hormonal compounds have given encouraging results: L 12717, 2-(4-chlorophenyl)-s-triazole[5,1-a]isoquinoline, and DL 204-IT, 2-(3-ethoxy-phenyl)-5,6-dihydro-s-triazole[5,1-a]isoquinoline.[50,51]

Malucidin This yeast extract has been used experimentally. Intravenous administration of a 3% solution during the first two trimesters of pregnancy resulted in resorption of fetuses, while abortion followed administration during the last trimester. Delivery began 23 ± 2 h after the injection with the same symptoms as in normal parturition.[155]

N-Desacetyl-thiocolchicine   NDTC has been used for induction of abortion.[145] Intravenous injection of a dose of 2 mg/kg bodyweight to bitches 40–53 days pregnant induces abortion within 15–40 h. Later, reproductive cycles and fertility are not influenced, but injection is followed by copious salivation, frequent vomiting and profuse diarrhoea lasting 2–3 days.[118]

THE MALE

Fertility is most often limited in females, but it can be brought about in the male either surgically or non-surgically.

Surgical methods

Castration

This operation is used less in dogs than cats for limitation of male reproduction, although it is the most effective method. Castration is more often used for controlling such unwanted male behaviour as wandering, fighting with other males, urine marking in the house and mounting other dogs or people. Prepubertal castration does not significantly reduce either mounting or the sexual responsiveness of the adult dog to oestrous bitches. Early castration results in loss of ability to lock with the bitch, probably due to underdevelopment of the penis.[4,11] Castration carried out after puberty is not completely effective in eradicating all male patterns of behaviour in adult dogs; only wandering seems to decline in almost all castrates, but other behavioural features seem to decline in only 50–70%.[63,65]

Behaviour alteration after castration

Nearly 50% of the alteration in behaviour takes place shortly after the operation, within 2 weeks, the remaining 50% of the effect developing gradually later.

Wandering can be eliminated in more than 90% of dogs. Urine marking is usually altered in about 50% either rapidly or gradually, but even if it is reduced or even eliminated indoors it may be unchanged out of doors, due to olfactory stimuli from other dogs.

Aggressive behaviour occurs in various forms, not all of which can be changed by castration. Thus, territorial and fear-induced aggression cannot be reduced, but intermale aggression can be expected to be altered in about 60% of dogs after castration.

Mounting of other dogs, humans or inanimate objects is often seen in puppies for a limited period, but this type of behaviour usually stop3 at puberty. In some dogs it may persist to become a serious problem in adulthood. Castration reduces mounting behaviour in up to two-thirds of dogs, and mounting humans is altered more frequently than mounting other dogs.

The differences between dogs in their response to castration are presumably due to individual differences in genetic makeup, or to varying environmental circumstances such as the presence or absence of the stimulation of other dogs. In dogs where castration does not alter the undesirable male behaviour, administration of long-acting progestogens such as MAP in a dose of about 10 mg/kg can be successful.

Mating behaviour may be retained for a long period after castration, although there is a decrease in the frequency of copulation and in the length of time erection is maintained as reflected by the duration of the copulatory lock. There is no difference in the libido between experienced and inexperienced males after castration,[62] but the production of prostatic fluid is reduced.[144]

Vasectomy

This method, much used in humans, is not widely employed in dogs, presumably because it does not eliminate undesirable behaviour Although vasectomy is a fairly simple and safe procedure, postoperative licking of the incision may lead to infection. Vasectomy is followed by reversible meiotic changes which result in spermatogenic arrest and a lowered frequency of mature spermatids, but after 15 weeks meiotic distribution has become normal.[88]

This method for limitation of fertility also implies a possibility of restoring fertility if vasovasostomy is performed. It is reported that although postvasovasostomy sperm counts never reach prevasectomy levels, the values are within the normal range.[59]

External vasectomy

This must be performed under general anaesthesia. Through a 2–3 cm incision in the skin and the tunica vaginalis a piece of either the vas deferens[14,108,149] or the epididymis[150] can be excised between two ligatures.

Ligation of epididymides causes more damage to the testes the closer the ligature is to them.[150] During the first 15 days after the operation the size of the testes is increased and the texture is more turgid; the size then decreases, the texture becomes soft and 1–3 months after ligation the seminiferous tubules undergo degenerative changes.

Vasectomy with ligation of the vas deferens results in degenerative changes in the seminiferous tubules during the first 4 months followed by regeneration, so that after 6 months the tubules appear normal.[149]

It is commonly believed, and thought to have been shown conclusively, that the first ejaculate after vasectomy is sperm-free because all ejaculated spermatozoa originate from the epididymides, and also that the volume decreases by 10–50%.[15] Pineda et al.,[113] however, found that ejaculates may contain spermatozoa as late as 21 days after bilateral vasectomy, indicating that although spermatozoa in the ejaculate originate from the epididymides, some may be present in the vasa deferentia. This is in agreement with other investigations where spermatozoa were present in the ejaculate for up to 60 days after vasectomy.[22] Furthermore, it has been found by others that the volume of the ejaculate does not change after vasectomy, whereas the total number of spermatozoa declines steadily in each successive ejaculate, and increased frequency of collection does not shorten the time necessary to reach azoospermia.[113] On the other hand, a decrease is reported to occur in the sperm count within 48 h of occlusion of the vas ductus deferens, accompanied by an apparent decrease in testicular volume.[158]

Internal vasectomy or vas occlusion
This is performed with a laparoscope inserted through a 1 cm incision near the umbilicus, and with accessory forceps through another small incision to grasp and cauterize each vas deferens intra-abdominally. The skin is finally closed with a single suture after withdrawal of the instruments.[157] This operation has been performed successfully in prepubertal dogs and followed by normal development into adulthood. In adult dogs it resulted in reduction of sperm numbers to zero within 48 h. It may become more widely used as a routine method as it is quick and requires only minimal surgical intervention.

Non-surgical methods

Chemical agents
Injection of various agents in the caudae of the epididymides in prepubertal and adult dogs can induce azoospermia because of scar tissue occluding the lumen of the vas deferens.[48]

Chlorhexidine
A single percutaneous bilateral intra-epididymal injection of 0.6 ml 1.5% chlorhexidine gluconate in 50% dimethyl sulphoxide (DMSO) causes a long-lasting and probably irreversible azoospermia in beagle dogs.[111,114] Clinically, no side-effects are seen other than a swelling of the scrotum lasting for 7 days. Treatment of prepubertal dogs results in ejaculates of seminal fluid at the same age as intact animals

but without spermatozoa. A single treatment given to adult dogs resulted in a decline in the number of spermatozoa in each successive ejaculate thereafter, and azoospermia on the 91st day after treatment. Two successive injections of the same dose 7 days apart caused azoospermia 35 days after the first injection.

Intra-epididymal injection of 0.5 ml of an aqueous solution of 3% chlorhexidine digluconate in 50% DMSO given into each cauda of the epididymides results in irreversible azoospermia within 35 to 42 days with no side-effects other than a transient swelling of the scrotum.[112]

α-Chlorohydrin (3-chloro-1,2-propandiol(α-chlorohydrin))

Following subcutaneous injection of 70 mg per kg bodyweight in the right thigh no spermatozoa were found in the epididymides or vasa deferentia after 33 days. This was followed by enlargement of the epididymides, and an increase in the diameter of the seminiferous tubules and the size of Leydig cell nuclei.[34]

Subcutaneous administration of 8 mg α-chlorohydrin for 30 days caused lesions in the testes with degenerative changes of the semen as well as the changes mentioned above.[35]

Danazol (17α-pregn-4-en-20-yno-(2,3-d)-isoxazol-17-ol)

This is an analogue of ethinyl testosterone. Injection of 2000 mg in one testis resulted in a marked increase in testicular size and a reduction in the diameter of the seminiferous tubules; no spermatozoa were found 25 days after treatment in either the epididymis or vas deferens ipsilateral to the treated testis, while normal numbers of spermatozoa were found in the control testis.[36]

KABI-1774 (2,6-cis-diphenylhexamethyl-cyclotetra-loxane)

Oral administration of 10 or 250 mg/kg per day results in behavioural changes starting, respectively, 14 and 7 days after the start of the treatment.[57] Only a small amount of the ejaculate is delivered in spite of normal pelvic movements and contractions of the penis, probably because of inhibition of prostatic secretion and impairment of transport of spermatozoa from the caudae epididymides. The result is abnormal motility, a decrease in the volume of ejaculate and total number of spermatozoa, and a high frequency of sperm tail abnormalities and distal cytoplasmic droplets.

Ultrasound

Ultrasound may be a useful method for male contraception since treatment with ultrasound $2W/cm^2$ for 15 min results in the arrest of spermiogenesis without altering either the behaviour or the testosterone level in blood.[41]

Immunization
This method of controlling reproduction is in the experimental stage, but may become a useful tool in the future. Injection of bovine LH to male dogs results in the formation of antibodies to the foreign hormone, the dogs are unable to ejaculate for approximately 6 weeks, and for up to 12 months reproductive function is impaired. As in castrated dogs the plasma androgen level decreases and there is atrophy of the testes, epididymides and prostate.[43]

REFERENCES

1. Anderson, R.K., Gilmore, E. and Schnelle, G.B. (1965) Utero-ovarian disorders associated with use of medroxyprogesterone in dogs. *J. Amer. vet. med. Assoc.*, **146**, 1311–1366.
2. Arbeiter, K. (1969) Hormone zur Behandlung der Hündin. *Wein. tierärztl. Monatsschr.* **55**, 587–591.
3. Austad, R., Lunde, A. and Sjaastad, Ø.V. (1976) Peripheral plasma levels of oestradiol-17β and progesterone in the bitch during the oestrous cycle, in normal pregnancy and after dexamethasone treatment. *J. Reprod. Fertil.*, **46**, 129–136.
4. Beach, F.A. (1970) Coital behaviour in dogs: VI. Long-term effects of castration upon mating in the male. *J. comp. physiol. Psychol.*, Monograph, **70**, (3), part 2, 1–32.
5. Beard, D.C. (1961) Hydroxyprogesterone acetate: use in estrogenic and progesterogenic states in the bitch. *Small Anim. Clin.*, **1**, 215–218.
6. Bell, E.T. and Christie, D.W. (1971) Erythrocytes and leucocytes in the vaginal smear of the beagle bitch. *Vet. Rec.*, **88**, 546–549.
7. Berchtold, M. (1970) Zum Problem der Trächtigkeitsunterbrechung bei der Hündin. In: *Tagung über Kleintierkrankheiten*, Wien (cited in W. Jöchle[71]).
8. Berzon, D.R. and DeHoff, J.B. (1974) Medical costs and other aspects of dog bites in Baltimore. *Public Health Rep.*, **89**, 377–381.
9. Bezard, P. (1981) *Prevention de l'oestrus chez la chienne par le mibolerone*. Thesis, Paris.
10. Bigbee, H.G. and Hennessey, P.W. (1977) Megestrol acetate for postponing estrus in first-heat bitches. *Vet. Med./small Anim. Clin.*, **72**, 1727–1730.
11. Le Boeuf, B.J. (1970) Copulatory and aggressive behaviour in the prepuberally castrated dog. *Horm. Behav.*, **1**, 127–136.
12. Brandt, H.-P. (1978) Die hormonelle Unterdrückung der Brunst. *Prakt. Tierarztl.* **59**, 92–94.
13. Brodey, R.S. and Fidler, I.J. (1966) Clinical and pathologic findings in bitches treated with progestational compounds. *J. Amer. vet. med. Assoc.*, **149**, 1406–1415.
14. Brueschke, E.E., Burns, M., Maness, J.H., Wingfield, J.R., Mayerhofer, K. and Zaneveld, L.J.D. (1974) Development of a reversible vas deferens occlusive device. I. Anatomical size of the human and dog vas deferens. *Fertil. Steril.*, **25**, 659–672.
15. Brueschke, E.E., Wingfield, J.R., Burns, M., and Zanevald, L.J.D. (1974) Development of a reversible vas deferens occlusive device. II. Effect of bilateral and unilateral vasectomy on semen characteristics in the dog. *Fertil. Steril.*, **25**, 673–686.
16. Bryan, H.S. (1960) Utility of 17α-acetoxyprogesterone in delaying estrus in the bitch. *Proc. Soc. exp. Biol. (NY)*, **105**, 23–26.
17. Bryan, H.S. (1973) Parenteral use of medroxyprogesterone acetate as an antifertility agent in the bitch. *Amer. J. vet. Res.*, **34**, 659–663.

18. Burke, T.J. (1977) Pregnancy prevention and termination. In: *Current Veterinary Therapy. VI. Small Anim. Pract.* 1241–1242. W.B. Saunders, Philadelphia, London, Toronto.
19. Burke, T.J. and Reynolds, H.A. (1975) Megestrol acetate for estrus postponement in the bitch. *J. Amer. vet. med. Assoc.*, **167**, 285–287.
20. Capel-Edwards, K., Hall, D.E., Fellowes, K.P., Vallance, D.K., Davies, M.J. and Robertson, W.B. (1973) Long-term administration of progesterone to the female beagle dog. *Toxicol. appl. Pharmacol.*, **24**, 474–488.
21. Chainey, D., McCoubrey, A. and Evans, J.M. (1970) The excretion of megestrol acetate by beagle bitches. *Vet. Rec.*, **86**, 287–288.
22. Chatterjee, S.N., Meenakshi, Sharma, R.N. and Kar, A.B. (1976) Semen characteristics of normal and vasectomized dogs. *Indian J. exp. Biol.*, **14**, 411–414.
23. Christie, D.W. and Bell, E.T. (1970) The use of progestagens to control reproductive function in the bitch. *Anim. Breed. Abstr.*, **38**(1), 1–21.
24. Clifton-Hadley, A. and Clifton-Hadley, R.S. (1978) Controlling oestrus in the dog. *Vet. Rec.*, **120**, 224.
25. Concannon, P.W. and Hansel, W. (1977) Prostaglandin F_2-induced luteolysis, hypothermia, and abortion in beagle bitches. *Prostaglandins*, **13**, 533–542.
26. Concannon, P., Altszuler, N., Hampshire, J., Butler, W.R. and Hansel, W. (1980) Growth hormone, prolactin, and cortisol in dogs developing mammary nodules and an acromegaly-like appearance during treatment with medroxyprogesterone acetate. *Endocrinology*, **106**, 1173–1177.
27. Concannon, P.W., Spraker, T.R., Casey, H.W. and Hansel, W. (1981) Gross and histopathologic effects of medroxyprogesterone acetate and progesterone on the mammary glands of adult beagle bitches. *Fertil. Steril.*, **36**, 373–387.
28. Coulden, L.W. (1964) Suppression of oestrus in the greyhound bitch. *Vet. Rec.*, **76**, 544.
29. Cox, J.E. (1970) Progestagens in bitches: a review. *J. small Anim. Pract.*, **11**, 759–778.
30. Crafts, R.C. (1948) The effects of estrogens on the bone marrow of adult female dogs. *Blood*, **3**, 276–285.
31. Craige, J.E. and Berger, C.J. (1965) Utero-ovarian disorders in dogs. *J. Amer. vet. med. Assoc.*, **147**, 316–318.
32. Curtis, E.M. and Grant, R.P. (1964) Masculinization of female pups by progestogens. *J. Amer. vet. med. Assoc.*, **144**, 395–398.
33. David, A., Edwards, K., Fellowes, K.P. and Plummer, J.M. (1963) Anti-ovulatory and other biological properties of megestrol acetate. *J. Reprod. Fertil.*, **5**, 331–346.
34. Dixit, V.P. and Lohiya, N.K. (1975) Chemical sterilization: effects of a single high dose of 3-chloro-1,2 propanediol on the testes and epididymides of dog. *Acta europ. fertil.*, **6**, 57–62.
35. Dixit, V.P., Lohiya, N.K. and Agrawal, M. (1975) Effects of α-chlorohydrin on the testes and epididymides of dog: a preliminary study. *Fertil. Steril.*, **26**, 781–785.
36. Dixit, V.P., Lohiya, N.K., Arya, M. and Agrawal, M. (1975) Chemical sterilization of male dogs after a single intra-testicular injection of danazol. *Folia biol. (Krakow)*, **23**, 305–310.
37. Döcke, F. (1975) Therapeutische Anwendung von Sexualhormonen bei Hund und Katze. In: *Veterinärmedizinische Endokrinologie*, 635–651. VEB Gustav Fischer Verlag, Jena.
38. Durr, J.L. (1976) Encourages ovariohysterectomy. Panel report. Preventing pregnancy after mismating in dogs. *Mod. vet. Pract.*, **57**, 1042.
39. El Etreby, M.F. (1979) Effect of cyprotrone acetate, levonorgestrel and progesterone on adrenal glands and reproductive organs in the beagle bitch. *Cell Tissue Res.*, **200**, 229–243.

40. Evans, J.M. (1972) Oestrus control in the bitch. *VII Int. Congr. Animal Reprod.*, Munich, Summaries, 203–204.
41. Fahim, M.S., Montie, J.E., Thompson, I.M. and Hall, D.G. (1976) Ultrasound as a new method of male contraception. *Fertil. Steril.*, **27**, 216.
42. Faulkner, L.C. (1975) Dimensions of the pet population problem. *J. Amer. vet. med. Assoc.*, **166**, 477–478.
43. Faulkner, L.C. (1975) An immunologic approach to population control in dogs. *J. Amer. vet. med. Assoc.*, **166**, 479–480.
44. Feldmann, B.M. (1974) The problem of urban dogs. *Science*, **185**, 903.
45. Ficus, H.J. (1977) Oestrus prevention with progestins. *Proc. 6th World Congr., Wld Small Anim. vet. Assoc.*, **55**.
46. Frank, D.W., Kirton, K.T., Murchison, T.E., Quinland, W.J., Coleman, M.E., Gilbertson, T.J., Feenstra, E.S. and Kimball, F.A. (1979) Mammary tumors and serum hormones in the bitch treated with medroxyprogesterone acetate or progesterone for four years. *Fertil. Steril.*, **31**, 340 and 346.
47. Franti, C.E. and Kraus, J.F. (1974) Aspects of pet ownership in Yolo county California. *J. Amer. vet. med. Assoc.*, **164**, 166–171.
48. Freeman, C. and Coffey, D.S. (1973) Sterility in male animals induced by injection of chemical agents into the vas deferens. *Fertil. Steril.*, **24**, 884–890.
49. Galliani, G. and Lerner, L.J. (1976) Pregnancy termination in dogs with novel nonhormonal compounds. *Amer. J. vet. Res.*, **37**, 263–268.
50. Galliani, G. and Omodei-Salé, A. (1982) Pregnancy termination in dogs with non-hormonal compounds: evaluation of selected derivatives. *J. small Anim. Pract.*, **23**, 295–300.
51. Galliani, G., Lerner, L.J., Caramel, C., Maraschin, R., Nani, S. and Nava, A. (1982) Pregnancy termination in dogs with novel non-hormonal compounds. Studies of 2-(3-ethoxy-phenyl)-5,6-dihydro-s-triozole[5,1-α]isoquinoline (DL 204-IT). *Arzneim.-Forsch./Drug Res.*, **32**, (I), 123–127.
52. Gee, R.W. (1965) Uterine disorders associated with use of medroxyprogesterone. *Aust. vet. J.*, **41**, 299–300.
53. Geil, R.G. and Lamar, J.K. (1977) FDA studies of estrogen, progestagens, and estrogen/progestagen combinations in the dog and monkey. *J. Toxicol. environm. Health*, **3**, 179–193.
54. Gerber, H.A., Jöchle, W. and Sulman, F.G. (1973) Control of reproduction and undesirable social and sexual behaviour in dogs and cats. *J. small Anim. Pract.*, **14**, 151–158.
55. Goyings, L.S., Sokolowski, J.H., Zimbelman, R.G. and Geng, S. (1977) Clinical, morphologic, and clinicopathologic findings in beagles treated for two years with melengestrol acetate. *Amer. J. vet. Res.*, **38**, 1923–1931.
56. Gräf, K.-J., El Etreby, M.F.A., Richter, K.-D., Günzel, P. and Neumann, F. (1975) The progestogenic potencies of different progestogens in the beagle bitch. *Contraception*, **12**, 529–540.
57. Gustafsson, B. and Jönsson, M. (1972) The effect on sexual behaviour and semen picture in dogs of a new antifertility compound KABI-1774. *Acta pharmacol. (Kbh.)*, **31**, Suppl. I.
58. Halnan, C.R.E. (1965) The use of norethisterone acetate for the control of the signs of oestrus in the bitch. *J. small Anim. Pract.*, **6**, 201–206.
59. Hamidinia, A., Beck, A.D. and Wright, N. (1983) Morphologic changes of the vas deferens after vasectomy and vasovasostomy in dogs. *Surg. Gynec. Obstet.*, **156**, 737–742.
60. Harris, D., Imperato, P.J. and Oken, B. (1974) Dog bites – an unrecognized epidemic. *Bull. NY Acad. Med.*, **50**, 981–1000.
61. Harris, T.W. and Wolchuk, N. (1963) The suppression of estrus in the dog and cat with long-term administration of synthetic progestational steroids. *Amer. J. vet. Res.*, **24**, 1003–1006.

62. Hart, B.L. (1968) Role of prior experience on the effects of castration on sexual behaviour of male dogs. *J. comp. physiol. Psychol.*, **66**, 719–725.
63. Hart, B.L. (1976) Behavioral effects of castration. *Canine Pract.*, **3**(3), 10–21.
64. Hill, R., Averkin, E., Brown, W., Gagne, W.E. and Segre, E. (1970) Progestational potency of chlormadinone acetate in the immature beagle bitch: Preliminary report. *Contraception*, **2**, 381–390.
65. Hopkins, S.G., Schubert, T.A. and Hart, B.L. (1976) Castration of adult male dogs: effects on roaming, aggression, urine marking, and mounting. *J. Amer. vet. med. Assoc.*, **168**, 1108–1110.
66. van der Horst, C.J.G. and Vogel, F. (1977) Some effects of prostaglandin on corpora lutea and on the uterus in the cycling dog. *T. Diergeneesk.*, **102**, 117–123.
67. Houpt, K.A., Coren, B., Hintz, H.F. and Hilderbrant, J.E. (1979) Effect of sex and reproductive status on sucrose preference, food intake, and body weight of dogs. *J. Amer. vet. med. Assoc.*, **174**, 1083–1085.
68. Jackson, P.S., Furr, B.J.A. and Hutchinson, F.G. (1982) A preliminary study of pregnancy termination in the bitch with slow-release formulations of prostaglandin analogues. *J. small Anim. Pract.*, **23**, 287–294.
69. Jöchle, W. (1974) Progress in small animal reproductive physiology, therapy of reproductive disorders, and pet population control. *Folia vet. lat. (Milano)*, **4**, 706–731.
70. Jöchle, W. (1974) Pet population control: chemical methods. *Canine Pract.*, **1**(3), 8–18.
71. Jöchle, W. (1975) Hormones in canine gynecology. A review. *Theriogenology*, **3**, 152–165.
72. Jöchle, W. (1978) Review: the pet population control scene. *Animal Reproduction Report. (Theriogenology Digest)* **no. 2**, 1–12.
73. Jöchle, W. (1980) Review: the pet population control scene (II). *Animal Reproduction Report. (Theriogenology Digest)* **no. 14**, 18–30.
74. Jöchle, W., Thomlinson, R.V. and Andersen, A.C. (1973) Prostaglandin effects on plasma progesterone levels in the pregnant and cycling dog (beagle). *Prostaglandins*, **3**, 209–217.
75. Jöchle, W., Lamond, D.R. and Andersen, A.C. (1975) Mestranol as an abortifacient in the bitch. *Theriogenology*, **4**, 1–9.
76. Joshua, J.O. (1965) The spaying of bitches. *Vet. Rec.*, **77**, 642–647.
77. Kamp, J.N. (1974) Prevention of pregnancy. In: *Current Veterinary Therapy. V. Small Anim. Pract.*, 925. W.B. Saunders, Philadelphia, London, Toronto.
78. Kennelly, J.J. (1969) The effect of mestranol on canine reproduction. *Biol. Reprod.*, **1**, 282–288.
79. Klitsgaard, J., Larsen, N.L. and Linnet, A. (1970) Regulation of heat in dog by injection of MAP. *Proc. XI Nord. vet. Congr., Bergen,* 251.
80. Klucker, D. (Cit. Wildt, D.E., Kinney, G.M. and Seager, S.W.J. (1977) Reproduction control in the dog and cat: an examination and evaluation of current and proposed methods. *J. Amer. Anim. Hosp. Assoc.*, **13**, 223–231.
81. Knecht, C.D. (1966) A brief survey of progestogen involvement in utero-ovarian disorders. *Illinois Vet.*, **9**, 3–10.
82. Krzeminski, L.F., Sokolowski, J.H., Dunn, G.H., VanRavenswaay, F. and Pineda, M. (1978) Serum concentrations of mibolerone in beagle bitches as influenced by time, dosage form, and geographic location. *Amer. J. vet. Res.*, **39**, 567–572.
83. LaDu, R.W. (1976) Uses ECP. Panel report. Preventing pregnancy after mismating in dogs. *Mod. vet. Pract.*, **57**, 1041–1042.
84. Lake, S.G. (1977) Controlling oestrus in the dog. *Vet. Rec.*, **101**, 530.
85. Lee, A.C. and Carlson, W.D. (1965) Reproduction in the beagle following irradiation of the ovaries. *Radiat. Res.*, **25**(210), Abstr. no. 104.

86. Lee, A.C. and Carlson, W.D. (1967) The histopathology of beagle ovaries after acute irradiation of the individual gonads. *Radiat. Res.*, **31**(653), Abstr. no. Ga-1.
87. Lowenstine, L.J., Ling, G.V. and Schalm, O.W. (1972) Exogenous estrogen toxicity in the dog. *Calif. Vet..*, **26**, 14–19.
88. MacDougall, M.K., McCowin, K., Derrick, F.C., Glover, W.L. and Jacobson, C.B. (1975) The effects of vasectomy on spermatogenesis in the dog, canis familiaris: a meiotic analysis. *Fertil. Steril.*, **26**, 786–790.
89. Magnusson, L.-E. & Ottander, G. (1969) Uppskjutande av löpning hos tik. *Svensk Vet.-Tidn.*, **21**, 31–34.
90. Mahi, C.A. and Yanagimachi, R. (1979) Prevention of *in vitro* fertilization of canine oocytes by anti-ovary antisera: a potential approach to fertility control in the bitch (1). *J. exp. Zool.*, **210**, 129–135.
91. Mandell, M.T. (1976) Never for breeding bitches. Panel report. Preventing pregnancy after mismating in dogs. *Mod. vet. Pract.*, **57**, 1042.
92. Mann, C.J. (1971) Some clinical aspects of problems associated with oestrus and with its control in the bitch. *J. small Anim. Pract.*, **12**, 391–397.
93. Mathieu, L., Rambaud, M. and Marguin, M. (1978) L'acetate de médroxyprogesterone dans le contrôle des chaleurs de la chieene. Étude clinique. *Anim. Compagnie*, **13**, 185–197.
94. Matteuzzi, A. (1969) On the use of slow progestogen in the suppression of heat in small animals. *Nuova Vet.*, **45**, 147–152.
95. Mellin, T.N., Orczyk, G.P., Hichens, M. and Behrman, H.R. (1976) Chemical inhibition of estrus in the beagle. *Theriogenology*, **5**, 165–174.
96. Moltzen, H. (1963) Hinausschiebung der Läufigkeit bei Hunden und Katzen mit Perlutex Leo. *Kleintier-Prax.*, **8**, 25–27.
97. Mullican, M. (1976) Estrone found effective. Panel report. Preventing pregnancy after mismating in dogs. *Mod. vet. Pract.*, **57**, 1041.
98. Murray, G.H. and Eden, E.L. (1952) Progesterone to delay estrum in bitches. *Vet. Med.*, **47**, 467–468.
99. Nelson, L.W. and Kelly, W.A. (1976) Progesterone-related gross and microscopic changes in female beagles. *Vet. Pathol.*, **13**, 143–156.
100. Nugent, T. (1976) Good results with Stilbestrol. Panel report. Preventing pregnancy after mismating in dogs. *Mod. vet. Pract.*, **57**, 1042.
101. Okin, Jr., R.E. (1976) Pyometra is complication. Panel report. Preventing pregnancy after mismating in dogs. *Mod. vet. Pract.*, **57**, 1041.
102. van Os, J.L. (1977) Oestrus control in bitches with proligestone, a new progestogen. *Proc. 6th World Congr., Wld Small Anim. vet. Assoc.*, 56–57.
103. van Os, J.L. (1982) *Oestrus control in the bitch with proligestone.* Thesis, Utrecht.
104. van Os, J.L. and Oldenkamp, E.P. (1978) Oestrus control in bitches with proligestone, a new progestational steroid. *J. small Anim. Pract.*, **19**, 521–529.
105. van Os, J.L. and Oldenkamp, E.P. (1980) Ist eine langfristige Östrusverhütung risikolos möglich? *Kleintier-Prax.*, **25**, 223–226.
106. van Os, J.L. and Oldenkamp, E.P. (1982) Oestrusregulation und Mammatumoren. *Kleintier-Prax.*, **27**, 79–86.
107. van Os, J.L., van Laar, P.H., Oldenkamp, E.P. and Verschoor, J.S.C. (1981) Oestrus control and the incidence of mammary nodules in bitches, a clinical study with two progestogens. *Vet. Quart.*, **3**, 46–56.
108. Oslund, R.M. (1924) Vasectomy on dogs. *Amer. J. Physiol.*, **70**, 111–117.
109. Pearson, H. (1973) The complications of ovariohysterectomy in the bitch. *J. small Anim. Pract.*, **14**, 257–266.
110. Pettit, G. (1965) Progesterone-induced pyometra in the bitch. *Anim. Hosp.*, **1**, 151–158.
111. Pineda, M.H. (1978) Chemical vasectomy in dogs. *Canine Pract.*, **5**(2), 34–46.
112. Pineda, M.H. and Hepler, D.I. (1981) Chemical vasectomy in dogs. Long-term study. *Theriogenology*, **16**, 1–11.

113. Pineda, M.H., Reimers, T.J. and Faulkner, L.C. (1976) Disappearance of spermatozoa from the ejaculates of vasectomized dogs. *J. Amer. vet. med. Assoc.*, **168**, 502–503.
114. Pineda, M.H., Reimers, T.J., Faulkner, L.C., Hopwood, M.L. and Seidel, G.E. (1977) Azoospermia in dogs induced by injection of sclerosing agents into the caudae of the epididymides. *Amer. J. vet. Res.*, **38**, 831–838.
115. Prole, J.H.B. (1974) The control of oestrus in racing greyhound bitches using norethisterone acetate. *J. small Anim. Pract.*, **15**, 213–219.
116. Prole, J.H.B. (1974) The effect of the use of norethisterone acetate to control oestrus in greyhound bitches on subsequent racing performance and fertility. *J. small Anim. Pract.*, **15**, 221–228.
117. LeRoux, P.H. and Van Der Walt, L.A. (1978) Ovarian autograft as an alternative to ovariectomy in bitches. *J. Amer. Anim. Hosp. Assoc.*, **14**, 418–419.
118. Ruckstuhl, B. (1976) Über die Brauchbarkeit des N-desacetyl-thiocolchicins zur Trächtigkeitsunterbrechung bei der Hündin. *Kleintier-Prax.*, **21**, 302–307.
119. Ruckstuhl, B. (1977) Probleme der Nidationsverhütung bei der Hündin. *Schweiz. Arch. Tierheilk.*, **119**, 57–65.
120. Ruckstuhl, B. (1978) Die Incontinentia urinae bei der Hündin als Spätfolge der Kastration. *Schweiz. Arch. Tierheilk.*, **120**, 143–148.
121. Runić, S., Miljković, M., Bogumil, R.J., Nahrwold, D. and Bardin, C.W. (1976) The *in vivo* metabolism of progestins. I. The metabolic clearance rates of progesterone and medroxyprogesterone acetate in the dog. *Endocrinology*, **99**, 108–113.
122. Rüsse, M. and Jöchle, W. (1963) Über die sexuelle Ruhigstellung weiblicher Hunde und Katzen bei normalem und gestörtem Zyklusgeschehen mit einem peroral wirksamen Gestagen. *Kleintier-Prax.*, **8**, 87–89.
123. Saxena, O.P. (1966) Ligature of fallopian tubes (salpingectomy) in bitches. *Indian vet. J.*, **43**, 83–84.
124. Schalm, O.W. (1978) Exogenous estrogen toxicity in the dog. *Canine Pract.*, **5**(5), 57–61.
125. Schneider, R. (1975) Observations on overpopulation of dogs and cats. *J. Amer. vet. med. Assoc.*, **167**, 281–284.
126. Schneider, R. and Vaida, M.L. (1975) Survey of canine and feline populations. Alameda and Contra Costa counties, California, 1970. *J. Amer. vet. med. Assoc.*, **166**, 481–486.
127. Sekeles, E., de Lange, A., Samuel, L. and Aharon, D.C. (1982) Oestrus control in bitches with chlormadinone acetate. *J. small Anim. Pract.*, **23**, 151–158.
128. Shivers, C.A., Sieg, P.M. and Kitchen, H. (1981) Pregnancy prevention in the dog: potential for an immunological approach. *J. Amer. Anim. Hosp. Assoc.*, **17**, 823–828.
129. Simmons, J.G. and Hamner, C.E. (1973) Inhibition of estrus in the dog with testosterone implants. *Amer. J. vet. Res.*, **34**, 1409–1419.
130. Sokolowski, J.H. (1971) The effects of ovariectomy on pregnancy maintenance in the bitch. *Lab. Anim. Sci.*, **21**, 5, 696–699.
131. Sokolowski, J.H. (1976) Efficacy and safety of mibolerone for estrous inhibition in the bitch. *VIII Int. Congr. Animal Reprod., Krakow*, vol. **III**, 507–510.
132. Sokolowski, J.H.L. (1976) Androgens as contraceptives for pet animals with specific reference to the use of mibolerone in the bitch. In: *Pharmacology in the Animal Health Sector*, ed. Davis, L.E. and Faulkner, L.C., 164–175. Colorado State University Press, Ft. Collins, Col.
133. Sokolowski, J.H. (1978) Evaluation of estrous activity in bitches treated with mibolerone and exposed to adult male dogs. *J. Amer. vet. med. Assoc.*, **173**, 983–984.
134. Sokolowski, J.H. and Geng, S. (1977) Effect of prostaglandin $F_{2\alpha}$-THAM in the bitch. *J. Amer. vet. Med. Assoc.*, **170**, 536–537.
135. Sokolowski, J.H. and Kasson, C.W. (1978) Effects of mibolerone on conception,

pregnancy, parturition, and offspring in the beagle. *Amer. J. vet. Res.*, **39**, 837–839.

136. Sokolowski, J.H. and VanRavenswaay, F. (1976) Effects of melengestrol acetate on reproduction in the beagle bitch. *Amer. J. vet. Res.*, **37**, 943–945.

137. Sokolowski, J.H. and Zimbelman, R.G. (1973) Canine reproduction: effects of a single injection of medroxyprogesterone acetate on the reproductive organs of the bitch. *Amer. J. vet. Res.*, **34**, 1493–1499.

138. Sokolowski, J.H. and Zimbelman, R.G. (1973) Canine reproduction: effects of a single injection of medroxyprogesterone acetate on the reproductive organs of intact and overiectomized bitches. *Amer. J. vet. Res.*, **34**, 1501–1503.

139. Sokolowski, J.H. and Zimbelman, R.G. (1974) Canine reproduction: effects of multiple treatments of medroxyprogesterone acetate on reproductive organs of the bitch. *Amer. J. vet. Res.*, **35**, 1285–1287.

140. Sokolowski, J.H. and Zimbelman, R.G. (1976) Evaluation of selected compounds for estrus control in the bitch. *Amer. J. vet. Res.*, **37**, 939–941.

141. Sokolowski, J.H., Medernach, R.W. and Helper, L.C. (1968) Exogenous hormone therapy to control the estrous cycle of the bitch. *J. Amer. vet. med. Assoc.*, **153**, 425–428.

142. Special report (1974) Conclusions and Recommendations from the National Conference on the Ecology of the Surplus Dog and Cat Problem. *J. Amer. vet. med. Assoc.*, **165**, 363–370.

143. Stabenfeldt, G.H. (1974) Physiologic, pathologic and therapeutic roles of progestins in domestic animals. *J. Amer. vet. med. Assoc.*, **164**, 311–317.

144. Stubbs, A.J. and Resnick, M.I. (1978) Protein electrophoretic patterns of canine prostatic fluid. Effect of hormonal manipulation. *Invest. Urol.*, **16**, 175–178.

145. Thiersch, J.B. (1967) Abortion in the bitch with *N*-desacetyl-thiocolchicine. *J. Amer. vet. med. Assoc.*, **151**, 1470–1473.

146. Turner, T. (1982) Oestrus control in the bitch. *Vet. Drug*, **13**, 4, 16–17.

147. Tyslowitz, R. and Dingemanse, E. (1941) Effect of large doses of estrogens on the blood picture of dogs. *Endocrinology*, **29**, 817–827.

148. Vandaele, W. (1977) Progrès récents dans la connaissance du cycle sexuel des chiennes. Précautions à prendre lors de l'emploi de progestagènes. *Ann. Méd. vét.*, **121**, 369–381.

149. Vare, A.M. and Bansal, P.C. (1973) Changes in the canine testes after bilateral vasectomy – an experimental study. *Fertil. Steril.*, **24**, 793–797.

150. Vare, A.M. and Bansal, P.C. (1974) The effects of ligation of cauda epididymidis on the dog testis. *Fertil. Steril.*, **25**, 256–260.

151. Vickery, B. and McRae, G. (1980) Effect of a synthetic prostaglandin analogue on pregnancy in beagle bitches. *Biol. Reprod.*, **22**, 438–442.

152. Vickery, B.H., McRae, G.I., Kent, J.S. and Tomlinson, R.V. (1980) Manipulation of duration of action of a synthetic prostaglandin analogue (TPT) assessed in the pregnant beagle bitch. *Prostaglandins and Med.* **5**, 93–100.

153. Weikel, Jr., J.H. and Nelson, L.W. (1977) Problems in evaluating chronic toxicity of contraceptive steroids in dogs. *J. Toxicol. environm. Health*, **3**, 167–177.

154. Wells, C.G. (1976) Good results with ECP. Panel report. Preventing pregnancy after mismating in dogs. *Mod. vet. Pract.*, **57**, 1042.

155. Whitney, L.F. (1960) Further studies on the effect of malucidin on pregnancy. *Vet. Med.*, **55**, 57–65.

156. Whitney, G.D. (1968) Prevention of pregnancy. In: *Current Veterinary Therapy. III. Small Anim. Pract.*, 679–680. W.B. Saunders, Philadelphia, London, Toronto.

157. Wildt, D.E. and Seager, S.W.J. (1977) Reproduction control in dogs. *Vet. Clin. N. Amer.*, **7**, 775–787.

158. Wildt, D.E., Seager, S.W.J. and Bridges, C.H. (1981) Sterilization of the male dog and cat by laparoscopic occlusion of the ductus deferens. *Amer. J. vet. Res.*, **42**, 1888–1897.

152   REPRODUCTION IN THE DOG

159. Withers, A.R. and Whitney, J.C. (1967) The response of the bitch to treatment with medroxyprogesterone acetate. *J. small Anim. Pract.*, **8**, 265–271.
160. Wright, L.J., Feinstein, A., Heap, R.B., Saunders, J.C., Bennett, R.C. and Wang, M.-Y. (1982) Progesterone monoclonal antibody blocks pregnancy in mice. *Nature (Lond.)*, **295**, 415–417.
161. Zimbelman, R.G. and Lauderdale, J.W. (1973) Failure of prepartum or neonatal steroid injections to cause infertility in heifers, gilts, and bitches. *Biol. Reprod.*, **8**, 388–391.

ADDITIONAL READING

Baker, B.A., Archbald, L.F., Clooney, L.L., Lotz, K. and Godke, R.A. (1980) Luteal function in the hysterectomized bitch following treatment with prostaglandin $F_{2\alpha}$ ($PGF_{2\alpha}$). *Theriogenology*, **14**, 195–205.
Barsanti, J.A., Edwards, P.D. and Losonsky, J. (1981) Testosterone responsive urinary incontinence in a castrated male dog. *J. Amer. Anim. Hosp. Assoc.*, **17**, 117–119.
Baumgärtner, W. and Posselt, H.J. (1982) Östrogeninduzierte hämorrhagische Diathese bei einer Hündin. *Kleintier-Prax.*, **27**, 419–420.
Beach, F.A. (1974) Effects of gonadal hormones on urinary behaviour in dogs. *Physiol. Behav.*, **12**, 1005–1013.
Berchtold, M. and Zindel-Grunder, S. (1981) Viszerale Transplantation von Ovargewebe zur Verhinderung von Nebenwirkungen nach der Kastration der Hündin. *Zuchthygiene*, **16**, 80.
Burke, T.J. (1982) Pharmacologic control of estrus in the bitch and queen. *Vet. Clin. N. Amer.*, **12**, 79–84.
Calixto, J.B., Aucélico, J.G. and Jurkiewicz, A. (1979) Relationship between modulation by estradiol, progesterone and calcium upon the pharmacological reactivity of uteri of dogs. *Res. Commun. chem. Pathol. Pharmacol.*, **25**, 447–460.
Cattin, D.G. (1980) Chemotherapeutic oestrus control in the bitch and latest advances. *J. S. Afr. vet. Assoc.*, **51**, 213–218.
Dorn, A.S. and Swist, R.A. (1977) Complications of canine ovariohysterectomy. *J. Amer. Anim. Hosp. Assoc.*, **13**, 720–724.
Dunbar, I.F. (1975) Behaviour of castrated animals. *Vet. Rec.*, **96**, 92–93.
Dürr, A. (1975) Pyometra nach Östrogenbehandlung. *Schweiz. Arch. Tierheilk.*, **117**, 349–354.
Eigenmann, J.E. and Venker-van Haagen, A.J. (1981) Progestagen-induced and spontaneous canine acromegaly due to reversible growth hormone overproduction: clinical picture and pathogenesis. *J. Amer. Anim. Hosp. Assoc.*, **17**, 813–822.
Ficus, H.-J. and Jöchle, W. (1975) Erwünschte und unerwünschte Gestagenwirkungen bei der Hündin. *Tierärztl. Prax.*, **3**, 231–241.
Greene, J.A. (1979) An alternative method for transfixation of the uterine stump. *Canine Pract.*, **6**, 26–29.
Gundlach, C.E. (1980) New products for use in the bitch for birth control. *XI. Int. Congr. Diseases of Cattle. Satellite Symp. Diseases of Small Animals, Tel Aviv*, 37–44.
Haase, F., Beier, S., Hartmann, D. and Elger, W. (1977). Development of a qualitative canine bioassay for gestagens. *Acta endocr, (Kbh.)*, **84**, Suppl. 208, 122–123.
Hall, A. and Dale, H.E. (1964) The effect of gonadotrophic hormones and progesterone on the estrous cycle of the female dog. *Vet. Med./small Anim. Clin.*, **59**, 852–854.
Hart, B.L. (1974) Environmental and hormonal influences on urine marking behaviour in the adult male dog. *Behav. Biol.*, **11**, 167–176.
Heywood, R. and Wadsworth, P.F. (1980) The experimental toxicology of estrogens. *Pharmacol. Ther.*, **8**, 125–142.
Julien, F. (1978) *La vasectomie chez le chien*. Thesis, Paris, 44 pp.

Kaplan, B. (1981) A technique of canine castration using anatomic structures for hemostasis. *Vet. Med./small Anim. Clin.*, **76**, 193–196.

Kumar, M.S.A., Chen, C.L. and Kalra, S.P. (1980) Distribution of luteinizing hormone releasing hormone in the canine hypothalamus: effect of castration and exogenous gonadal steroids. *Amer. J. vet. Res.*, **41**, 1304–1309.

Kunin, S. and Terry, M. (1980) A complication following ovariohysterectomy in a dog. *Vet. Med./small Anim. Clin.*, **75**, 1000–1001.

Lessey, B.A. and Gorell, T.A. (1980) A cytoplasmic estradiol receptor in the immature beagle uterus. *J. Steroid Biochem.*, **13**, 211–217.

Lessey, B.A. and Gorell, T.A. (1981) Nuclear progesterone receptors in the beagle uterus. *J. Steroid Biochem.*, **14**, 585–591.

Mahi-Brown, C.A., Yanagimachi, R. and Hoffmann, J.C. (1983) Ovulation failure in bitches actively immunized with porcine zona pellucida. *Biol. Reprod.*, **28**, Suppl. 1, 158.

Mailhac, J.M., Barraud, F., Valon, F. and Chaffaux, S. (1980) Maîtrise de la reproduction chez les carnivores. Les interventions chirurgicales. *Le Points Vét.*, **10**, 17–39.

Medleau, L., Johnson, C.A., Perry, R.L. and Dulisch, M.L. (1983) Female pseudohermaphroditism associated with mibolerone administration in a dog. *J. Amer. Anim. Hosp. Assoc.*, **19**, 213–215.

Meesters, J.P.M. (1965) Hysterectomy and ovariotomy in the dog from the left flank. *T. Diergeneesk.*, **90**, 96–97.

Nelson, L.W., Botta, J.A., Jr. and Weikel, J.H., Jr. (1973) Estrogenic activity of norethindrone in the immature female beagle. *Res. Commun. chem. Pathol. Pharmacol.*, **5**, 879–883.

Oettel, M., Albrecht, G. and Stölzner, W. (1972) Endokrinologische und klinische Befunde bei der Anwendung von Östriol in der Kleintierpraxis. *Arch. exp. Vet.-Med.*, **26**, 1053–1060.

Pearson, H. (1973) The complications of ovariohysterectomy in the bitch. *J. small Anim. Pract.*, **14**, 257–266.

Ratcliffe, J. (1971) Vaginal bleeding in a spayed bitch: a case report. *J. small Anim. Pract.*, **12**, 169.

Reed, R.A. and Thornton, J.R. (1982) Stilboestrol toxicity in a dog. *Aust. vet. J.*, **58**, 217–218.

Schütt, I. (1978) Abort und Nidationsverhütung beim Hund. *Kleintier-Prax.*, **23**, 319–324.

Schulman, J. (1982) Canine urinary incontinence due to atony of the bladder and urethral incompetence. *Vet. Med./small Anim. Clin.*, **77**, 1495–1497.

Shille, V.M. (1982) Mismating and termination of pregnancy. *Vet. Clin. N. Amer.*, **12**, 99–106.

Sokolowski, J.H. (1974) Pharmacologic control of fertility in small domestic animals. *41st Ann. Meet. Amer. Anim. Hosp. Assoc.*, **475**.

Solti, F., Holló, I., Iskum, M., Nagy, J. and Ruzsa, P. (1965) Über die Wirkung der weiblichen Sexualhormone auf den Kreislauf und auf die Durchblutung der Extremitäten beim Hund. *Acta med. Acad. Sci. hung.*, **21**, 337–341.

Verstegen, J., De Coster, R. and Brasseur, M. (1981) Influence des oestrogènes sur les constantes sanguines chez la chienne. *Ann. Méd. vét.*, **125**, 397–403.

de Vries, H. and Daykin, P.W. (1976) The development of a pharmaceutical preparation for controlling oestrus in the bitch. *VIII Int. Congr. Anim. Reprod., Krakow*, 514–517.

Wagner, R. (1968) Präventive und inhibitive Zyklusbeeinflussung bei der Hündin. *Kleintier-Prax.*, **13**, 133–135.

Weiger, G. (1979) Ein Beitrag zur Kastration der Hündin. *Kleintier-Prax.*, **24**, 339–340.

Yamamoto, T., Honjo, H., Yamamoto, T., Oshima, K., Otsubo, K., Okoda, H. and Shibata, K. (1979) The metabolic fate of estradiol benzoate in female dog. *J. Steroid Biochem.*, **11**, 1287–1294.

Chapter 8
Pregnancy

DEVELOPMENT OF THE UTERUS

The ova develop to the stage of morula in the oviducts, enter the uterus as morulae of 16 cells or more 8–12 days after breeding and become blastocysts shortly thereafter.[18] After an average unattached period of 1 week, implantation takes place, 17–22 days after ovulation and breeding.

About 24–48 h after ovulation the fertilized ova have been transported into the middle section of the oviducts where cleavage begins after 72 h, and at 96, 120, 144, 168 and 192 h the ova are at the 2 cell, 2–5 cell, 8 cell, 8–16 cell and 16 cell stages, respectively, in the distal section.[39] At 204–216 h the ova are transported into the uterus as morulae. Implantations take place with an almost equal distribution of fetuses in the two horns. In Figures 8.1 to 8.3 the fetal membranes and the characteristic placenta zonaria are shown.

After breeding at 21 days the liquid-filled fetal membranes form

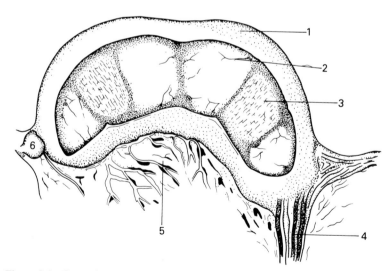

Figure 8.1 Opened uterus showing two intact fetal membranes, with the typical zonary placenta, in the dog. (1) Uterus. (2) Chorion without villi. (3) Chorion with villi. (4) Uterine cervix. (5) Uterine broad ligament. (6) Ovary. (From Nielsen[29])

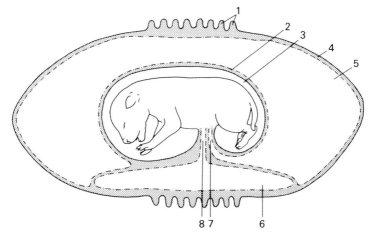

Figure 8.2 Diagram of fetal membranes in dog, longitudinal section. (1) Chorionic villi. (2) Allantoamnion. (3) Amniotic cavity. (4) Allantochorion. (5) Allantoic sac. (6) Yolk sac. (7) Allantoic stalk. (8) Yolk stalk. (From Nielsen[29])

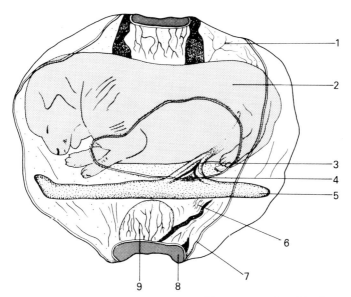

Figure 8.3 Opened fetal membranes in the dog. (1) Allantochorion. (2) Allantoamnion. (3) Umbilical cord. (4) Yolk sac. (5) Apex of yolk sac. (6) Allantois covering the yolk sac. (7) Cut edge of allantochorion. (8) Marginal haematoma. (9) Zonary placenta. (From Nielson[29])

distinct separated ovoid swellings on the uterus with an average diameter of 12–15 mm; at 28 days the uterine swellings become more spherical in form, with a diameter of 25 mm and the same consistency as at the 3-week stage; at 35 days the fetal membranes and the uterus have grown so much that the full and now folded uterus is of uniform diameter throughout its length. Due to the weight of the fetuses the uterus is bent downwards, forming a bow from the anterior border of the pelvis to the posterior sublumbar region; at 49 days after breeding, the uterus occupies the abdominal cavity from the pelvis to the liver, and then dorsally and caudally.

Some variations in fetal size may occur from litter to litter. In an investigation of 4 litters it was found that the average length of a fetus at the 30-day pregnancy stage was 8, 16, 17 and 20 mm, respectively, and at the 48th day the fetal length varied from 96 to 130 mm and the weight from 43 to 128 g.[2] In Figure 8.4 the growth curve and the representative developmental stages of the dog embryo are shown.

Figure 8.4   (*Opposite*) Growth curve and representative developmental stages of the dog. (After Evans and Sack[11])

Days of gestation		
	15	Primitive streak complete
	16	Neural tube forming; somite formation begins
	17	Three primary brain vesicles present; neural tube closed to level of 8th somite
	18	Allantois forming
	20	Cephalic flexure present; first and second branchial arches present; torsion of caudal part of embryo; approximately 30 somites
	22	Embryo C-shaped; amnion complete; third branchial arch present; *forelimb* and *hindlimb buds* forming; olfactory pits forming
	25	Fourth branchial arch present; limb buds cylindrical; mammary ridge present; acoustic meatus present
	25–28	*Hand plate* present; pinna present as ridge; shallow grooves between forelimb digits; eyes pigmented
	30–32	Eyelids forming; *tactile hair follicles* on upper lip and above eyes; external genitalia consist of genital tubercle and lateral swellings; grooves between digits; pinna partly covers acoustic meatus; five pairs of mammary primordia present; *intestines herniate* into umbilical cord
	33	*Palate fused*
	35	Eyelids partly cover eyes; pinna covers acoustic meatus; digits separated distally; external genitalia differentiated
	38	Tactile hairs appear
	40	*Eyelids fused*; intestines returned to abdominal cavity; claws formed
	43	Hair follicles present on body; digits widely spread
	45	*Body hair* forming; colour markings appear
	53	Hair covering complete; digital pads present
	57–63	Birth

A relationship has been reported between the three parameters: gestational age, fetal weight and fetal crown–rump length. The gestational age in days in beagles is calculated: 28.360 + 1.8811 × (crown–rump length in cm) −0.0097129 × (weight in g) with an accuracy of ± 4 days.[28]

LENGTH OF PREGNANCY

It is difficult to specify exactly the length of pregnancy, because (1) the bitch may allow mating over a period of several days and mating may even occur before ovulation; (2) ovulations also occur over a

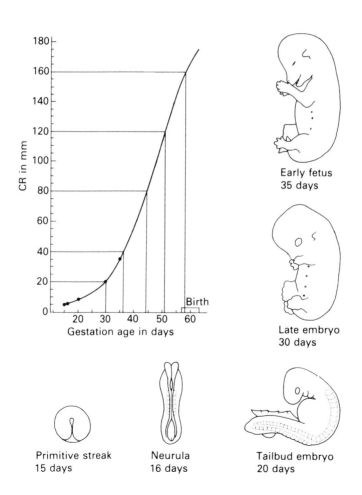

Early fetus
35 days

Late embryo
30 days

Primitive streak
15 days

Neurula
16 days

Tailbud embryo
20 days

period of time; (3) recently ovulated eggs cannot be fertilized but require some time, 2–5 days, for the meiotic cleavage and maturation before fertilization is possible; (4) and spermatozoa may survive several days in the reproductive tract of the bitch. The age of the bitch and the number of fetuses also have some influence. Thus there is a positive correlation between the duration and the number of the pregnancy;[30] on the other hand, there is a negative correlation between the litter size and the duration of the pregnancy.[19,31] The chances of delivering live and surviving puppies diminish with increasing time from the average duration of pregnancy.

The duration of the gestation period averages 63 ± 2 days in most breeds (with variations from 54 to 72 days) calculated from the first day of acceptance (Table 8.1). Prediction of the most probable time for parturition is thus possible when the first day of acceptance is known (Table 8.2). If the day of parturition and the length of pregnancy are to be calculated from when the cell content in the vaginal smear changes from predominance of superficial cells to predominance of intermedium and basal cells, parturition starts 57 days later and the duration of the pregnancy averages 60 days, since the day of fertilization is 3 days before this change in the vaginal smear no matter when breeding occurs.[19]

During pregnancy the bodyweight of the dam increases. On average the weight gain from oestrus to parturition is 36% (range 20–55%),[7] and the increase is most pronounced during the latter part of the period (Figure 8.5).

Table 8.1 Duration of pregnancy in various dog breeds. (After Krzyzanowski et al[22])

Breed	Duration of pregnancy, days	
	Average	Variation
Boxer	63.5	56–71
Doberman	62.8	58–71
Great Dane	62.6	59–69
Fox terrier, wire-haired	62.6	55–72
Dachshund, smooth-haired	62.5	55–71
Scottish collie	62.4	56–72
Cocker spaniel	62.4	56–69
German short-haired pointer	62.3	57–71
German wire-haired pointer	62.2	56–71
Jack Russell (ratter)	62.1	55–70
Alsatian	62.1	54–72
Poodle, medium-size	61.8	54–68
Poodle, miniature	61.6	57–69
Poodle, large	61.5	54–70
Pekinese	61.4	54–72

Table 8.2 Pregnancy table based on a pregnancy length of 63 days. In leap years a correction of −1 day is to be made in pregnancies including the 29 February, and breeding of a bitch on 27 December will result in parturition on 29 February.

Day	January	February	March	April	May	June	July	August	September	October	November	December
1	0305	0405	0503	0603	0703	0803	0902	1003	1103	1203	0103	0202
2	0306	0406	0504	0604	0704	0804	0903	1004	1104	1204	0104	0203
3	0307	0407	0505	0605	0705	0805	0904	1005	1105	1205	0105	0204
4	0308	0408	0506	0606	0706	0806	0905	1006	1106	1206	0106	0205
5	0309	0409	0507	0607	0707	0807	0906	1007	1107	1207	0107	0206
6	0310	0410	0508	0608	0708	0808	0907	1008	1108	1208	0108	0207
7	0311	0411	0509	0609	0709	0809	0908	1009	1109	1209	0109	0208
8	0312	0412	0510	0610	0710	0810	0909	1010	1110	1210	0110	0209
9	0313	0413	0511	0611	0711	0811	0910	1011	1111	1211	0111	0210
10	0314	0414	0512	0612	0712	0812	0911	1012	1112	1212	0112	0211
11	0315	0415	0513	0613	0713	0813	0912	1013	1113	1213	0113	0212
12	0316	0416	0514	0614	0714	0814	0913	1014	1114	1214	0114	0213
13	0317	0417	0515	0615	0715	0815	0914	1015	1115	1215	0115	0214
14	0318	0418	0516	0616	0716	0816	0915	1016	1116	1216	0116	0215
15	0319	0419	0517	0617	0717	0817	0916	1017	1117	1217	0117	0216
16	0320	0420	0518	0618	0718	0818	0917	1018	1118	1218	0118	0217
17	0321	0421	0519	0619	0719	0819	0918	1019	1119	1219	0119	0218
18	0322	0422	0520	0620	0720	0820	0919	1020	1120	1220	0120	0219
19	0323	0423	0521	0621	0721	0821	0920	1021	1121	1221	0121	0220
20	0324	0424	0522	0622	0722	0822	0921	1022	1122	1222	0122	0221
21	0325	0425	0523	0623	0723	0823	0922	1023	1123	1223	0123	0222
22	0326	0426	0524	0624	0724	0824	0923	1024	1124	1224	0124	0223
23	0327	0427	0525	0625	0725	0825	0924	1025	1125	1225	0125	0224
24	0328	0428	0526	0626	0726	0826	0925	1026	1126	1226	0126	0225
25	0329	0429	0527	0627	0727	0827	0926	1027	1127	1227	0127	0226
26	0330	0430	0528	0628	0728	0828	0927	1028	1128	1228	0128	0227
27	0331	0501	0529	0629	0729	0829	0928	1029	1129	1229	0129	0228
28	0401	0502	0530	0630	0730	0830	0929	1030	1130	1230	0130	0301
29	0402	0502	0531	0701	0731	0831	0930	1031	1201	1231	0131	0302
30	0403		0601	0702	0801	0901	1001	1101	1202	0101	0201	0303
31	0404		0602		0802		1002	1102		0102		0304

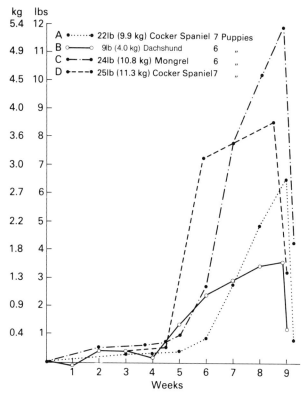

Figure 8.5 Increase in bodyweight during pregnancy. (After Arthur[3])

The change in the shape of the body, which is due to increasing abdominal contents, is visible about the 50–55th day of pregnancy, and at the same time fetal movements become visible in the flank of the bitch. In the first half of pregnancy the physiological behaviour is almost unchanged, but in the last part the bitch may show symptoms of abdominal pain from fetal growth causing the uterus to elongate and change position. The appetite changes, and instead of one large meal daily, frequent small meals are preferred.

During the second half of pregnancy mammary growth occurs; the nipples become enlarged and firm, and late in pregnancy a serous secretion may be present in them. The alteration in the mammary glands is more obvious in the primiparous bitch; in multiparous bitches no mammary enlargement is apparent until the last 3–4 days of pregnancy. In the primiparous bitch lactation begins 24 h prior to parturition, whereas in the multiparous bitch milk appears several days before parturition.

ANTENATAL EXAMINATIONS

Examinations in relation to pregnancy may be necessary before breeding, during pregnancy, and shortly before as well as after parturition.

Examination before breeding is especially indicated in bitches prior to the first pregnancy, but also in bitches having previously had a complicated parturition, to decide whether pregnancy and parturition are advisable or should be avoided.

Before breeding, the owner should be informed of the physiological events and the risks of dystocia following crossbreeding. The bitch should have reached a size and stage of development corresponding to the norm of the breed in question, and only those without anomalies of the pelvis and the birth canal should be used for breeding. Excited and nervous dogs as well as those with hereditary abnormalities should be excluded from breeding. Old or overweight bitches should not be allowed to undergo a pregnancy since this may be too great a burden physically. Mating is not advisable with male dogs known to have produced disproportionately large fetuses.

Optimal feeding is implicit in normal pregnancy. The owner should be instructed to weigh the bitch regularly during the gestation period to avoid overfeeding, and the bitch should be exercised daily to avoid obesity, which leads to a risk of uterine inertia. The feed should consist of one-quarter animal and three-quarter vegetable products with sufficient calcium and phosphorus. In Table 8.3 the requirements of pregnant and lactating bitches are shown, but individual differences exist, and the requirements also depend on the activity of the bitch, the coat and the environmental temperature. The energy requirement is increased to 35–40 g dry matter per kg bodyweight during the first 6–7 weeks and increases to 45 g dry matter per kg in the last 2 weeks (Table 8.4).

Pregnancy and pathological conditions associated with distension

Table 8.3 Nutritional requirements of pregnant and lactating bitches. (After Hedhammar[15])

Recommended dietary composition, dry basis	Food consumption as compared to non-pregnant status (100%)
Protein 20–40%	Pregnancy (weeks):
Fat 10–20%	1–3 100%
Calcium 1.1%	4–6 100–125%
Phosphorus 0.9%	7–9 125–150%
Vitamin A 5000–10 000 IU	Lactation (weeks):
Vitamin D 500–1000 IU	1–2 150–200%
Vitamin E 50 IU	3–4 200–300%

Table 8.4   The daily requirements of dry matter and digestible protein in the feed. (After Jensen[21])

Bitch	Dry matter in the feed, g/day	Per cent digestible protein in the feed, dry matter
Mature bitch		
light exercise	$35 \times$ kg bodyweight$^{0.75}$	10
intensified exercise	$45 \times$ kg bodyweight$^{0.75}$	10
Pregnant bitch		
1st to 5th week	$35 \times$ kg bodyweight$^{0.75}$	10
6th to 7th week	$40 \times$ kg bodyweight$^{0.75}$	12
8th to 9th week	$45 \times$ kg bodyweight$^{0.75}$	15

of the abdomen, such as pseudopregnancy, ascites or tumours, must be distinguished by appropriate examination.

Just prior to parturition the bitch should be examined to estimate her physical condition and the number of fetuses and their viability, as judged by fetal movements and auscultation of the fetal heart.

More frequent examinations may be indicated in bitches affected by diseases that may influence pregnancy.

DIAGNOSIS OF PREGNANCY

The interoestrous period is longer than gestation, so the non-recurrence of oestrus cannot be used as an indication of pregnancy, nor can increased bodyweight or udder development, since these may be due to wrong feeding or pseudopregnancy.

Various methods, palpation as well as special apparatus, may be used for diagnosing pregnancy. Some may give information on the presence of fetuses and others on their viability. Each method has a more or less limited period of use, and most increase in accuracy with the progression of pregnancy. The owner should be informed that a positive diagnosis can be related only to the current state of the bitch and that it is no assurance of the birth of living puppies. Fetal death and abortion may occur at any time during the pregnancy. A frequency of up to 11% has been reported, estimated in relation to the total number of corpora lutea.[2]

Diagnostic methods

Abdominal palpation
During the first 2–3 weeks of pregnancy it is not possible to feel any change in the dimensions of the uterus.

On the 21st day after breeding, pregnancy diagnosis is possible unless the bitch is nervous, has a tense abdominal wall or is too fat, thus making deep abdominal palpation impossible. The uterine swellings are distinct with a diameter of 12–15 mm. Such early diagnosis is rather difficult in bitches pregnant with only one fetus. The 28th day after breeding is the optimal time for pregnancy diagnosis by palpation. The uterine spherical swellings have an average diameter of 25 mm and are very distinctly separated (as in the pregnant queen, see Figure 18.2, page 276.) The accuracy of detecting pregnancy by palpation has been found to be 87% and for non-pregnancy 73%.[1]

The 35th day after breeding is characterized by the swellings being no longer separated but confluent; pregnancy diagnosis is difficult or impossible at this stage, although the size of the uterus has increased. The diameter of the uterine horns is uniform, thus making it difficult to distinguish between pregnancy and pyometra.

From the 56th day after breeding it is often possible to palpate the fetuses, and a fetus may be felt rectally if the forequarters of the bitch are elevated.

Ultrasound
This can be used either by the Doppler, A-scanning or B-scanning methods. No matter which method is used, several locations along the lateral surface of the nipple line should be tested using an adequate amount of lubricant jelly placed between the skin and the probe to ensure proper transmission of the sound waves.

The Doppler method
The Doppler method is based on the appearance of movements in certain organs, i.e. pulsation in a uterine artery, in the naval artery or in the fetal heart, this pulsation reflecting the ultrasound signal, which is received and transformed to an auditory signal by the instrument. Fetal movements may also be diagnosed. This method can be used to detect fetal heartbeats from the 29–35th day of pregnancy.[16,20,26,32,37] The accuracy of detection increases as pregnancy progresses, and is 85% and 100% in the periods 36–42 days and 43 days to term, respectively; in non-pregnancy it is 100%.[1] The detection of an abdominal arterial pulse in small bitches (2–3 kg) could result in a false diagnosis.[34]

A-scanning
A-scanning is based upon the presence of fetal liquid, which reflects the ultrasonic waves. These reflected signals will then appear on a screen indicating the depth from which they are reflected. This method may be used successfully from as early as the 18–20th day of pregnancy.[38] Even though implantation does not take place until about this time, there should be sufficient fluid in the uterus. In the

optimal period, from 32 to 62 days after mating, the accuracy of detecting pregnancy in bitches was found to be 90% and of non-pregnancy 85%.[1] Care should be taken not to place the probe too far caudally, as signals may then be obtained from the urinary bladder.

B-scanning
B-scanning is claimed to be superior to A-scanning and Doppler methods because of the particular advantage that it also indicates the presence of dead fetuses.[36] This examination can be performed after the 18–19th day after mating;[23,24] the period between the 28th and 35th days of gestation seems the most suitable time for counting fetuses, and details of the fetal bodies are clearly visible after the 40th day of pregnancy.[4]

X-ray
This method can be used after the 49th day, when the calcification of the fetal skeleton is sufficient to show contrast. In a few bitches the fetal vertebrae and the ribs are obvious by X-ray examination as early as the 40th day.

The uterine swellings should be visible by X-ray from the 30–35th day of pregnancy; CO_2 is insufflated into the peritoneal cavity in amounts of 200–800 ml according to the size of the bitch.[25]

Auscultation
Late in the pregnancy it may be possible to hear the fetal heartbeat through the abdominal wall of the dam using a stethoscope.[25]

Fetal electrocardiography
This is not yet used in small animals, but the method has been used in cattle and mares for the diagnosis of pregnancies with living twins and for monitoring the health of the fetus by applying electrodes to the skin of the dam.[5,6,30] In bitches it has been tried experimentally using surgically positioned electrodes.[42]

Other examinations

Vaginal smears
These cannot be used for pregnancy diagnosis, but from about the 55–57th day of pregnancy crystallization patterns appear, which are more pronounced as parturition approaches, except in bitches exhibiting uterine atony at whelping.[27]

Haematological examination
During pregnancy the composition of the blood alters in the bitch, but none of these parameters is used for pregnancy diagnosis.[8,9]

In cases of pseudopregnancy the parameters are within the normal range for non-pregnant bitches, even though the bitch may behave as if pregnant and even lactate. A haematological examination may thus be useful for differentiating between pseudopregnancy and normal pregnancy.[9,25]

Erythrocytes   Erythrocyte numbers decrease from day 21 until parturition, as do the haemoglobin percentage and the cell volume. In 70% of bitches examined less than 5 million/μl erythrocytes were found during the last week of pregnancy.[8] At the same time the cell volume was less than 40% and the haemoglobin percentage less than 14. The greatest decrease occurred in young bitches and those inadequately fed.

Haematocrit   This decreases continuously from day 20 to a minimum of 30 just prior to parturition. In non-pregnant bitches examined at the same time after oestrus the haematocrit was found to be 45.[7] After parturition the haematocrit increases again, but even in the postpartum period the value is lower than in non-pregnant bitches.

Sedimentation rate   This increases from day 21, reaching a maximum at parturition.

Blood platelets   These increase in number from the 21st–28th day of pregnancy, and just prior to parturition the number may be 500 000/μl.

Leukocytes   These increase in number from day 21, reach a maximum around day 49, then decrease. The increase is caused mainly by an increasing number of neutrophils. A total count of more than 30 000/μl is to be considered abnormal.[8]

Coagulation factors   Factors VII, VIII, IX and XI increase in concentration in the first 4–6 weeks of the pregnancy, then decrease until parturition.[13]

Fibrinogen   This increases two to three-fold during pregnancy.[14]

Serum creatinine level   This decreases 25–33% 21 days after breeding, averaging 0.8 mg/dl (range 0.6–0.9) in multigravid bitches compared with 1.1 mg/dl (range 0.9–1.2) in primigravid bitches.[12]

Serum gammaglobulin (IgG)   This decreases 40–45% 21 days after breeding, averaging 648 mg/dl (range 440–1220) in multigravid bitches compared with 1108 mg/dl (range 840–1460) in primigravid bitches.[12]

Hormone assay

Until recently it has not been possible to use hormone assay for pregnancy diagnosis in the bitch as in other animal species, since no significant differences have been found in the concentrations of P and oestrogens in pregnant and non-pregnant bitches in metoestrum. But the fact that during implantation the blastocysts produce and secrete oestrogens which are luteotrophic and antiluteolytic and therefore prevent expulsion of blastocysts may be of use in pregnancy diagnosis. Thus, it is reported that measurement of the urinary levels of total oestrogen between days 14 and 25 after first service, optimal between days 19 and 22, can be used for detection of pregnancy. In non-pregnant bitches the average concentration is 3.0–4.5 μg/ml compared to concentrations between 6.0 and 9.3 μg/ml in pregnant bitches.[33]

Chromosome examination

In humans this is used for examination of cells in amniotic fluid to diagnose hereditary syndromes at an early stage of pregnancy, and it also gives information on the sex of the fetus. In bitches the amount of fetal fluid just prior to parturition is 8–30 ml, and of allantoic fluid 16–50 ml per fetus,[35] but chromosome examination is possible in spite of these limited amounts of fluid. Samples for sex determination have been taken from fetuses at the 40th day of pregnancy.[17]

Laparotomy

Laparotomy may give information on the state of the uterus as well as other abdominal organs, though this is more easily obtained by laparoscopy.[10,40,41]

Laparoscopy

Laparoscopy may be performed under general anaesthesia and with an established pneumoperitoneum by the use of a 180° laparoscope, diameter 5 mm. In bitches weighing more than 10 kg a 10 mm diameter laparoscope may be used, and in bitches whose bodyweight is less than 1.5 kg, a diameter of 2–7 mm is preferable.

The anaesthetized bitch is placed on its back with the posterior part elevated 30°, and pneumoperitoneum is established by injecting an air mixture containing 5% CO_2 through the abdominal wall 4–8 cm from the midline with a Verre's cannula. The amount of air depends on the size of the bitch: less than 0.5 l is needed for bitches with a bodyweight below 2.5 kg, 0.5 l for bitches weighing 2.5–4.5 kg, 1 l for 4.5–14 kg, and 1–2 l for > 14 kg.

An incision is then made 1–2 cm from the midline and 2–4 cm posterior to the umbilicus. A trochar and cannula are introduced at an angle of 30°, and the trochar is then replaced by the laparoscope introduced through the cannula. Other instruments such as biopsy forceps may be introduced through another incision 6–8 cm from the midline on the opposite side of the abdomen. After examination the

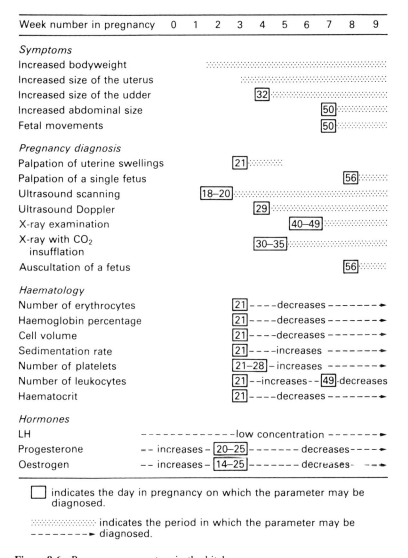

Week number in pregnancy	0	1	2	3	4	5	6	7	8	9

Symptoms
Increased bodyweight
Increased size of the uterus
Increased size of the udder — 32
Increased abdominal size — 50
Fetal movements — 50

Pregnancy diagnosis
Palpation of uterine swellings — 21
Palpation of a single fetus — 56
Ultrasound scanning — 18–20
Ultrasound Doppler — 29
X-ray examination — 40–49
X-ray with CO_2 insufflation — 30–35
Auscultation of a fetus — 56

Haematology
Number of erythrocytes — 21 – – – –decreases – – – – – – – ►
Haemoglobin percentage — 21 – – – –decreases – – – – – – – ►
Cell volume — 21 – – – –decreases – – – – – – – ►
Sedimentation rate — 21 – – – –increases – – – – – – – ►
Number of platelets — 21–28 – increases – – – – – – – ►
Number of leukocytes — 21 – –increases – – 49 –decreases
Haematocrit — 21 – – – –decreases – – – – – – – ►

Hormones
LH — – – – – – – – – – – –low concentration – – – – – – – ►
Progesterone — – – increases – 20–25 – – – – – – – decreases – – – – ►
Oestrogen — – – increases – 14–25 – – – – – – – decreases – – – ►

☐ indicates the day in pregnancy on which the parameter may be diagnosed.

⸬⸬⸬⸬⸬ indicates the period in which the parameter may be
– – – – – – – ► diagnosed.

Figure 8.6 Pregnancy parameters in the bitch.

laparoscope and instruments used are removed and the air escapes through the cannula before it is removed. The incisions are closed by a single suture in the skin.

This type of examination may be used for the diagnosis of early pregnancy, pyometra and ovarian cysts, for biopsies or samples of fetal fluid, for intraperitoneal insemination in cases of reproductive disturbances, or for tubal ligation for reproductive control. The ovaries in the bursa cannot be inspected until the bursa inlet is enlarged by a cutting and burning instrument. It does not interfere with the reproductive capacity even if more than one investigation is performed.

SUMMARY

The symptoms of pregnancy, methods of pregnancy diagnosis, haematology and hormone concentrations during pregnancy are summarized in Figure 8.6.

REFERENCES

1. Allen, W.E. and Meredith, M.J. (1981) Detection of pregnancy in the bitch: a study of abdominal palpation, A-mode ultrasound and Doppler ultrasound techniques. *J. small Anim. Pract.*, **22**, 609–622.
2. Andersen, A.C. and Simpson, M.E. (1973) *The Ovary and Reproductive Cycle of the Dog (Beagle)*. Geron-X, Inc., Los Altos, California.
3. Arthur, G.H. (1975) *Veterinary Reproduction and Obstetrics*, 4th ed. Baillière Tindall, London.
4. Bondestam, S., Alitalo, I. and Kärkkäinen, M. (1983) Real-time ultrasound pregnancy diagnosis in the bitch. *J. small Anim. Pract.*, **24**, 145–151.
5. Bosc, M.J. and Chupin, D. (1975) Essai d'application de l'électrocardiographie au diagnostic de gestation uni- ou multifoetale chez la vache. *Ann. Zootech.*, **24**, 117–123.
6. Colles, C.M., Parkes, R.D. and May, C.J. (1978) Foetal electrocardiography in the mare. *Equine vet. J.*, **10**, 32–37.
7. Concannon, P.W., Powers, M.E., Holder, W. and Hansel, W. (1977) Pregnancy and parturition in the bitch. *Biol. Reprod.*, **16**, 517–526.
8. Doxey, D.L. (1966) Cellular changes in the blood as an aid to diagnosis. *J. small Anim. Pract.*, **7**, 77–89.
9. Doxey, D.L. (1966) Some conditions associated with variations in circulating oestrogens. Blood picture alterations. *J. small Anim. Pract.*, **7**, 375–385.
10. Dukelow, W.R. (1978) Laparoscopic research techniques in mammalian embryology. In: *Methods in Mammalian Reproduction*, ed. Daniel, Jr., J.C. pp. 437–460. Academic Press, New York, San Francisco, London.
11. Evans, H.E. and W.O. Sack (1973) Prenatal development of domestic and laboratory mammals: growth curves, external features and selected references. *Zbl. Vet.-Med., Reihe C*, **2**, 11–45.
12. Fisher, T.M. and Fisher, D.R. (1981) Serum assay for canine pregnancy testing. *Mod. vet. Pract.*, **62**, 6, 466.
13. Gentry, P.A. and Liptrap, R.M. (1977) Plasma levels of specific coagulation factors and oestrogens in the bitch during pregnancy. *J. small Anim. Pract.*, **18**, 267–275.

14. Gentry, P.A. and Liptrap, R.M. (1981) Influence of progesterone and pregnancy on canine fibrinogen values. *J. small Anim. Pract.*, **22**, 185–194.
15. Hedhammar, Å. (1981) Personal communication.
16. Helper, L.C. (1970) Diagnosis of pregnancy in the bitch with an ultrasonic Doppler instrument. *J. Amer. vet. med. Assoc.*, **156**, 60–62.
17. Höhn, H. (1974) Darstellung von Chromosomen aus Zellen der Amnionflüssigkeit beim Hund. *Dtsch. tierärztl. Wochenschr.*, **81**, 90–93.
18. Holst, P.A. and Phemister, R.D. (1971) The prenatal development of the dog: preimplantation events. *Biol. Reprod.*, **5**, 194–206.
19. Holst, P.A. and Phemister, R.D. (1974) Onset of diestrus in the beagle bitch: definition and significance. *Amer. J. vet. Res.*, **35**, 401–406.
20. Jackson, P.G.G. and Nicholson, J.M. (1979) The use of ultrasound to monitor fetal life in a pregnant bitch. *Vet. Rec.*, **104**, 36.
21. Jensen, K. (1974) Feeding of dogs. *Medlemsbl. danske Dyrlægeforen.*, **57**, 745–755.
22. Krzyzanowski, J., Malinowski, E. and Studnicki, W. (1975) Examination on the period of pregnancy in dogs of some breeds. *Med. Weteryn.*, **31**, 373–374.
23. Laiblin, Ch. (1983) Trächtigkeitsfeststellung bei der Hündin mit einem zweidimensionalen Impulsechoverfahren. *Zuchthygiene*, **18**, 126.
24. Laiblin, Ch., Schmidt, S. and Dudenhausen, J.W. (1982) Erste Erfahrungen mit dem ADR-Real-Time-Scanner zur Trächtigkeitsdiagnose bei Schaf, Schwein, Hund und Katze. *Berl. Münch. tierärztl. Wochenschr.*, **95**, 473–476.
25. Laing, J.A. (1970) *Fertility and Infertility in the Domestic Animals. Aetiology, Diagnosis and Treatment*, 2nd ed. Baillière Tindall and Cassell, London.
26. Lamm, A.M. (1970) Pregnancy diagnosis in the dog using ultrasound. *Proc. 11th Nord.-vet. Congr., Bergen*, p. 250.
27. Nava, G.A. (1969) The Papanicolaou phenomenon in the last ten days of pregnancy in the bitch. *Veterinaria (Milano)*, **18**, 373–379. (*Anim. Breed. Abstr.* (1970), **38**, p. 490, no. 2971.)
28. Nelson, N.S. and Cooper, J. (1975) The growing conceptus of the domestic cat. *Growth*, **39**, 435–451.
29. Nielsen, E.H. (1965) *Veterinær embryologi – embryogenesis og placenta*. Medical Book Company, Copenhagen.
30. Parkes, R.D. and Colles, C.M. (1977) Fetal electrocardiography in the mare as a practical aid to diagnosing singleton and twin pregnancy. *Vet. Rec.*, **100**, 25–26.
31. Pearson, M. and Pearson, K. (1931) On the relation of the duration of pregnancy to size of litter and other characters in bitches. *Biometrika*, **22**, 309–323.
32. Preu, K.P. (1975) Zur Trächtigkeitsdiagnose an der Hündin mittels Ultraschall. *Kleintier-Prax.*, **20**, 195–199.
33. Richkind, M. (1983) Possible use of early morning urine for detection of pregnancy in dogs. *Vet. Med./small Anim. Clin.*, **78**, 1067–1068.
34. Rižnar, S. and Makek, Z. (1977) Use of ultrasound in bitches for pregnancy diagnostic and obstetric purposes. *Vet. Arh. (Zagreb)*, **47**, 179–182.
35. Roberts, S.J. (1971) *Veterinary Obstetrics and Genital Diseases (Theriogenology)*, 2nd ed. Cornell University Press, Ithaca, New York.
36. Schmid, G. (1978) Die Trächtigkeitsdiagnose mit dem B-Scan-Verfahren beim kleinen Haustier. *Zuchthygiene*, **13**, 87. Abstr. Tagung über Physiologie der Fortpflanzung der Haustiere, 16–17 February 1978, Giessen.
37. Sereda, J. and Kowakcyt, S. (1978) Pregnancy diagnosis in the bitch using ultrasound. *Med. Weteryn.*, **2**, 86–88. (*Anim. Breed. Abstr.* (1979), **47**, p. 44, no. 431).
38. Smith, D.M. and Kirk, G.R. (1975) Detection of pregnancy in the dog. *J. Amer. Anim. Hosp. Assoc.*, **11**, 201–203.
39. Tsutsui, T. (1975) Studies on the reproduction in the dog. V. On cleavage and transport of fertilized ova in the oviduct. *Jap. J. Anim. Reprod.*, **21**, 70–75.

40. Wildt, D.E., Kinney, G.M. and Seager, S.W.J. (1977) Laparoscopy for direct observation of internal organs of the domestic cat and dog. *Amer. J. vet. Res.*, **38**, 1429–1432.
41. Wildt, D.E., Levinson, C.J. and Seager, S.W.J. (1977) Laparoscopic exposure and sequential observation of the ovary of the cycling bitch. *Anat. Rec.*, **189**, 443–450.
42. Williams, H.B., Rankin, J.S. and Hamlin, R.L. (1978) An invasive procedure for in utero fetal electrocardiology: recording of fetal electrocardiograms by applying electrodes through the uterus. *Amer. J. vet. Res.*, **39**, 183–184.

ADDITIONAL READING

Arbeiter, K. (1981) Trächtigkeitsdiagnose bei Hund und Katze. *Tierärztl. Prax.*, **9**, 367–373.
Bernstine, R.L., Litt, B.D. (1970) Doppler signal and interval measurements in the fetal dog. *Amer. J. Obstet. Gynec.*, **106**, 893–898.
Boulcott, S.R. (1967) The feeding behaviour of adult dogs under conditions of hospitalization. *Brit. vet. J.*, **123**, 498–507.
Ficus, H.J. (1973) Röntgendiagnostik von Organerkrankungen in der Kleintierpraxis (1). *Tierärztl. Prax.*, **1**, 81–94.
Fisher, T.M. and Fisher, D.R. (1982) The effects of pregnancy and other factors on the canine ocular fundus. *Vet. Med./small Anim. Clin.*, **77**, 71–72.
Kehrer, A. (1973) Zur Entwicklung und Ausbildung des Chorions der Placenta zonaria bei Katze, Hund und Fuchs. *Z. Anat. Entwickl.-Gesch.*, **143**, 25–42.
Monty, Jr., D.E., Wilson, O. and Stone, J.M. (1979) Thyroid studies in pregnant and newborn beagles, using ^{125}I. *Amer. J. vet. Res.*, **40**, 1249–1256.
Moore, J.A. and Kelly, J.H. (1979) Vomiting and shock in advanced pregnancy (a case report). *Vet. Med./small Anim. Clin.*, **74**, 670–674.
Romsos, D.R., Palmer, H.J., Muiruri, K.L. and Bennink, M.R. (1981) Influence of a low carbohydrate diet on performance of pregnant and lactating dogs. *J. Nutr.*, **111**, 678–689.
Royal, L., Fernay, J. and Tainturier, D. (1979) Mise au point sur les possibilités actuelles de diagnostic de la gestation chez les carnivores domestiques. *Rev. Méd. vét.*, **130**, 859–890.
Weinberg, S.R., Bakarich, A.C., Ledney, G.D., McGarry, M.P. and MacVittie, T.J. (1983) Hemopoiesis in pregnant beagles following low-dose total-body irradiation and surgery. *Experientia*, **39**, 864–866.
Wildt, D.E. (1980) Laparoscopy in the dog and cat. In: *Animal Laparoscopy*, ed. Harrison, R.M. and Wildt, D.E. Williams and Wilkins, Baltimore/London, Chapter 3.

Chapter 9
Parturition and Newborn Puppies

PREPARATION FOR PARTURITION

The bitch should be installed in the whelping box at least one week before parturition, so that she can become accustomed to it and to the surroundings in which the whelping is to take place. The box should be placed in a quiet warm place with fresh air without draughts and away from other dogs. It should be raised some distance above the floor and of a size which permits the bitch to move around without difficulty. The bottom may be covered with newspapers for cheap, effective and easy cleaning. In one side of the box there should be an entrance, and along the inside of the box there should be a rail to prevent the puppies being laid on.

SIGNS OF IMPENDING PARTURITION

Various symptoms and changes in the behaviour of the bitch are obvious before parturition (Table 9.1). The last 2–3 days prior to parturition the bitch should be fed less, and constipating food must be avoided. In this period the bitch rests most of the time, and just before parturition she collects paper, clothes, etc., and starts making a nest. Some bitches may crawl underneath chairs and other furniture. Usually they become restless, change position, and normally seek quiet, dark places. The vulval region becomes oedematous, and there may be a slight viscous discharge.

Relaxation of the pelvic and abdominal muscles is said to be the most consistent and reliable physical sign of impending parturition under laboratory conditions but the presence or absence of milk production is not a reliable sign.[42] In some bitches there may be milk production for more than 2 weeks prior to parturition, whereas in others the onset of lactation is not observed until just prior to parturition.

Before parturition the body temperature decreases from 38 °C to 37 °C (Table 9.2). The cause of this is not known, but it is presumably the result of a decrease in activity, lowered food intake and also the decreasing concentration of P. The temperature increases again at or after parturition and is maintained for at least 4 days at levels above those recorded prior to the transient prepartum hypothermia. This may be due to an overcompensation for the previous decrease.[8]

Table 9.1 Symptoms observed preceding parturition in 86 primiparous and 73 multiparous bitches. (After Naaktgeboren[52])

	Interval from observation of symptom to expulsion of first fetus		Observed in bitches (%)
	Average	Variation	
Growth of the nipples	4 w	7 w–2 w	6.3
Oedema of the vulva	2 w	2.5 w–1 d	4.4
Fetal movements	2 w	3 w–1 d	12.6
Distension of the abdomen	11 d	3 w–1 d	10.7
Laziness	10 d	3 w–1 d	12.6
Alopecia of the abdominal wall	4–5 d		3.2
Nestbuilding	3.5 d	3 w–30 min	45.9
Vomiting	3.5 d	2 w–45 min	13.2
Hollowing of the croup	2.5 d	1 w–8 h	14.9
Milk in the nipples	2 d	1 w–15 min	19.3
Frequent urination	2 d	1 w–1.5 h	16.9
Loose faeces	2 d		4.4
Increased drinking	1 d		1.3
Restlessness	30 h	2 w–1 h	57.7
Scraping	24 h	1 w–30 min	32.0
Decreased appetite	20 h	1 w–3 h	37.1
Seeking protection near the owner	20 h	2 d–1 h	15.1
Circling in the nest	20 h	2 d–1 h	5.7
Lowered body temperature	19 h	30 h–8 h	22.6
Shivering	18 h	1 d–15 min	7.5
Lying flat on the floor	18 h		2.5
Bloody or mucoid vaginal discharge	12 h	4 w–1 min	34.6
Accelerated respiration	12 h	2 d–15 min	30.8
Licking of the vulval region	12 h	2 d–1 min	13.8
Whining and howling	10 h	4 d–2 min	21.3
Intensive swallowing movements	3 h		4.4
Hiding	1 h		1.3
Curving the tail	30 min		3.2
Refusing to go outside			2.5
Changed expression of the eyes			1.3
Pressing against the nest wall			1.3

The concentration of corticoids in the plasma increases in the last few days before parturition from 23 ng/ml to 37 ng/ml, then drops to 15 ng/ml at parturition.[8] The concentration of cortisol in the serum changes in parallel with the concentration of corticoid in total plasma, and it increases from a mean level of 23 ± 1 ng/ml to 63 ± 7 ng/ml at 8–24 h prepartum and then decreases to 19 ± 4 ng/ml 8–12 h postpartum.

The concentration of P falls below 1 ng/ml before parturition[3,7–9,14,25,34,35,53,63]

Serum prolactin levels average 40 ± 7 ng/ml in the last week of

Table 9.2   Changes in the body temperature and the concentration of progesterone before and after parturition. (After Concannon et al.[8])

Hours before/after parturition	−108	−84	−60	−48	−36	−24	−12	+8
Temperature (°C)	38.0	37.9	37.9	37.8	37.6	37.2	37.0	38.4
Progesterone (ng/ml)	6.3	6.4	5.9	5.5	4.4	2.1	1.6	0.9

pregnancy and increase during the last 16–56 h prepartum, reaching peak levels of 117 ± 24 ng/ml 8–32 h prepartum.[9] The concentration declines post partum to 37 ± 8 ng/ml followed by an increase due to the suckling stimulus. During the first week post partum the concentration increases to 86 ± 19 ng/ml followed by a slow decrease during the following 4 weeks to 43 ± 6 ng/ml, followed by an abrupt fall to 13 ± 2 ng/ml after weaning.

Exact knowledge of the mechanism regulating parturition in the bitch is incomplete, but if the birth process is more or less of the same nature as in farm animals parturition may be explained as shown in Figure 9.1. Near the end of the gestation period and after maturation of the fetal endocrine organs, a still unknown factor stimulates the fetal hypothalamus to release corticotropic releasing hormone, which acts on the anterior pituitary gland; this causes it to release adrenocorticotropic hormone, producing hypertrophia and hyperfunction of the fetal adrenal glands, and the release of corticosteroid hormone, which stimulates production and release of P from the placenta. The concentration of P which has declined gradually during the later part of the pregnancy, will, because of luteolysis induced by prostaglandin, decline sharply just before parturition, causing alteration of the P:oestrogen ratio. This increases the myometrial excitability and conductivity. Prostaglandin induces production of relaxin in the ovaries and/or placenta, and this, in combination with oestrogen, may be responsible for the relaxation of the pelvic ligaments and the reproductive tract.

Uterine contractions are induced directly by prostaglandin, by oxytocin released from the anterior pituitary gland by prostaglandin, and through impulses from the cervix and vagina due to distension of the reproductive tract from the fetus and fetal membranes, impulses which also cause abdominal straining and thereby contribute to fetal expulsion.

PARTURITION

First-stage labour

The duration of this stage during which the cervix relaxes and dilates averages 4 h, but may last 6–12 h, and even up to 36 h in a primi-

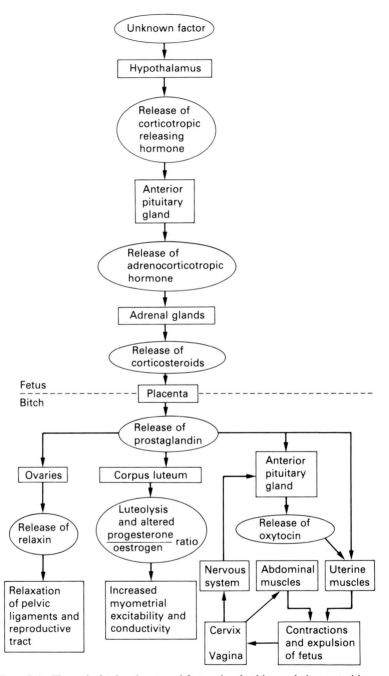

Figure 9.1 The main fetal and maternal factors involved in regulating parturition.

gravid nervous bitch.[20] The bitch is uneasy, apprehensive and frightened, refuses food, may shiver, pant and vomit, and glances anxiously at her flank. Some bitches are calm while others are excited, some pet bitches prefer the presence of the owner, whereas others, especially those in colonies, seek solitude. Weak uterine contractions occur intermittently, often accompanied by urination.

Second-stage labour

This stage follows the first and is characterized by visible straining with the bitch perhaps in a sitting position with her back pressed against the box wall. In between contractions the bitch licks the vulval region. The allantochorion ruptures in the pelvis, and some fluid is visible in the vulva; the amnion then passes into the pelvis, thus preventing the total escape of allantoic fluid. The liquid-filled membrane is pressed through the cervix and the vagina and becomes visible in the vulva, where it ruptures spontaneously or by the bitch licking it. At this stage the head of the first fetus is engaging the pelvis of the bitch, and the head and shoulders are then brought into the pelvis by two or three vigorous contractions. After one or two more efforts the head and shoulders are delivered and the puppy slips out easily. The first fetus to be expelled comes from the uterine horn containing most fetuses,[24,60] and the next puppy is usually expelled from the contralateral horn.[71] The fetus rotates around its longitudinal axis, and the head and the limbs extend before engaging the pelvic inlet. The fetus itself presumably takes an active part in the procedure, since a fetus dying prior to rotation, remains unrotated.

After the start of regular contractions the first fetus cannot be expected to survive more than about 6 h in the uterus, and each of the subsequent fetuses only about 2 h.[20]

Normally the bitch is silent during this phase, but some bitches, especially primigravidae and particularly during the birth of the first puppy, may cry in pain as the head of the fetus is forced through the vulva.[19] A very few bitches may even cry during the whole of second-stage labour, independent of its duration or difficulty.

The respiratory rate is increased from the normal 16–20 per min, and becomes either superficial, accelerating to 100–175 per min, or deeper, increasing to 40–60 per min. While the respiration rate is slow, the bitch lies down with partly closed eyes. From time to time she sniffs the whelping box and her genitalia, paddles or pants and shivers. These symptoms, leading to expulsion of a puppy, are repeated.

Third-stage labour

This stage covers the expulsion of the fetal membranes and lasts 5–15 min, but more puppies may be delivered before the fetal membranes are expelled, and all puppies may even be born before the expulsion of all fetal membranes. If the placenta is not delivered immediately but the navel cord is present in the vulva, the bitch may grasp the cord and pull out the membrane. Most bitches lie down during delivery and this position is often maintained during the whole expulsion period. Even if a bitch leaves the box for a short period for defecation or to drink, she returns and resumes parturition in the same position.

The second and third stages are then repeated but the interval between the deliveries is extremely variable, from 5–10 min to 20–60 min (average 30 min). A bitch pregnant with a large litter may give birth to the first puppies at the rate described, but the last puppies may follow after a rest of up to 4 h. The intervals increase with advanced parturition, and the time for delivery of a normal litter of 5 averages only 3 h, but it may last 10–24 h, and in such cases it is very difficult to distinguish between normal rest and secondary uterine inertia.

If labour is to be normal it is essential for the bitch to be accustomed to her surroundings. Parturition may be arrested in kennel dogs taken indoors even if the expulsion of a fetus has begun but labour is resumed as soon as the bitch is replaced in familiar surroundings without human contact. Later in parturition the bitch will accept and permit examination without any interference in the process of expulsion. In some breeds, especially hunt, border and sealyham terriers, any kind of intervention is resented, especially forceps delivery and caesarean section. Some bitches may be so disturbed that they kill all the puppies delivered under such circumstances.[20] On the other hand, some bitches will whelp only if the owner is present throughout the whole procedure.

Behaviour at birth

The presence of other dogs or strange humans and noise amongst the puppies may lead to inertia of varying duration. When the disturbance ceases, activity is recommenced, but a pause of as long as 15–60 min is not unusual and, if the disturbance is repeated, the delivery may be extended for up to 6 h.[6] If the disturbance occurs in a pause between the expulsion of two fetuses, labour may cease or weaken, stopping the expulsion of the next fetus. If the bitch is disturbed during an expulsion, this is arrested, and several periods with uterine contractions may be seen before expulsion occurs.

After delivery of a puppy the bitch is interested in it for a short period and licks the fetal fluid from its head and mouth. The fetal membranes may be ingested but this may cause the bitch to vomit. The bitch then licks her own body before attending again to the puppies. The umbilical cord is bitten at varying distances from the body, but when delivery is finished, they are all bitten to the same length, 1.5–3 cm. The puppies are cleaned and arranged along the bitch's body and she lies resting and protecting them until the next expulsion.

Stillborn puppies are treated in the same way as normal living ones so long as they are warm. Some bitches reject cold, dead puppies, whereas others exhibit a very pronounced parental instinct and take care of both dead and living puppies.

Most bitches, including primigravidae, act as described, but sometimes delivery takes place outside the whelping box and sometimes the bitch neglects the puppies until the final one is born. The period of getting accustomed to the box and the surroundings, mentioned above, minimizes the risk of deviations from normal behaviour. The presence of the owner may have a calming influence on a nervous bitch.

Involution of the uterus

During the first 2–3 days post partum the vaginal discharge is red and watery in consistency. The amount decreases, with the colour becoming red-brown during the following week, and 2–3 weeks after parturition the discharge is sparse with a light red to red-brown colour and a mucoid consistency.

As early as 2–4 days after parturition the size of the uterus is reduced to a diameter of 1–2 cm.

CONCEPTION RATES AND LITTER SIZES

It is of the greatest importance to decide how long a bitch can be used for breeding without a significant drop in her performance. Strasser and Schumacher[66] state that those bitches whose performance is below average after the fifth year of life or after their sixth litter should be excluded from breeding. Bitches with an above-average performance may be left to breed until there is a distinct drop in performance or until the end of the eighth year of life.

Conception rate

Various factors, such as the number of matings per cycle, the mating time in oestrus and the duration of oestrus, are important for assessing the conception rate.

Number of matings per cycle
The conception rate increases if more than one mating is permitted in a cycle (Table 9.3).

Table 9.3 Conception rates in relation to breed, number of matings per cycle and time in oestrus of first mating. (After Rowlands[57])

Day in oestrus of first mating	Airedale		Foxhound	F × A*	Mongrel		Conception rate (%)
	One mating	Two matings	Two matings	Two matings	One mating	Two matings	
1	1:1**	7:8	5:8	2:4	—	10:12	78
2	1:1	7:7	1:1	0:2	—	2:2	85
3	1:2	16:16	1:1	6:6	3:4	8:9	92
4	2:4	1:1	2:2	1:1	—	3:3	82
5	1:2	1:1	—	—	—	0:1	50
6	—	—	—	—	0:1	1:1	
7	—	—	—	—	—	—	
Breed		88	75	69		82	
Conception rate (%)	60	96			60	86	82

*F × A = foxhound–airedale cross.
**Number of bitches conceived : number of cycles.
Difference between conception rates in airedales and mongrels = 6% ± 9%.
Difference between conception rates following single and double mating in these two breeds = 32% ± 13%.

Table 9.4   Conception rate in relation to duration of oestrus. (After Rowlands[57])

Duration of oestrus (days)	Number of cycles	Conception rate (%)
5–7	11	100
8–10	26	100
11–13	30	90
14–16	5	60
17–19	4	50

Mating time in oestrus

The best conception rate is achieved if the first of two matings is performed before or at the fourth day of oestrus (Table 9.3). Holst and Phemister[30] found the best conception rate, if more than one mating was permitted, to be when the first took place at the initial acceptance of the male and remating was on the third day or later.

Duration of oestrus

In bitches mated twice, the first mating being before the fifth day of oestrus, it has been found that conception occurred in all bitches where oestrus did not exceed 10 days, whereas a significant decrease occurred when oestrus was of longer duration (Table 9.4).

Cycle number

There seems to be a slight increase in the conception rate in bitches mated in their third cycle (Table 9.5).

Breed

Even if breed differences in conception rate do not exist, crossbreeding may have some influence. In Table 9.3 it can be seen that the conception rate does not differ significantly between the two breeds, foxhound and airedale, but crossbred foxhound–airedales have a lower conception rate than the two pure breeds.

Table 9.5   Conception rates in succeeding cycles. (After Rowlands[57])

Order of cycle	Number of cycles	Conception rate (%)
First	31	84
Second	18	88
Third	19	95
Fourth	9	89
Fifth to seventh	9	78

Litter size

The number of puppies born is not equal to the number of ova fertilized, since embryonic mortality may occur in the bitch due to endometrial dysfunction or infection. There seems to be some connection between embryonic mortality and age. Intrauterine death is obvious when a litter is made up of fully developed puppies together with fetuses which have died at various stages of development[19] or when more implantation places than fetuses have been found at caesarean section.[69] By comparing ovulation and fertilization rates it can be shown that the number of fetuses was 1 fewer than the number of corpora lutea in 25.4% of cases, 2 fewer in 7.5% and 3–5 fewer in 8%.[70]

By studying reproduction records it was found that it is not possible to judge individual breeding performance by the number of puppies per litter. Account must be taken of the number of puppies per bitch per year and the breeding loss.[62]

The biggest litter reported consisted of 23 puppies,[23] and even though small litters may occur in all breeds there are great variations in the litter size in dogs of different breeds. Various factors may play a part. The average litter size is shown in Table 9.6.

Body size of the dam

There is a positive correlation between body size and litter size,[56,68] but the total weight of a litter is fairly constant in all breeds because the birthweight of puppies is highest in small breeds.

Age of the dam and the sire

Younger bitches produce larger litters. Thus in the Hungarian shepherd breed the largest litters are produced up to 3 years of age, the litter size declining gradually thereafter.[62] In the beagle the number of puppies per bitch per year is within an optimum range up to the fifth year of age, that is on average the fifth to sixth litter.[66]

The age of the sire, if under 8 years, appears to have little effect on litter size, whereas older male dogs produce rather small litters and the litter size tends to be lowest when both the dam and the sire are more than 8 years old.[62]

Genetic conditions

An inherited predisposition to big or small litters is found in both male and female dogs. The sire has a significant influence on the litter size in some breeds, such as airedale terrier and cocker spaniel,[43] and both the sire and the dam influence the litter size in bernese and appensellen farm dogs.[70]

Table 9.6 Litter size in various breeds of dog. (After Lyngset and Lyngset[44] and Robinson[56])

Breed	Mean litter size	Breed	Mean litter size
Airedale terrier	7.6	Hamilton stoever	5.7
Alsatian	8.0	Hungarian sheepdog	6.7
Appensellen sennerhunde	8.0	Hygen stoever	5.9
Australian terrier	5.0	Irish setter	7.2
Basenjii	5.5	Irish terrier	6.1
Beagle	5.6	Karelian bear dog	5.3
Bedlington terrier	5.6	Kerry blue terrier	4.7
Bernese mountain dog	5.8	King Charles spaniel	3.0
Bloodhound	10.1	Labrador retriever	7.8
Boston terrier	3.6	Lakeland terrier	3.3
Boxer	6.4	Lappland dog	4.8
Breton	6.2	Luzern stoever	4.9
Bulldog	5.9	Manchester terrier	4.7
Bull terrier	6.2	Mastiff	7.7
Cairn terrier	3.6	Newfoundland	6.3
Chow	4.6	Norwegian buhund	5.1
Cocker spaniel	4.8	Norwegian elkhound, grey	6.0
Collie	7.9	Norwegian elkhound, black	4.8
Dachshund, smoothhair	4.8	Pappillion	5.0
Dachshund, longhair	3.1	Pekinese	10.0
Dachshund, wirehair	4.5	Pinscher, miniature	3.4
Dalmatian	5.8	Pointer	6.7
Dandie Dinmont terrier	5.3	Pomeranian	2.0
Doberman	7.6	Poodle, miniature	4.3
Drever retriever	5.2	Poodle, dwarf	4.8
Dunker stoever	6.9	Poodle, standard	6.4
English foxhound	7.3	Puffin dog	2.8
English setter	6.3	Retriever	5.2
English springer spaniel	6.0	Rottweiler	7.5
English white terrier	4.4	Samoyed	6.0
Entlebucher sennerhunde	5.5	St Bernard	8.5
Finnish stoever	5.9	Schiller stoever	5.7
Fox terrier, smoothhair	4.1	Schnauzer, giant	8.7
Fox terrier, wirehair	3.9	Schnauzer, miniature	4.7
French bulldog	5.8	Schnauzer, standard	5.1
German shorthair pointer	7.6	Scottish terrier	4.9
German wirehair pointer	8.1	Shetland sheepdog	4.0
Golden retriever	8.1	Shih Tzu	3.4
Gordon setter	7.5	Siberian husky	5.9
Greenland dog	5.1	Swiss running hound	5.4
Greyhound	6.8	Welsh corgi, pembroke	5.5
Griffon bruxellois	4.0	Welsh terrier	4.0
Grosse schweizer sennerhunde	7.9	West highland white terrier	3.7
Halden stoever	6.2	Whippet	4.4

Litter number

In some investigations litters after the third and fourth were found to decrease in size, and the frequency of stillbirths increased,[37-40,54] but in others the litter size was found to be independent of the cycle number.[57]

Feeding and surroundings

These factors also influence litter size, thus a good diet and optimal environment have a tendency to produce bigger litters.[37-40]

Sex ratio

The average percentage of male births is 50.6%, but the sex ratio is genetically based, and the percentage of male puppies born per litter varies from breed to breed with more males born in large breeds and fewer in small breeds (Table 9.7).

Normal development

Table 9.8 gives data on the normal development in newborn and growing puppies.

Table 9.7   Sex ratio within breeds, the frequency of male puppies born is above the average in large breeds and below in small breeds. (After Tedor and Reif[68])

Breed	Percentage of males born per litter
Doberman pincher	53
Boxer	51.7
Weimaraners	51.7
Brittany spaniel	51.7
Golden retriever	51.6
Fox terrier	51.5
Collie	51.4
Norwegian elkhound	51.3
Labrador retriever	51.1
German shepherd dog	50.8
Average for breeds investigated	*50.6*
Poodle	50.4
Beagle	50.3
Yorkshire terrier	49.6
Chihuahua	49.6
Dachshund	49.4
Miniature schnauzer	49.2

Table 9.8   Data (mean) on the newborn and growing puppy.

Eyes closed until	10–12 days of age
Ears closed until	13–15 days of age
Ability to stand at	10 days of age
Walks with steady gait at	21 days of age
Voluntary control of urination and defecation at	16–21 days of age
Rectal temperature: newborn–1 week	35–37.2 °C
2 weeks–3 weeks	37.2–37.8 °C
after 4 weeks	37.5–39.0 °C

Sleeps about 90% of the time
Occasional muscle contractions during sleep in the first 4 weeks ('activated sleep')
After the third day the flexor dominance involving the body muscles and neck muscles
is replaced by extensor dominance

NURSING CARE OF NEWBORN PUPPIES

The normal rectal temperature of a newborn puppy is 35.5–36.0 °C, falling to 29.5 °C within the first minutes of extrauterine life during the drying period.[10,28] The fall is quickly reversed when the puppy is dry and comfortable, and the temperature rises over the first 7 days to approximately 38.0 °C.

The puppies should be protected against draught and cold, and the temperatures in the surroundings should be as follows:

first week	29–32 °C
second week	26–29 °C
third week	23–26 °C
fourth week	23 °C

If the temperature is too high, the puppies lie still and apathetic, and if it is too low, they whine; activity is raised by moderate cold.[55] Puppies having a rectal temperature of between 35.5 °C and 21 °C still have a firm muscle tonus and a normal red mucosa in spite of a cold skin surface, but the respiratory frequency is raised from 20 to 40 and the heart frequency to 200/min. The puppies squeal and whine at each respiratory movement, but the whining and the activity cease as soon as they are placed alongside the bitch or another warm object. The rectal temperature decreases to 15.5 °C after 2 h at 10 °C, resulting in apathetic puppies with uncoordinated movements, a respiratory rate of 20–25 and a heartbeat of less than 50 per min. Perfectly healthy puppies may survive 12 h in cold surroundings if they are rewarmed at 26.5–29.5 °C.[11]

The bitch remains in the nest during the first few days following the parturition, only leaving for feeding, defecation and urination. Some bitches must be removed forcibly for the purpose of cleaning the box.

After the first 2 days the bitch may leave the box for longer periods, and after 2 weeks she may leave the puppies for up to 2–3 h at a time. Some bitches may be aggressive towards other dogs or even the owner during delivery and the nursing period.

The bitch stimulates the puppies to urinate and defecate by licking their perineal regions. She often rejects a newborn puppy which fails to thrive and it is pushed out of the box or even killed or buried outside.

When the puppies start to move about, the bitch will supervise them. Some will try to keep the puppies in the nest, and bring back puppies that are moved or move away by themselves. With increasing age the puppies are left to themselves in the whelping box and older ones that fall out of the box are seldom retrieved even if they cry.

The lactation period lasts on average 6 weeks starting with a colostral secretion for 1–3 days, followed by increasing secretion of normal milk for the next 2–3 weeks. The amount increases until the fifth and sixth week, then decreases. The amount depends on the body size of the bitch and the number of puppies, but the change from colostrum to normal milk may be delayed in litters below average size.[58]

Feeding the bitch and the puppies

In the lactation period the bitch must be fed to the optimum level to produce enough milk. Food intake begins to increase after parturition and reaches a maximum 3–4 weeks later (Table 9.9). In the latter part of the pregnancy and in the second, third and fourth weeks of lactation, the energy requirement is increased by 20%, 50%, 75% and 100%, respectively. The diet must also contain sufficient amounts of calcium, phosphorus and vitamins.

Immediately after parturition the weight of the bitch should not deviate by more than 5–10% from its average non-pregnant weight. Feeding should be moderate and not forced since this may interfere with milk production and cause 'milk eczema' amongst the puppies. If this condition develops, the ration of protein and fat must be cut down and the bitch given 1–2 teaspoons of sodium bicarbonate.[27]

Table 9.9 The daily requirements of dry matter and digestible protein in the feed. (After Jensen[33])

Lactation period	Dry matter in the feed (g per day)	Percentage of digestible protein in the feed, dry matter
First week	$40 \times$ kg bodyweight$^{0.75}$	15
Second week	$55 \times$ kg bodyweight$^{0.75}$	15
Third to fourth weeks	$70 \times$ kg bodyweight$^{0.75}$	15

If the puppies have sufficient milk, they should be pear-shaped with the heavier part downward, and heavy and firm on handling. The chance of survival is related to the growth pattern. In puppies gaining weight from the beginning of nursing and those with an initial weight loss not exceeding 10% of the birthweight followed by a weight gain, the prognosis of survival is good, whereas it is poor in those with a weight loss exceeding 10% of the birthweight unless suitable therapy is instituted.[72]

Inadequate amounts of milk may be due to the bitch dying, pathological conditions such as metritis, mastitis or the underdeveloped udder which occurs in cases of premature parturition.

Insufficient intake of milk may be seen in premature and underdeveloped puppies, if the teats are big or if there is an extreme oedema of the udder. Pain caused by mastitis may make the bitch prevent the puppies from sucking.

It is important to ensure sufficient colostrum on the first day of life even though it is not of vital importance, as the placenta of the bitch, to a certain extent, permits passage of globulins from the mother to the puppies.

When the puppies are 4 weeks old, the bitch or the male dog, 10–60 min after being fed, may go direct to the puppies and vomit, or go into a corner and whine, thus causing the puppies to leave what they are doing and rush to the dog, gathering around its head. It then vomits, and the vomit is immediately ingested by the puppies.[36] Such behaviour is most obvious when the bitch has been away from the puppies for several hours and returns immediately after feeding.[47] If the puppies are receiving supplementary feed they will ingest only a little of the vomit, even though they may be interested in it, and afterwards the bitch will eat what has not been eaten by the puppies.

Milk production decreases about 5 weeks after parturition, and the periods of suckling become less frequent as the puppies become more active and violent. Some bitches punish puppies trying to approach the feeding bowl.

At the age of 4 weeks the puppies are given supplementary food starting with a mixture of milk and chopped meat or egg yolk. The amount should be about 2 g per day and be raised gradually. When lactation ends, feeding is offered 4 to 5 times daily until 3 months of age, then 3 times daily until the age of 6 months, and then once or twice daily.

Artificial feeding

Newborn puppies may be fed successfully with various commercial substitutes.[4,31,46,65] A good substitute can be made by mixing 0.8 l of cows' milk, 0.2 l of cream containing 12% of fat, one egg yolk, 6 g bone meal, 2000 IU vitamin A and 500 IU vitamin D[5] while heating the mixture to 40 °C, when 4 g citric acid is added to coagulate the casein. The puppies are fed individually according to their age. Milk substitute should be given in daily amounts of 15–20%, 22–25%, 30–32% and 35–40% of the bodyweight to puppies of 3, 7, 14 and 21 days of age, respectively. If a puppy is too weak to suck, it must be fed by a tube introduced in the oesophagus to the level of the eighth rib. The tube should be of soft rubber and 3–4 mm in diameter. If this is not available, a sterile male dog catheter can be used. The distance from the nose to the last rib should be marked to ensure the correct placement of the tube or catheter. The milk substitute is given every 8 h and after each feed urination and defecation reflexes must be stimulated by a light massage or by washing the abdomen and perineal region.

Artificially raised puppies have a lesser weight gain during the first 3 weeks of life, but thereafter the weight increases to normal. Too much feeding may cause diarrhoea; in such cases the substitute should be diluted and the dosage lowered.

Puppies suckled by their dams double their bodyweight in 1 week, but in artificially reared puppies this does not happen until after 2 weeks.[59]

PATHOLOGICAL CONDITIONS AND MORTALITY IN NEWBORN PUPPIES

During the pregnancy, parturition and in the postpartum period various pathological conditions of known and unknown aetiology may occur and lead to the death of the fetus or puppy. Those occurring early in the pregnancy may be followed by death and resorption of the conceptus, whereas later abnormalities may lead to lesions of different fetal organs.

Various factors, such as congenital abnormalities, infections, damage during parturition and managerial factors, may lead to prenatal or neonatal death or death during the first days or weeks of life among puppies apparently born healthy – the so-called fading puppy syndrome (Table 9.10).

Mortality rates

A mortality rate of 25–34% may occur. Rowlands[57] reports a frequency of stillborn puppies of 4.6% and a mortality of 34% within the first 8 weeks of life, and furthermore that 82% of the deaths occurred within the first week and 12.5% within the second week. Death immediately after delivery may be caused by dystocia; about 8% of puppies born alive die shortly after from this.[2]

Table 9.10 Possible factors involved in prenatal or neonatal deaths or the fading puppy syndrome[*]

Factor	Origin
Congenital defects	Genetic
	Teratogenic
	Maternal malnutrition
Infections	Virus
	adenovirus
	parvovirus
	herpesvirus
	Bacteria
	Brucella canis
	staphylococci
	Pseudomonas aeruginosa
	Escherichia coli
	beta-haemolytic streptococci
	Fungi
	Endoparasites
	roundworms
	hookworms
	Ectoparasites
	mites
Management	Mismanagement of
	prepartum period
	whelping
	nursing bitch
	suckling puppies
	Thermoregulation

[*]Based on: Malherbe[45], Yutuc[74], Davies and Skulsky[12], Jackson[32], Ablett and Baker[1], Fox[17,18], Hime[29], McKelvie and Andersen[49], Drake and McCarthy[13], Gowing[22], Spalding *et al.*[64], Scothorn *et al.*[61], Evans[15], Wright and Cornwell[73], Leipold[41], Hashimoto *et al.*[26], Farrow and Malik[16].

In a survey covering 392 puppies, all of which died within the first 12 weeks of life, it was found that three-quarters of the deaths occurred during the first 2 weeks after delivery.[67] The main reasons were asphyxia, bacterial or viral infections, prematurity and inadequate maternal care. In 13% of the deaths more than one factor was involved.

There are no differences between male and female puppies in the number of stillbirths.[21] The frequency of stillborn puppies is independent of the litter size, whereas postpartum mortality occurs far less in smaller litters than in larger ones (Table 9.11). Each extra puppy in a litter increases the mortality rate for the puppies by 3.48% in the first week and by 3.43% within 6 weeks.[21]

In spite of a higher number of male puppies at birth there are

Table 9.11   Death rate in relation to litter size. (After Rowlands[57])

Litter size	Number of litters	Total	Mortality (%)		Average litter loss
			Stillbirth	Postnatal	
1–7	41	29.6	4.7	24.9	1.41
8–13	52	43.2	4.6	38.6	3.56
Difference (%)		13.6	0.1	13.7	
Standard error		±3.75		±3.6	

Table 9.12   Mortality in relation to sex. (After Rowlands[57])

	Number of males	Number of females	Difference (% ± SE)
Number of puppies born	383	328	
Number of puppies born (%)	54	46	8 ± 3.75
Total death rate (%)	42.5	34.4	8.1 ± 3.64
Stillbirths (%)	4.7	4.6	0.1
Postnatal deaths (%)	37.9	29.9	8 ± 3.54
Number of puppies weaned	215	210	

Table 9.13   Death rates of puppies born in succeeding litters (After Rowlands[57])

Order of litter	Number of litters	Total	Mortality (%)	
			Stillbirth	Postnatal
First	43	30.0	2.3	27.7
Second	24	41.9	5.9	36.0
Third	15	43.0	4.1	38.9
Fourth to sixth	11	54.8	10.8	44.0

almost equal numbers of each sex at weaning because of a significantly greater mortality among males (Table 9.12).

The number of stillbirths and the postpartum mortality rate increase with increasing age of the dam (Table 9.13).

Congenital abnormalities

Congenital defects may be of genetic or teratogenic origin, or due to prenatal malnutrition of the bitch; they may include various organs and, if severe, eventually be incompatible with life. In order to avoid these defects the breeding programme must exclude bitches as well as male dogs known or suspected to be carriers of genetic defects. Medication during the pregnancy should be avoided if possible, since this may result in fetal defects; for example, teratogenic defects in dogs have been reported following administration of excessive amounts of vitamin A or D, and in other species from administration of corticosteroids or various kinds of antibiotic.[51] (See Chapter 3, page 55, the risk of masculinization of female puppies after administration of P to the pregnant bitch.)

The diet of the bitch should be complete during pregnancy and lactation. Prenatal malnutrition may result in the birth of abnormally small puppies which appear physically immature, be unable to suck effectively and are subject to hypoglycaemia, hypothermia, stunted growth and death. Signs of immaturity and diminished chances of survival are a birthweight more than 25% below average, lack of normal growth of hair coat and irregular respiration. The diagnosis of immaturity may be confirmed by necropsy since such puppies have an abnormal liver:brain ratio. A liver:brain ratio of 1.5:1 is adequate for survival, and a ratio of 2–3:1 is necessary for strong, vigorous puppies.[50]

Viral or bacterial infections

Viral or bacterial infections may cause infertility, fetal or postnatal death among puppies. Intrauterine infections with canine herpesvirus may occur possibly leading to postnatal death. Other viruses may also be involved in neonatal diseases, and infections with *Escherichia coli* or beta-haemolytic streptococci may cause deaths in puppies. Loss may be caused also by endoparasites or ectoparasites.

Other problems

Other problems include: mismanagement during pregnancy, whelping and during the postpartum and growing periods, for example, improper feeding of the pregnant bitch, wrong whelping procedure whether it is too much or too little interference, improper feeding of the lactating bitch, undermothering or overmothering of the puppies or the wrong environmental temperature. A significant proportion of deaths occurring from delivery to weaning is the result of direct trauma, such as dystocia or excessive licking causing navel hernia, evisceration and cannibalism. Death may also occur among nervous and inexperienced dogs by the bitch overlaying or treading on the puppies. The bitch should therefore be fed according to her pregnant or lactational status (pages 161 and 184) and parturition should be managed as described above (page 173). Sufficient assistance and supervision must be offered during parturition and lactation, the temperature kept as near optimal as possible (see nursing care, page 183), and the puppies provided with sufficient colostrum and dewormed. Disinfection, sterilization, chemotherapy, use of a kennel-specific vaccine, and paraimmunization, which activates a number of non-specific host-responses, may be used to minimize the mortality.[48]

Perinatal death

In all cases of perinatal death the fetuses, the puppies and possibly also the placenta should be examined macroscopically, histopathologically and microbiologically. The examination should include blood, stomach contents, material from the abdominal cavity and visceral tissue from the fetus or puppy, and blood and vaginal swabs from the bitch. The findings should be correlated with all available information on mating, pregnancy, parturition, the postpartum period and, if possible, with any similar information about the sibs of the bitch.

During the neonatal period and the following 6 weeks of life various pathological conditions may occur (Table 9.14). Signs of illness may include a weight loss of more than 10%, disappearance of movements during sleep, a decrease in body tonus and muscle strength and a continuation of flexor dominance after the third day of life.

Therapy

A puppy with symptoms of immaturity should have supplementary feeds with 5% glucose every 2–3 h, and, if the condition is unchanged after 20–24 h, feeding should also include milk substitutes. The puppy should be kept at a temperature as for hypothermia, mentioned below, and in a relative humidity of 85–90%.

Table 9.14 Time when pathological conditions (excluding congenital defects) occur in puppies.

Pathological condition	Age in days
Hypoxia	immediately after birth
Dehydration	0–20
Hypothermia	0–20
Diarrhoea	0–20
Haemolytic syndrome	1–4
Umbilical infection	1–4
Septicaemia	1–40
Toxic milk syndrome	3–14
Neonatal dermatitis	4–10
Puppy viraemia	5–21
Neonatal ophthalmia	0–10

A sick puppy should be given intensive care including rehydration and protection against hypothermia. But it must be emphasized that prophylaxis such as navel hygiene and disinfection, initiation of respiration, provision of optimal environmental temperature and humidity, and ingestion of sufficient colostrum are of primary importance.

Dehydration
Dehydration is treated by parenteral administration of a mixture of equal parts of Ringer's lactate solution and 5% glucose, followed by 5–10% glucose orally until hydration is re-established.[51]

Hypothermic puppies
These should be maintained at an ambient temperature of 30–32 °C for the first 4 days of life, and thereafter the temperature may be decreased to 27 °C at the first or second week, and to 23 °C at the fourth or fifth week.

Diarrhoea
Diarrhoea during the first weeks of life is often related to the food eaten, and is characterized by green or possibly white faeces which are due to depletion of digestive enzymes. Treatment comprises rehydration and correction of the diet.

Haemorrhagic syndrome
This is characterized by bleeding from the nose and mouth, by haemorrhage in internal tissues, intraperitoneally and intrapulmonarily, and by haematuria. The therapy consists of administration of 0.01–0.1 mg of vitamin K_1, possibly supplemented by plasma transfusion, adminstered prophylactically to the pregnant bitch during the latter half of the pregnancy in a dose of 1–5 mg.

Umbilical infection
Umbilical infection, most commonly caused by streptococci, should primarily be avoided prophylactically but, when manifest, treatment with antibiotics should be established.

Septicaemia
This may be caused by many micro-organisms with symptoms such as discomfort, crying, abdominal distension, accelerated respiration, weakness, incoordination and dehydration. Therapy comprises administration of antibiotics and fluid.

The toxic milk syndrome
The toxic milk syndrome, in which there is abdominal distension and straining, may be caused by toxins in the milk due to subinvolution of

the uterus or metritis. The puppies should be removed and artificially fed and the bitch treated in accordance with the basic illness.

Neonatal dermatitis

Neonatal dermatitis, due to insufficient cleansing of the puppies by the bitch and drying up of the fetal fluid as well as infection by haemolytic staphylococci, is located on the head and neck, and is treated by antibiotics and cleaning with mild shampoos.

Puppy viraemia

This occurs with clinical symptoms similar to septicaemia. The treatment comprises attempts at arresting replication of the virus, for example by exposure of the affected litter to environmental temperatures of 37–38 °C for 3 h, then 32–35 °C for 21 h and a prophylactic supply of fluid to avoid dehydration.[51]

Neonatal ophthalmia

This is due to acute purulent conjunctivitis, and treatment consists of antibiotics and of cleaning the eyes.

REFERENCES

1. Ablett, R.E. and Baker, L.A. (1960–61) Canine virus hepatitis serum neutralizing antibody levels of brood bitches having a history of puppy loss. *J. small Anim. Pract.*, **1**, 174–178.
2. Andersen, A.C. (1957) Puppy production in the weaning age. *J. Amer. vet. med. Assoc.*, **130**, 151–158.
3. Austad, R., Lunde, A. and Sjaastad, Ø.V. (1976) Peripheral plasma levels of oestradiol-17 and progesterone in the bitch during the oestrous cycle, in normal pregnancy and after dexamethasone treatment. *J. Reprod. Fertil.*, **46**, 129–136.
4. Baines, F.M. (1981) Milk substitutes and the hand rearing of orphan puppies and kittens, *J. small Anim. Pract.*, **22**, 555–578.
5. Björck, G., Olsson, B. and Dyrendahl, S. (1957) Artificial raising of puppies. *Nord. Vet.-Med.*, **9**, 285–304.
6. Bleicher, N. (1962) Behavior of the bitch during parturition. *J. Amer. vet. med. Assoc.*, **40**, 1076–1082.
7. Concannon, P.W., Hansel, W. and Visek, W.J. (1975) The ovarian cycle of the bitch: plasma estrogen, LH and progesterone. *Biol. Reprod.*, **13**, 112–121.
8. Concannon, P.W., Powers, M.E., Holder, W. & Hansel, W. (1977) Pregnancy and parturition in the bitch. *Biol. Reprod.*, **16**, 517–526.
9. Concannon, P.W., Butler, W.R., Hansel, W., Knight, P.J. and Hamilton, J.M. (1978) Parturition and lactation in the bitch: Serum progesterone, cortisol and prolactin. *Biol. Reprod.*, **19**, 1113–1118.
10. Crighton, G.W. (1961) Rectal temperatures in new-born normal puppies. *Vet. Rec.*, **73**, 646–649.
11. Crighton, G.W. (1969) Thermal regulation in the newborn dog. *Mod. vet. Pract.*, **50**(2), 35–46.
12. Davies, M.E. and Skulski, G. (1956) A study of beta-haemolytic streptococci in the fading puppy in relation to canine virus hepatitis infection in the dam. *Brit. vet. J.*, **112**, 404–416.

13. Drake, J.C. and McCarthy, M.O.J. (1964) Some hypotheses on neonatal mortality in pups. *Vet. Rec.*, **76**, 1283–1287.
14. Edqvist, L.-E., Johansson, E.D.B., Kasström, H., Olsson, S.-E. and Richkind, M. (1975) Blood plasma levels of progesterone and oestradiol in the dog during the oestrous cycle and pregnancy. *Acta endocrinol. (Kbh.)*, **78**, 554–564.
15. Evans, J.M. (1968) Neonatal disease in puppies associated with bacteria and toxoplasma. *J. small Anim. Pract.*, **9**, 453–461.
16. Farrow, B.R.H. and Malik, R. (1981) Hereditary myotonia in the chow chow. *J. small Anim. Pract.*, **22**, 451–465.
17. Fox, M.W. (1963) Neonatal mortality in the dog. *J. Amer. vet. med. Assoc.*, **143**, 1219–1223.
18. Fox, M.W. (1965) The pathophysiology of neonatal mortality in the dog. *J. small Anim. Pract.*, **6**, 243–254.
19. Freak, M.J. (1948) The whelping bitch. *Vet. Rec.*, **60**, 295–306.
20. Freak, M.J. (1962) Abnormal conditions associated with pregnancy and parturition in the bitch. *Vet. Rec.*, **74**, 1323–1339.
21. Gaines, F.P. and VanVleck, L.D. (1976) The influence of beagle sires on gestation length, litter size, birth weight, and livability. *Carnivore Genet.* **3**, 75–79.
22. Gowing, G.M. (1964) Advanced demodectic dermatitis in 4-day-old dachshund puppies. *Mod. vet. Pract.*, **45**, 13, 70.
23. *Guinness Book of Records* (1978) 24th ed. Guinness Superlatives Ltd, Enfield.
24. Günther, S. (1954) Klinische Beobachtungen über die Geburtsfunktion bei der Hündin. *Arch. exp. Vet.-Med.*, **8**, 748–752.
25. Hadley, J.C. (1975) Total unconjugated oestrogen and progesterone concentrations in peripheral blood during pregnancy in the dog. *J. Reprod. Fertil.*, **44**, 453–460.
26. Hashimoto, A., Hirai, K., Okada, K. and Fujimoto, Y. (1979) Pathology of the placenta and newborn pups with suspected intrauterine infection of canine herpes-virus. *Amer. J. vet. Res.*, **40**, 1236–1240.
27. Hedhammar, Ä. (1970) Utfodring av valpar och unghundar. *Svensk Vet.-Tidn.*, **22**, 603–612.
28. Hime, J.M. (1961) Hypothermia in the newborn puppy. *Vet. Rec.*, **73**, 727.
29. Hime, J.M. (1963) An attempt to simulate the 'fading syndrome' in puppies by means of an experimentally produced haemolytic disease of the newborn. *Vet. Rec.*, **75**, 692–694.
30. Holst, P.A. and Phemister, R.D. (1974) Onset of diestrus in the beagle bitch. Definition and significance. *Amer. J. vet. Res.*, **35**, 401–406.
31. Ingmand, E.B. (1971) A practitioner's approach to canine nutrition. *Vet. Med./small Anim. Clin.*, **66**, 557–560.
32. Jackson, C. (1958) Preliminary note on the pathology of canine neonatal mortality. *Vet. Rec.*, **70**, 288–289.
33. Jensen, K. (1974) Feeding of dogs. *Medlemsbl. danske Dyrlaegeforen.*, **57**, 745–755.
34. Jöchle, W., Tomlinson, R.V. and Andersen, A.C. (1973) Prostaglandin effects on plasma progesterone levels in the pregnant and cycling dog (Beagle). *Prostaglandins*, **3**, 209–217.
35. Jones, G.E., Boyns, A.R., Bell, E.T., Christie, D.W. and Parkes, M.F. (1973) Immunoreactive luteinizing hormone and progesterone during pregnancy and following gonadotrophin administration in beagle bitches. *Acta endocrinol. (Kbh.)* **72**, 573–581.
36. Joshua, J.O. (1972) Unusual behaviour in the bitch. *Vet. Rec.*, **90**, 162–163.
37. Kaiser, G. (1971) Die Reproduktionsleistung der Haushunde in ihrer Beziehung zur Körpergrösse und zum Gewicht der Rassen. I. *Z. Tierzücht. Züchtungsbiol.*, **88**, 118–168.
38. Kaiser, G. (1971) Die Reproduktionsleistung der Haushunde in ihrer Beziehung

zur Körpergrösse und zum Gewicht der Rassen. II. *Z. Tierzücht. Züchtungsbiol.*, **88**, 241–253.

39. Kaiser, G. (1971) Die Reproduktionsleistung der Haushunde in ihrer Beziehung zur Körpergrösse und zum Gewicht der Rassen. III. *Z. Tierzücht. Züchtungsbiol.*, **88**, 316–340.

40. Kaiser, G. and Huber, W. (1969) Beziehungen zwischen Körpergrösse und Wurfgrösse beim Haushund. *Rev. Suisse Zool.*, **76**, 656–673.

41. Leipold, H.W. (1977) Nature and causes of congenital defects of dogs. *Vet. Clin. N. Amer.*, **8**, 47–77.

42. Long, D., Mezza, R. and Krakowa, S. (1978) Signs of impending parturition in the laboratory bitch. *Lab. Anim. Sci.*, **28**, 178–181.

43. Lyngset, A. (1973) The influence of the male dog on litter size. *Nord. Vet.-Med.*, **25**, 186–191.

44. Lyngset, A. and Lyngset, O. (1970) Litter size in dogs. *Nord. Vet.-Med.*, **22**, 186–191.

45. Malherbe, W.D. (1944) Joint-ill in puppies, *J. S. Afr. vet. med. Assoc.*, **15**, 73–74.

46. Mapletoft, R.J., Schutte, A.P., Coubrough, R.I. and Kuhne, R.J. (1974) The perinatal period of dogs. Nutrition and management in the hand-rearing of puppies. *J. S. Afr. vet. Assoc.*, **45**, 183–189.

47. Martins, T. (1949) Disgorging of food to the puppies by the lactating dog. *Physiol. Zool.*, **22**, 169–172.

48. Mayr-Bibrack, B. (1982) Neues über Ursachen und Bekämpfungsmöglichkeiten des infektiösen Welpensterbens. *Kleintier-Prax.*, **27**, 3–10.

49. McKelvie, D.H. and Andersen, A.C. (1963) Neonatal deaths in relation to the total production of experimental beagles to the weaning age. *Lab. Anim. Care*, **13**, 725–730.

50. Mosier, J.E. (1974) Pediatric nutrition and diet. *J. Amer. Anim. Hosp. Assoc.*, **10**, 78–83.

51. Mosier, J.E. (1978) The puppy from birth to six weeks. *Vet. Clin. N. Amer.*, **8**, 79–100.

52. Naaktgeboren, C. (1971) *Die Geburt bei Haushunden und Wildhunden.* A. Ziemsen Verlag, Wittenberg Lutherstadt.

53. Parkes, M.F., Bell, E.T. and Christie, D.W. (1972) Plasma progesterone levels during pregnancy in the beagle bitch. *Brit. vet. J.*, **128**, 15–16.

54. Pearson, M. and Pearson, K. (1931) On the relation of the duration of pregnancy to size of litter and other characters in bitches. *Biometrika*, **22**, 309–323.

55. Prescott, C.W. (1972) Neonatal diseases in dogs and cats. *Aust. vet. J.*, **48**, 611–618.

56. Robinson, R. (1973) Relationship between litter size and weight of dam in the dog. *Vet. Rec.*, **92**, 221–223.

57. Rowlands, I.W. (1950) Some observations on the breeding of dogs. *Proc. Conf. Soc. Study Fertil., London*, **Vol. II**, 40–55.

58. Rüse, I. (1961) Die Laktation der Hündin. *Zbl. Vet.-Med.*, **VIII**, 252–281.

59. Rüse, I. (1966) Über die mutterlose Aufzucht von Hundewelpen. *Zbl. Vet.-Med.*, **130**, 125–131.

60. Schultze, W. (1955) Zum Geburtsvorgang bei der Hündin. *Mh. Vet.-Med.*, **10**, 533–539.

61. Scothorn, M.W., Koutz, F.R. and Groves, H.F. (1965) Prenatal *Toxocara canis* infection in pups. *J. Amer. vet. med. Assoc.*, **146**, 45–48.

62. Siert-Roth, U. (1958) Litter size and sex ratio in Hungarian shepherd dogs. *Zool. Gart.*, **22**, 204–208. (*Anim. Breed. Abstr.*, **27**, 344, no. 1548).

63. Smith, M. and McDonald, L.E. (1974) Serum levels of luteinizing hormone and progesterone during the estrous cycle, pseudopregnancy and pregnancy in the dog. *Endocrinology*, **94**, 404–412.

64. Spalding, V.T., Rudd, H.K., Langman, B.A. and Rodgers, S.E. (1964) Isolation of

C.V.H. from puppies showing the 'fading puppy' syndrome. *Vet. Rec.*, **76**, 1402–1403.
65. Spreull, J.S.A. (1949) Some observations on breeding, whelping and rearing. *Vet. Rec.*, **61**, 579–582.
66. Strasser, H. and Schumacher, W. (1968) Breeding dogs for experimental purposes. II. Assessment of 8-year breeding records for two beagle strains. *J. small Anim. Pract.*, **9**, 603–612.
67. Suter, M. (1977) *Peri- und postnatale Todesursachen beim Hund.* Thesis, Zürich, 39 pp.
68. Tedor, J.B. and Reif, J.S. (1978) Natal patterns among registered dogs in the United States. *J. Amer. vet. med. Assoc.*, **172**, 1179–1185.
69. Tröger, C.P. (1968) Embryonaler Fruchttod bei der Hündin. *Berl. Münch. tierärztl. Wochenschr.*, **81**, 245–248.
70. Tsutsui, T. (1975) Studies on the reproduction in the dog. VI. Ovulation rate and transuterine migration of the fertilized ova. *Jap. J. Anim. Reprod.*, **21**, 98–101.
71. van der Weiden, G.C., Taverne, M.A.M., Okkens, A.C. and Fontijne, P. (1981) The intra-uterine position of canine foetuses and their sequence of expulsion at birth. *J. small Anim. Pract.*, **22**, 503–510.
72. Wilsman, N.J. and Van Sickle, D.C. (1973) Weight change patterns as a basis for predicting survival of newborn pointer pups. *J. Amer. vet. med. Assoc.*, **163**, 971–975.
73. Wright, N.G. and Cornwell, H.J.C. (1968) Viral induced neonatal disease in puppies. *J. small Anim. Pract.*, **9**, 449–452.
74. Yutuc, L.M. (1949) Prenatal inspection of dogs with ascarids, *Toxocara canis*, and hookworms, *Ancylostoma caninum. J. Parasitol.*, **35**, 358–360.

ADDITIONAL READING

Aguirre, G. (1973) Hereditary retinal diseases in small animals. *Vet. Clin. N. Amer.*, **3**, 515–528.
Bain, A.D., Gauld, I.K., Watkins, J.E. and Withers, A.R. (1965) A simple technique for the demonstration, post-mortem, of canine chromosomes. *J. small Anim. Pract.*, **6**, 127–129.
Beaver, B.V. (1982) Somatosensory development in puppies. *Vet. Med./small Anim. Clin.*, **77**, 39–41.
Bibrack, B. (1974) Die Bedeutung von Virusinfektionen beim Welpensterben und beim Zwingerhusten. *Kleintier-Prax.*, **19**, 223–229.
Blunden, A.S. (1983) The 'fading puppy complex': an assessment of a paraimmunity inducer as a means of control. *Vet. Rec.*, **113**, 201.
Bouw, J. (1982) Breeding and health impairment in dogs. *T. Diergeneesk.*, **107**, 12–20.
Breazile, J.E. (1978) Neurologic and behavioral development in the puppy. *Vet. Clin. N. Amer.*, **8**, 31–45.
Clark, G.A. (1932) Some foetal blood-pressure reactions. *J. Physiol. (Lond.)*, **74**, 391–400.
Dunbar, I., Ranson, E. and Buehler, M. (1981) Pup retrieval and maternal attraction to canine amniotic fluids. *Behav. Processes*, **6**, 249–260.
Edney, A.T.B. (Ed.) (1982) *Dog and Cat Nutrition.* Pergamon Press, Oxford, New York, Toronto, Sidney, Paris, Frankfurt.
Gelatt, K.N. (1973) Pediatric ophthalmology in small animal practice. *Vet. Clin. N. Amer.*, **3**, 321–333.
Gerber, J.G., Hubbard, W.C. and Nies, A.S. (1979) Uterine vein prostaglandin levels in late pregnant dogs. *Prostaglandins*, **17**, 623–627.
Griffith, I.R., Duncan, I.D., McCulloch, M. and Harvey, M.J.A. (1981) Shaking pups: a disorder of central myelination in the spaniel dog. *J. Neurol. Sci.*, **50**, 423–433.

Harkness, J.E. and McCormick, L.F. (1981) Swimming-puppy syndrome in a litter of German shepherd pups. *Vet. Med./small Anim. Clin.*, **76**, 817–821.

Jensen, C. and Ederstrom, H.E. (1955) Development of temperature regulation in the dog. *Amer. J. Physiol.*, **183**, 340–344.

King, J.W.B. (1978) Survival of single pups. *Vet. Rec.*, **103**, 433.

Kirk, R.W. (1965) Canine pediatrics. *J. Amer. vet. med. Assoc.*, **147**, 1475–1479.

Kirk, R.W. and Sifferman, R.L. (1978) Preventive pediatric programs for the growing puppy. *Vet. Clin. N. Amer.*, **8**, 129–142.

Kronfeld, D.S. (1977) Feeding for breeding. *Pure-bred Dogs Amer. Kennel Gaz.*, **94**(7), 34–41.

Larsson, S. and Vesterlund, S. (1975) Förhindrande av kolibacillos hos nyfödda valpar. *Svensk Vet.-Tidn.*, **27**, 65–66.

Meier, K.U., Teute, H.-W. and Daerr, H.-Ch. (1973) Die Aufzucht von Beagle-Welpen mit WELPI®-LAC. *Kleintier-Prax.*, **18**, 32–39.

Meutstege, F.J. (1982) Hip dysplasia in dogs. *T. Diergeneesk.*, **107**, 26–29.

Meyer, H., Mundt, H.-C., Thomèe, A. and Rauchfuss, R. (1981) Zum Mineralstoffwechsel von Saugwelpen sowie graviden und laktierenden Hündinnen. *Kleintier-Prax.*, **26**, 115–120.

Miller, Jr., J.A. and Miller, F.S. (1960) Mutual antagonism between hypothermia and asphyxia in neonatal puppies. *Anat. Rec.*, **136**, 357.

Mosier, J.E. (1977) Nutritional recommendations for gestation and lactation in the dog. *Vet. Clin. N. Amer.*, **7**, 683–692.

Mosier, J.E. (1979) Nutritional considerations in canine reproduction. *Norden News*, **54**, 3, 28–30.

Naaktgeboren, C. and Bontekoe, E.H.M. (1976) Vergleichend-geburtskundliche Betrachtungen und experimentelle Untersuchungen über psychosomatische Störungen der Schwangerschaft und des Geburtsablaufes. *Z. Tierzücht. Züchtungsbiol.*, **93**, 264–320.

Narfström, K. (1981) Cataract in the West Highland white terrier. *J. small Anim. Pract.*, **22**, 467–471.

Newberne, P.M. (1974) A survey and some thoughts on the current state of pet nutrition. *J. Amer. Anim. Hosp. Assoc.*, **10**, 111–121.

Patterson, D.F. (1971) Canine congenital heart disease: epidemiology and etiological hypotheses. *J. small Anim. Pract.*, **12**, 263–287.

Richardson, R.C. (1978) Diseases of the growing puppy. *Vet. Clin. N. Amer.*, **8**, 101–128.

Ries, R.L. and Harris, L.R. (1982) Urolithiasis in a puppy. *Mod. vet. Pract.*, **63**, 8, 646–647.

Riser, W.H., Haskins, M.E., Jezyk, P.F. and Patterson, D.F. (1980) Pseudoachondroplastic dystocia in miniature poodles: clinical, radiologic, and pathologic features. *J. Amer. vet. med. Assoc.*, **176**, 335–341.

Romsos, D.R., Palmer, H.J., Muiruri, K.L. and Bennink, M.R. (1981) Influence of a low carbohydrate diet on performance of pregnant and lactating dogs. *J. Nutr.*, **111**, 678–689.

Sheffy, B.E. (1978) Nutrition and nutritional disorders. *Vet. Clin. N. Amer.*, **8**, 7–29.

Slappendel, R. J. (1982) Hereditary features of haemophilia and their consequences in dog-breeding. *T. Diergeneesk.*, **107**, 23–25.

Strasser, H. (1965) Zur künstlichen Aufzucht von Hundewelpen. *Die Blauen Hefte*, no. **28**, 26–28.

Thiel, U. (1980) Myocarditis in puppies. *Vet. Rec.*, **106**, 43.

Tillmann, H. (1952) Physiologie und Pathophysiologie der Gebärmutter unter den Bedingungen der Multiparität. *Berl. Münch. tierärztl. Wochenschr.*, **65**, 101–105.

Venker-van Haagen, A.J. (1982) Laryngeal paralysis in Bouviers Belge des Flandres and breeding advice to prevent this condition. *T. Diergeneesk.*, **107**, 21–22.

Chapter 10
Dystocia, Obstetric and Postparturient Problems

When the bitch is examined at a late stage of pregnancy, the owner should be given a basic description of the bitch's behaviour, the various stages of parturition, the discharge and the colour of the fetal fluid, and be told when and how professional aid should be sought.

DYSTOCIA

Various symptoms may indicate dystocia:

1. Obvious abnormal fetal presentation, position or posture – for example, only one leg appearing in the vagina.
2. Strong and persistent labour for 20–30 min without expulsion of the fetus.
3. Weak and infrequent unproductive labour for 2–3 h.
4. A lapse of more than 4 h after the expulsion of a puppy (the absence of labour signs may be because all fetuses have been expelled).
5. Prolonged pregnancy, pathological vaginal discharge and signs of intoxication.

The interval between the expulsion of puppies should be used only as a guide and not as a definite rule. The time required for the expulsion of a puppy varies from the normal 10–15 min to several hours in bitches with weak and infrequent uterine contractions.

Dystocia can be classified as maternal or fetal (Tables 10.1 and 10.2). The most frequently occurring forms are associated with inertia and discrepancy between the size of the fetus and the birth canal.

Maternal dystocia

Inertia
This may occur in various forms. It is found in older bitches or following painful and pathological conditions such as hernia and rupture of the diaphragm. In the bulldog the abdominal muscles may be so relaxed and weak that contractions are insufficient to guide the fetus to the pelvic inlet.[27]

Table 10.1 Causes of maternal dystocia.

Birth canal (constriction)	Insufficient dilatation	Vulva	Congenital defect
			Fibrosis
			Immaturity
		Vagina	Congenital defect
			Fibrosis
			Neoplasia or cysts
			Cystocele
			Abscess
			Prolapse
		Cervix	Inertia
			Hormonal imbalance
			Fibrosis
			Congenital defect
		Uterus	Torsion
			Congenital defect
			Inguinal hernia
	Inadequate pelvis	Pelvis	Immaturity
			Fracture
			Breed
			Diet
			Development
			Disease
			Neoplasia
Forces (expulsive defect)	Uterine	Primary inertia	Myometrial defect
			Intrinsic 'weakness'
			Overstretching

Toxic degeneration
Fatty infiltration
Senility
Dietetic deficiencies
Infections of the uterus
Systemic illness
Breed
Heredity

Chemical deficiency
Oestrogen/progesterone
Oxytocin
Prostaglandins
Relaxin
Calcium
Glucose

Premature birth

Environmental disturbance

Absence of adequate fetal fluid

Fetal/maternal dystocia

Secondary inertia

Rupture of uterus

Torsion of uterus

Trauma of uterus

Abdominal

Age
Debility
Pain
Herniation of uterus
Ruptured diaphragm
Perforated trachea (anaesthesia)

Table 10.2   Causes of fetal dystocia.

Hormone deficiency	Adrenocorticosteroid	
Oversize	Absolute	Small litter Breed Prolonged gestation
	Developmental defect	Duplication Ascites Anasarca Hydrocephalus Hydrops amnii
Faulty disposition	Presentation	(Posterior) Transverse Bicornual Simultaneous
	Position	Ventral Lateral

	Posture	Anterior
		  carpal flexion
		  elbow flexion
		  shoulder flexion
		  head flexion lateral
		  head flexion upward
		  head flexion downward
		Posterior
		  hock flexion
		  hip flexion

Fetal death		

Primary inertia

This is characterized by failure of the uterine muscles to expel fetuses of normal size through an unobstructed birth canal. Uterine contractions are weak, infrequent and unproductive. The bitch may not show any sign of labour, or may be slow in the expulsion of the puppies, or atony may appear after expulsion of some of them.

Primary inertia occurs in all breeds and at all ages, but some, such as scottish terrier, aberdeen and border terriers, dachshund and chihuahua, are more susceptible. It occurs rather frequently in inbred bitches and especially in older bitches who are overweight and under-exercised. Overdistension of the uterus by a large number of fetuses, one single oversized fetus or hydrops amnii may be the cause. Or it may be hereditary, or of mental origin, or occur in association with hypocalcaemia or systemic endocrine imbalance, or be due to premature parturition.

Secondary inertia

This is caused by obstruction, of fetal or maternal origin, leading to cessation of contraction of the uterine muscles.

Torsion of the uterus

This condition, in which none of the fetuses or only some are expelled (depending on the site of torsion), is rather rare.

Inguinal hernia

Inguinal hernia, with a part of the pregnant uterus including a fetus in the inguinal canal, may occur.[47,54,65]

Rupture of the uterus

Rupture of the uterus with displacement of a fetus into the abdominal cavity and complicated with peritonitis may cause dystocia.[10,26,39,45,48,51]

Abnormality of the uterus

An abnormality of the uterus in which the body and the first few cm of both horns form a thin fibrous tube about 1 cm in diameter has been described as one form of maternal dystocia.[22]

Insufficient dilatation of the cervix

This may lead to transverse presentation of a fetus.

Dorsoventral bands of tissue in the vagina have been described as preventing expulsion of fetuses in bulldogs[57] and bull terriers.[33]

Neoplasms

These are not rare in the bitch, but seldom cause dystocia which prevents the expulsion of fetuses.

Abnormalities of vagina and vulva

Abnormalities, congenital or acquired, of vagina or vulva may also cause dystocia.

Prolapse of the vagina

This does occur in bitches especially during oestrus, but it rarely causes dystocia.[2,28,53,62]

Fetopelvic discrepancy in size

The dimensions of pelvis and fetus in certain breeds are given in Tables 10.3 and 10.4. The pelvic inlet increases, though not proportionally, with the size of the breed.[64] The fetal weight is 1–2% of the dam's weight in large breeds and 4–8% in smaller breeds.[36,44]

Fetopelvic disproportion is a fairly common type of dystocia and in origin may be maternal (relative fetal oversize), fetal (absolute fetal oversize) or both. The pelvis may be too narrow because of immaturity, retarded development, healed fractures or congenital causes.

Absolute fetal oversize causes fetal dystocia in primigravid bitches with a single fetus. Oversized fetuses have been reported as appear-

Table 10.3   Pelvic dimensions in the bitch. (After Arthur[8] and Rasbech[49])

	Pelvic inlet	
Breed	Sacropubic (mm)	Iliac (mm)
Pekinese	31–37	38–39
Manchester terrier	32	30
Pomeranian	40	32
Mongrel	42	35
Sealyham terrier	44	49
Terrier	45–49	38–40
Dachshund	48	47
Bull terrier	48	47
Airedale terrier	58	50
Doberman	63	63
Alsatian	65	60

ing frequently in large litters, the first fetus most often being big.[20] In corgis because of extreme variations in the size of the puppies, both absolute and relative fetal oversize may occur.[8] Crossbreeding may result in malproportion of the fetus, and the heads especially may be too big. The same problem occurs in brachycephalic breeds and sealyham and scottish terriers where the fetuses have large heads and the dams narrow pelves.[8]

The thorax may be too large resulting in arrest of expulsion when the head has passed the pelvis. The pelvis of the fetus is seldom so big as to cause problems, but dystocia has been described as being caused by the fetal abdominal organs being pressed back, producing a balloon-like distension of the posterior part of the fetus, which was too big to pass through the pelvis.[36]

Table 10.4   Dimension of puppies at birth. (After Arthur[8] and Larsen[36])

	Head		Chest	
Breed	Lateral (mm)	Dorsoventral (mm)	Across shoulders (mm)	Dorsoventral (mm)
Mongrel	28–32	30–32	32–38	28–48
Terrier	29–30	32–34	32–34	38
Dachshund	30	32	29	35
Spaniel	32–38	32–38	32–42	42–48
Pincher, miniature	34	32	40	47
Doberman	38	35	43	44

Fetal dystocia

Monsters are relatively rarely responsible for dystocia, but fetal hydrocephalus, anasarca, ascites, chondrodystrophy and double monsters have been described as causes.[8,12,19,63]

In bulldogs dystocia has been described as being caused by a stunted tail pressed against the pelvic inlet.[36]

Fetal obstruction

This is not uncommon when fetuses from opposite horns enter the birth canal at the same time. The fetuses then lie anterior to the pelvic inlet.[63]

Abnormal presentation

Transverse presentation may occur if there is only one fetus which is too big in relation to the birth canal, or if the cervix does not dilate sufficiently before labour starts, or closes before expulsion of all the fetuses.

Abnormal positions or postures

These may occur in either anterior or posterior presentation.

Deviation of the head

Deviation of the head to one side or ventrally is rather common in spite of the relatively short length of the neck, and is often combined with inertia (Figures 10.1 and 10.2). Lateral deviation of the head often occurs in the last fetus.[8,27] Fetuses with dorsal deviation of the head have been described as being expelled spontaneously in bitches of larger breeds.[36]

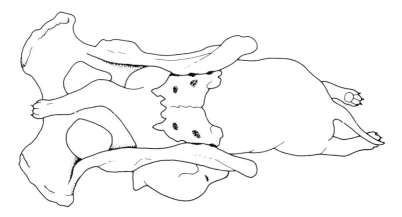

Figure 10.1   Lateral deviation of the head (shoulder presentation). (After Arthur[8])

Figure 10.2 Vertex posture ('butt' presentation) with bilateral shoulder flexion. (After Arthur[8])

Flexion of anterior or posterior limbs
This occurs in dead or oversized fetuses. Small fetuses may be expelled with flexed forelimbs. Carpal flexion, incomplete extension of the elbow and breech presentation (Figure 10.3) are reported to be rather rare,[12,63] but some researchers have found that breech presentation occurs quite frequently.[8,36]

Figure 10.3 Bilateral hip flexion posture (breech presentation). (After Arthur[8])

OBSTETRIC CARE

Obstetric examination

A careful obstetric examination should include the history; general clinical examination; abdominal palpation; abdominal auscultation, ultrasonic examination; inspection of the vulva; inspection and exploration of the vagina; and (if indicated) radiographic examination.

History

Since a good case history is of major importance, the most detailed information should be sought. The date of breeding must be known, to decide whether the gestation period has been correctly estimated. The breed and size of the male dog and the course of the pregnancy should also be known, since various diseases associated with these may influence parturition and the viability of the fetuses. Previous whelping experience including information on dystocia and any injuries should be sought together with the time of onset of parturition, the frequency and intensity of expulsive efforts, number of puppies born, time intervals between deliveries and details of any assistance given.

Clinical examination

General clinical examination
This should include inspection of visible mucous membranes. The frequency of respiration and pulse should be recorded and the rectal temperature taken. Other examinations should be conducted as circumstances indicate.

Abdominal palpation
This is carried out to estimate the size and tone of the uterus and to palpate the fetuses and their size and position in relation to the pelvic inlet. Fetal movements may be felt through the abdominal wall.

Abdominal auscultation
The sound of rapid fetal heartbeats may reveal fetal viability.

Ultrasonic examination
Ultrasonic examination by the use of Doppler apparatus may give information on the viability of the fetuses.

Inspection of the vulva
This gives information on discharge, oedema and any injuries.

The vagina

Inspection of the vagina may give information on the state of cervix dilatation, and the presence of fetal membranes or fetal extremities in the cervical orifice.

Exploration of the vagina can be performed by one or two lubricated fingers after thorough cleaning of the vulva and the perineal region. The presence of congenital abnormalities, the dimensions of the soft birth canal and the degree of dilatation of the cervix may be assessed. By pressing the finger dorsally the force of the contraction reflex may be estimated. If a fetus can be palpated, its viability, the presentation, the position and the posture are assessed. A fetus may survive several hours after discharge of the fetal fluid,[49] but not more than 20 min after entering the birth canal.[13] The viability may be assessed if the mouth of the fetus can be reached and sucking movements felt by inserting the finger into the mouth. Normal anterior presentation is easy to diagnose by palpating the nose and the head of the fetus.

Deviation of the head laterally is characterized by only one leg being palpable and can be differentiated from posterior presentation after gently drawing the fetus into the pelvis. Ventral deviation of the head in the nape posture is easier to diagnose because the ears and the occiput may be felt, whereas the more pronounced case of complete neck flexion, the breast–head posture, is more difficult to diagnose because only a mass of muscles covered with skin is felt.

Normal posterior presentation is easily diagnosed by palpating the two hindlegs and the tail.

Partial or complete breech presentation with one or both hindlegs flexed is diagnosed by palpating the tail and one or neither of the hindlegs.

Bicornual or transverse presentation can usually only be diagnosed by caesarean section, since normally the fetus cannot be reached by the palpating finger.

Radiographic examination

This may be indicated to determine whether parturition has finished or to evaluate the number and location of retained fetuses.

Obstetric aid

To secure a successful delivery of live, undamaged puppies without harming the dam, obstetric aid may be carried out in various ways: extraction of the fetus; episiotomy; embryotomy; medication; or surgical delivery, that is, caesarean section or hysterectomy.

Extraction of the fetus
Delivery by extraction is indicated in fresh, non-protracted cases, such as malpresentation of a fetus that may be corrected per vaginam, and in cases of slight relative oversize that can be overcome by traction. In inertia which does not respond to medical treatment, delivery by traction may be successful.

The bitch is best placed on a table in a standing position and the hindquarters cleaned. The fetus may be extracted either with the fingers or by forceps, using sufficient amounts of lubricant.

Digital extraction
Digital correction and extraction may be carried out by one or two fingers inserted in the vagina, the other hand grasping the fetus through the abdominal wall and leading it towards the pelvic inlet (Figure 10.4). A fetus in anterior presentation is fixed just behind the head, and in posterior presentation anterior to the pelvis. A fetus, especially if alive, should never be extracted by fixing a leg because of the risk of damaging it. The direction of traction should be in the natural direction of expulsion, that is initially dorsocaudally guiding the shoulder of a fetus in anterior presentation or the pelvis of a fetus in posterior presentation into the pelvis by the use of semirotatory movements. Thereafter the direction is altered to caudoventral in order to adjust the fetus as easily as possible to the natural direction of expulsion. Traction should be synchronized with uterine contractions and should not be too strong at first, because the birth canal needs some time to adjust its size to that of the fetus.

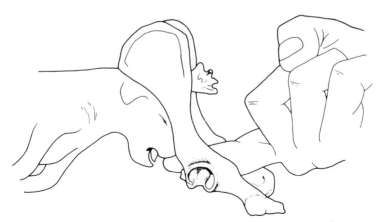

Figure 10.4 Correction of vertex posture with the finger. (After Arthur[8])

Forceps extraction

Instrumental delivery can be carried out by means of various kinds of special forceps, such as Hobday's and Palsson's, but ordinary forceps such as Ramley's sponge-holding forceps can also be used. Forceps without a ratchet are preferable as there is then no temptation to close the forceps completely. A forceps delivery should *never* be undertaken if the fetus is totally out of reach of the finger. The forceps must be used in such a way so as not to grasp the vaginal wall. This can be ensured if they are always applied with a rotatory movement before being closed and only between uterine contractions (Figure 10.5).

In anterior presentation the forceps, guided by a finger, should grasp the upper jaw, the lower jaw or the whole nose. When the head has been drawn through the pelvis, the fingers take over. Extraction can also be tried by gripping the folds of skin on each side of the head with the forceps – Palsson's forceps are made to grip behind the head like two fingers. Snare forceps can be used with the snare placed above the neck of the fetus and the jaws ventral to the neck. Traction can then be applied by the forceps to the free ends of the snare with this held tightly in position by the forceps (Figure 10.6).

In anterior presentation traction may also be applied to the puppy's head using a Hobday's vectis placed over the head and behind the occiput, and with a finger under the head (Figure 10.7).

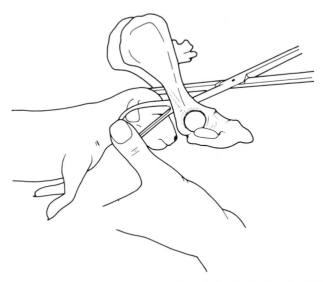

Figure 10.5 Delivery of a puppy with retention of the forelimbs using Hobday's forceps, while fixing the position of the fetus through the abdominal wall with the left hand, the forceps are applied to the skull with the right. (After Arthur[8])

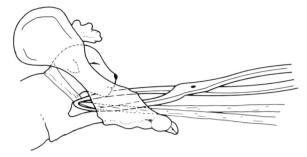

Figure 10.6   Robert's snare forceps applied to fetal neck. (After Arthur[8])

An ovariectomy hook fixed into the soft area between the mandibles may be used to apply traction to the fetus forming – with the middle finger in the mouth of the fetus – pincers on the symphysis of the lower jaw.[18]

In posterior presentation forceps may be used to grasp folds of the skin lateral to the thigh or the pelvis. The fetus may be fixed anterior to the pelvis by the use of Palsson's forceps. If the posterior part of the fetus has already passed the pelvic inlet, Hobday's vectis can be used for delivery placed over the dorsal part of the fetal pelvis behind the tuber coxae and with the index finger pressing upwards in front of the fetal pelvis.

Forced traction may be used for delivery of a dead fetus with a firm grip across the fetal cranium or pelvis and with the bitch completely anaesthetized.

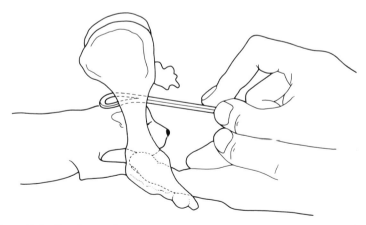

Figure 10.7   Traction applied to puppy's head using a guide and the finger. (After Arthur[8])

Abnormal position or posture
Abnormal position or posture should be corrected before traction is applied. Manipulation is helped if the position of the fetus is fixed by gripping it with the left hand through the abdominal wall. Deviation of the head is corrected simply by inserting a finger into the mouth of the fetus. Flexion of limbs either in anterior or posterior presentation is corrected by curling a finger around the retained limb and drawing it upwards and backwards into the maternal pelvis.

In those cases where two fetuses from opposite horns obstruct one another, one is retained while the other is extracted. If there is one fetus in anterior and one in posterior presentation, the latter should be extracted first because the pelvis is more easily brought into the birth canal than the head.[36]

Dystocia due to dorsoventral cord formation in the vagina making digital delivery impossible may be overcome either by cutting the cord which is seldom accompanied by serious bleeding, or by caesarean section.

In cases of narrowness of the vulva a finger inserted into the rectum of the dam may be used to press the head of the fetus towards the vulva while attempts are made to widen the vulval opening sufficiently for delivery with the fingers of the other hand.

Episiotomy
This may be indicated in cases of narrowness of the vulva. The incision is made in the upper commissure of the vulva either extending towards the anus or in a dorsolateral direction.[12,14]

Embryotomy
This method requires a certain amount of skill. It may be used where there is a dead fetus in the birth canal or one or two fetuses are retained in the uterus,[12,36,49,60] but caesarean section is usually preferred in such cases.

Medical treatment
The major indication for medication is inertia, and the treatment consists of the use of drugs initiating and supporting uterine contractions, so long as the cervix is dilated and possible maternal and fetal dystocia is corrected.

Uterine spasm, which may result in the fetus being forced towards the roof of the uterus, may be avoided by prior injection of calcium solution, such as calcium borogluconate 5–15 ml subcutaneously or intramuscularly, or calcium Sandoz 3–6 ml subcutaneously.[56] The injection of calcium is often sufficient for contractions to begin since the inertia may be associated with subclinical eclampsia.

Oxytocin
This is recommended in uterine inertia in a dose of 2–5 IU intramuscularly. To estimate the correct dose the milk letdown reflex may be used.[55] If the bitch lactates after injection, the dosage is correct, and if parturition does not proceed in spite of this, caesarean section may be indicated. In cases of slow fetal expulsion the injection may be repeated after 2 h if milk letdown is not evident, and the repeated injections should be separated by 2 h until there is either milk letdown or expulsion of a puppy. Oxytocin has only a short-term effect, and may cause parturition of the placenta leading to fetal death. If used late in parturition it is advisable to be ready to assist delivery without delay if natural delivery is not completed within 15–20 min after injection.

Ergometrine maleate
This has a more prolonged effect and may also be used in cases of inertia. The oral dose is 0.25–0.5 mg, and the effect is obtained within 10–30 min.[27] The intramuscular dose is 0.2–0.5 mg, acting within 10–15 min and lasting up to $1\frac{1}{2}$ h.[13]

Caesarean section
It has been said that if there are difficulties in expelling the first and second fetus, caesarean section is the preferred treatment, but if only one or two puppies remain in the uterus, other forms of delivery may be used.[32]
   Several kinds of dystocia may indicate surgical delivery: discrepancy between fetal and pelvic size; malpresentation not amenable to vaginal delivery; inertia which does not respond to medical treatment; fetal monsters; malformation of the birth canal; torsion of the uterus; and inguinal metrocele.
   Timing of the operation is of vital importance for survival of the puppies. It has been found that 82.7% of the puppies survived the first 24 h when caesarean section was carried out within 24 h of contractions starting, whereas only 30% survived when the operation was performed 28–50 h after contractions began.[65]

Anaesthesia
Various kinds of anaesthetics have been used, each with some advantages and disadvantages, and the choice depends partly on the condition of the bitch and partly on the routine of the operator. The transplacental transfer of the anaesthetic may also be considered as it may depress fetal viability and reduce the chance of survival. The use of anaesthetics for caesarean section in the dog and cat has been reviewed by Dodman.[21] Most of the drugs used may act toxically, since metabolization and excretion occur via the liver which, in the

fetus, is deficient in some enzymes. If the physical condition of the animal is poor, the operation may be performed under local analgesia combined with sedatives. If all the fetuses are known to be dead, barbiturates are useful.[24] When the fetuses are alive, short-acting barbiturates as well as repeated injections of ultrashort-acting barbiturates (such as thiobarbiturates) are contraindicated because of their depressive effect on respiration, and the fact that they are only slowly metabolized by the fetal liver. Ultrashort-acting barbiturates may be used initially and anaesthesia maintained by inhalation without deleterious effect on the fetus, causing only a short postoperative unconsciousness and resulting in speedy establishment of postnatal care. The same effect is obtained by using etorphine hydrochloride and methotrimeprazine; the latter may even be injected into the puppies to shorten the depressive effect of etorphine hydrochloride.

Total anaesthesia
Total anaesthesia may depress respiration and cause death, because respiration may already be inhibited by the presence of dead fetuses which fill the uterus and abdominal cavity and press against the diaphragm. This can cause regurgitation of stomach contents. Intubation and/or administration of morphine is therefore indicated.[32]

The gaseous anaesthetics are fairly safe for the fetuses because of their rapid excretion via the fetal respiratory tract.[42] In patients amenable to masking, induction may be established with halothane (4%) in oxygen (2 l/min) and nitrous oxide (3 l/min), but premedication may be necessary in larger bitches before applying the mask.[31] The amount of halothane should be kept at a minimum during maintenance. Minimum neonatal depression is achieved by the use of a relaxant.

Epidural anaesthesia
Epidural anaesthesia is suitable for caesarean sections in skilled hands, since there is no placental transmission and the fetuses are not affected by the anaesthetic. It is also a safe technique for debilitated, intoxicated and aged bitches. A dosage of approx. 0.5 ml 2% procaine or 2% lidocaine per kg of bodyweight provides analgesia postcrior to the umbilicus, but a reduced dosage should be used in obese and aged animals because of the risk of migration of the drug more anteriorly.[31]

Regardless of the type of anaesthetic the time between the induction and surgery should be as short as possible because of the detrimental effect of prolonged anaesthesia on the fetuses.

Surgical technique
To reduce haemorrhage when the placenta is separated and to hasten involution it is advisable to administer 5–10 IU oxytocin intramuscularly at the beginning of the operation.[57]

Midline incision with the patient lying on her back is best, since this allows better exposure of the uterus, limits haemorrhage and leaves a less noticeable scar.[24,46,52,57] Hysterectomy is also facilitated through the midline incision. A flank incision may be preferred, since this does not interfere with lactation and probably provides somewhat quicker healing.[8,38,59] The tendency to slower healing of midline incisions may be due to the weight of the udder.

The uterus is exposed and the surrounding viscera packed off to prevent fetal fluid escaping into the abdominal cavity. It is possible to reach all fetuses through one uterine incision, but if there is more than one fetus in each horn and the uterus is greatly distended, an incision should be made in each horn.[8] If future breeding is planned, the incision should be made in the body, as this does not interfere with later implantation, as might be the case after incision in the horns. After opening the amnion each fetus is removed from the uterus, the umbilical cord clamped with artery forceps and cut with scissors about 2 cm from the abdominal wall. The cord is checked for haemorrhage when the forceps are released. The placentae are removed and the uterine incision is closed by a double continuous Lembert suture using chromic catgut or another resorbable suture. The uterus is cleaned, the towel packs removed and the uterus returned into the peritoneal cavity. The laparotomy incisions are closed with standard sutures for large abdominal incisions. Systemic or local treatment with antibiotics into the peritoneum, and fluid therapy, may be given if indicated.

Good postoperative care of the bitch is important. She should be kept in a warm environment, not be allowed to jump over the side of the whelping box, and her movements should be generally restricted for the first 12 h after the operation. The puppies should be dried and care taken that respiration has begun and is established. They should be kept warm until the dam regains consciousness and then placed near her as this contact may accelerate the beginning of uterine involution.

Requests to spay a bitch following caesarean section should be refused because of the increased risk of surgical shock.

A bitch can undergo a number of pregnancies after caesarean section, and several caesarean sections can be performed on the same bitch, but after about the third litter the litter size normally decreases, and difficulties increase with the number of operations performed.

Postoperative complications
There may be some bleeding from the uterus resulting in bloody vaginal discharge. This can be stopped by intramuscular administration of 2–10 IU oxytocin or 0.2–0.5 mg ergometrine.

Other complications, such as collapse, peritonitis or breakdown of the abdominal wound, may occur. Death may take place up to 10 days after the operation and its causes may be classified into four groups[8] according to time.

Within 24 h of the operation The main causes are advanced toxaemia present at the time of operation and circulatory disturbances. It has been found that 13.3% of bitches died within 24 h after caesarean section compared with 2–4% of other patients subjected to a similar anaesthetic.[42] This is probably due to abnormalities of respiration and blood circulation during pregnancy.

36–72 h after the operation This may be due to peritonitis caused by an intrauterine infection.

3–6 days after the operation This may be caused by toxaemia following bacterial infection causing necrosis of the abdominal wound extending to the peritoneum.

5–10 days after the operation This may be due to breakdown of the abdominal wound with prolapse of viscera sometimes complicated by peritonitis.

Hysterectomy
This is indicated in cases with a grossly infected and gangrenous uterus. Supportive therapy should be given including antibiotics and fluids in sufficient amounts.

Prevention of dystocia

The breeding of excitable and neurotic bitches, or those with deformed pelves or congenital abnormalities of the birth canal, should be avoided. The bitch and the sire should be of the same breed, but in breeds with a predisposition to dystocia, such as brachycephalics, this may decrease with 'outbreeding'.

The diet should be balanced, the pregnant bitch well exercised and not allowed to become overweight so that she is kept in a state of good general health.

Examination during pregnancy and information from the owner concerning previous pregnancies and parturitions are all of importance, as are familiar surroundings and efficient but minimal supervision for whelping.

POSTPARTURIENT PROBLEMS

Postpartum haemorrhage

Postpartum haemorrhage can be caused by injuries during delivery, placental necrosis or subinvolution of placental sites. When nontraumatic, the condition is characterized by periodic vaginal discharge of fresh, coagulated blood.

Subcutaneous injection of medroxyprogesterone acetate (MAP), 2 mg/kg, results in reduction of haemorrhage within 24 h; after 2 days the discharge is blood-tinged and it stops the next day.[5,6]

Injection of 10–30 mg chlormadinone acetate (CAP) can also be used successfully in cases of non-traumatic postnatal haemorrhage.[4]

Agalactia

Oxytocin can be used for letdown of milk but is effective only if milk production is normal. It can be administered intravenously, intramuscularly or subcutaneously in doses of 0.2–1 IU.[3] It may also be administered by inhalation three to four times daily.[34]

Insufficient lactation

Prolactin (PRL), 2 IU daily, is sufficient for maintenance of lactation.[61]

Retention of placenta

The diagnosis can be confirmed by X-ray examination in the form of a double hysterogram[50] (Figure 10.8). Then 4 ml of contrast medium (for example, hypaque) is introduced into the uterus via the cervix and a pneumoperitoneum established with CO_2. Oxytocin can be used in a dosage of 1–5 IU, repeated if necessary after 6 h.[3]

Delayed involution

Injection of 5–10 IU of oxytocin has been claimed to be effective in causing involution,[29] and diethylstilboestrol (DES), 1–2 mg, given orally for 7–10 days has proved effective.[25]

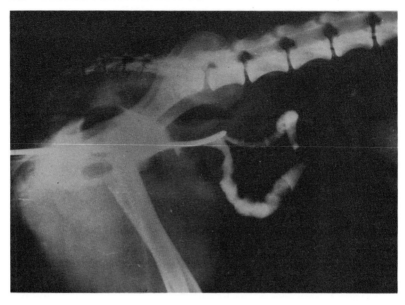

Figure 10.8 Hysterogram of acute metritis, 3 weeks post partum, the normally clear-cut areas of placentation are disrupted by metritis and only the terminal section of the dependent horn appears normal. (After Cobb[17])

Subinvolution of placental sites

In some bitches[1,11,16] the normal serosanguineous discharge lasts for more than the normal 4–6 weeks because of the condition without otherwise affecting the animal. The rectal temperature is normal as is the haemogram unless anaemia is present.

It is of great importance to institute prophylactic treatment against secondary endometritis. Complete regeneration of the endometrium may then follow spontaneously.

Prophylactic administration of oxytocin in the early postpartum period may also be of some value.

Prolapse of the uterus

Prolapse[29,30,57] may occur during and after parturition either in an incomplete or complete form with the uterus protruding from the vulva. In incomplete prolapse there is straining and abdominal pain. Vaginal exploration provides the diagnosis. The uterus is replaced and involution stimulated by injection of 5–10 IU oxytocin. If the uterus is damaged or if there is necrosis or severe haemorrhage, ovariohysterectomy should be performed.

Acute metritis

Prolonged labour, uterine subinvolution, retained placental or fetal tissue, and unskilled obstetrical intervention can lead to acute metritis.[16,37]

The main symptom is a copious purulent-to-sanguineopurulent vaginal discharge with hyperthermia, or in advanced cases with hypothermia. The bitch is depressed, weak and may show dehydration, tenesmus and hyperpnoea. The uterus is palpably enlarged. By the use of contrast material radiography may possibly be used to differentiate this condition from retained macerated fetuses and placenta (see Figure 10.8). Sometimes laparotomy may be necessary for a definite diagnosis.

Intrauterine lavage and manual manipulation of the uterine horns by laparotomy and administration of antibiotics have been used with success.[23] Besides antibiotic treatment and institution of drainage, various hormones and other drugs are used.

Ovariohysterectomy

This is the most effective treatment, but if it is to be avoided, adequate antibiotic therapy must be instituted after sensitivity testing.

Diethylstilboestrol (DES)

DES 1–2 mg orally for 5–7 days[25] or a single intramuscular injection of 0.5 mg/kg has been used in an effort to dilate the cervix.[41]

Oxytocin

This can be used in an attempt to stimulate myometrial activity in cases where the cervix is open, 1.0–2 IU/kg up to 40 IU, possibly repeated 1–2 h later.[15,37,41]

Prostaglandins

The use of $PGF_{2\alpha}$ once or twice a day in doses of 25–100 μg/kg has been tried with encouraging results[16] and a dose of 250 μg/kg subcutaneously is reported to have been successful in spite of the earlier reported adverse side-effects of the drug.[58]

Non-hormonal compounds

A non-hormonal compound, 1,4-bis-(methan-sulphonyl-oxy)-butan, in a dose of 3 mg/kg orally, given once or followed 4–6 days later by half the initial dose, combined with antibiotic treatment, has resulted in the uterus returning to normal size within 4–6 weeks, discharge being arrested within a few days, and a normal heat period 6–10 months later without deleterious side-effects in most of the bitches treated.[7]

To prevent acute metritis, cleanliness before, during and after whelping together with routine postpartum administration of oxytocin are to be recommended.

Puerperal tetany

This condition[9,16,30,35,40,43] may occur late in the pregnancy or, more often, during early lactation, with panting, whining, stiffness of gait and hypersalivation, symptoms of nervousness, followed later by tonic or clonic muscular spasms which may become more frequent and prolonged if no treatment is given.

Intravenous administration of 5–20 ml of a 10–20% solution of calcium gluconate followed by subcutaneous or oral administration of calcium salts will arrest the condition. It is essential to feed the puppies artificially for at least 24 h, and in cases of relapse longer or even permanent artificial feeding may be necessary.

The condition is more frequent in small and medium-sized bitches, and prophylactic oral administration of calcium to bitches weighing less than 5 kg with more than three puppies, and to those weighing 5–15 kg with more than five puppies, is recommended.

REFERENCES

1. Al-Bassam, M.A., Thomson, R.G. and O'Donnell, L. (1981) Involution abnormalities in the postpartum uterus of the bitch. *Vet. Pathol.*, **18**, 208–218.
2. Alexander, J.E. and Lennox, W.J. (1961) Vaginal prolapse in a bitch. *Canad. vet. J.*, **2**, 428–430.
3. Arbeiter, K. (1968) Hormone zur Behandlung der Hündin. *Wien. tierärztl. Monatsschr.*, **55**, 587–591.
4. Arbeiter, K. (1974) Personal communication (cited in Jöchle, W. (1975) Hormones in canine gynecology – a review. *Theriogenology*, **3**, 152–165.
5. Arbeiter, K. (1975) The use of progestins in the treatment of persistent uterine hemorrhage in the postpartum bitch and cow: a clinical report. *Theriogenology*, **4**, 11–13.
6. Arbeiter, K. (1976) Postpartale Metrorrhagien bei der Hündin. *Kleintier-Prax.*, **21**, 5–7.
7. Arbeiter, K. (1976) Über ein neues Prinzip zur Behandlung von Gynäkopathien der Hündin. *Prakt. Tierarzt*, **57**, 295–300.
8. Arthur, G.H. (1975) *Veterinary Reproduction and Obstetrics*, 4th ed. Baillière Tindall, London.
9. Austad, R. and Bjerkas, E. (1976) Eclampsia in the bitch. *J. small Anim. Pract.*, **17**, 793–798.
10. Banks, P.N. (1963) Uterine rupture in the bitch. *J. small Anim. Pract.*, **4**, 345–347.
11. Beck, A.M. and McEntee, K. (1966) Subinvolution of placental sites in a postpartum bitch. A case report. *Cornell Vet.*, **56**, 269–277.
12. Benesch, E. (1957) *Lehrbuch der tierärztlichen Geburtshilfe und Gynäkologie. 2. Auflage.* Urban and Schwarzenberg, München, Berlin, Wien.
13. Bennett, D. (1974) Canine dystocia – a review of the literature. *J. small Anim. Pract.*, **15**, 101–117.

14. Bölcsházy, K. (1955) Handling of dystocia in dogs by means of episiotomy. *Acta vet. Acad. Sci. Hung.*, **5**, 275–280.
15. Burke, Th.J. (1974) Acute metritis. In: *Current Veterinary Therapy. V. Small Anim. Pract.*, 923–924. W.B. Saunders, Philadelphia, London, Toronto.
16. Burke, Th. J. (1977) Post-parturient problems in the bitch. *Vet. Clin. N. Amer.* 7, 693–698.
17. Cobb, L.M. (1959) The radiographic outline of the genital system of the bitch. *Vet. Rec.*, **71**, 66–68.
18. Collins, D.R. (1966) A simple obstetrical technic for assisting with fetal delivery. *Vet. Med./small Anim. Clin.*, **61**, 455–458.
19. Curwen, P. (1949) Anasarca as a cause of dystocia in the bitch. *Vet. Rec.*, **61**, 572.
20. Dall, J.A. (1962) General discussion, pp. 1337–1338. In: Freak, M.J.[27]
21. Dodman, N.H. (1979) Anaesthesia for caesarean section in the dog and cat: a review. *J. small Anim. Pract.*, **20**, 449–460.
22. Dover, P.M. (1965) An unusual case of dystocia in a bitch. *N.Z. vet. J.*, **13**, 201.
23. Durfee, Ph.T. (1968) Surgical treatment of postparturient metritis in the bitch. *J. Amer. vet. med. Assoc.*, **153**, 40–42.
24. Eriksen, E.. and Hansen, F. (1978) Personal communication.
25. Ewing, G. (1973) Treatment of selected reproductive problems in the dog. *Proc. Ann. AVSSBS Conf.*, 90–94 (cited in Jöchle, W. (1975) Hormones in canine gynecology – a review. *Theriogenology*, 3, 152–165).
26. Ficus, H.J. and Hollenberg, U. (1971) Bildbericht über eine Geburtsstörung bei einem deutschen Schäferhund. *Kleintier-Prax.*, **16**, 28–29.
27. Freak, M.J. (1962) Abnormal conditions associated with pregnancy and parturition in the bitch. *Vet. Rec.*, **74**, 1323–1339.
28. Gehring, W. (1964) Prolapsus vaginae bei der Hündin. *Kleintier-Prax.*, **9**, 93–97.
29. Greiner, T.P. (1974) Genital emergencies. In: *Current Veterinary Therapy. V. Small Anim. Pract.*, 909–915. W.B. Saunders Co., Philadelphia, London, Toronto.
30. Hall, M.A. and Swenberg, L.N. (1977) Genital emergencies. In: *Current Veterinary Therapy. VI. Small Anim. Pract.*, 1216–1227. W.B. Saunders, Philadelphia, London, Toronto.
31. Hartsfield, S.M. (1979) Obstetrical anesthesia in small animals. *Calif. Vet.*, **33**, 18–24.
32. Heath, J.S. (1963) Indications and complications in caesarean section in the bitch. *J. small Anim. Pract.*, **4**, 289–292.
33. Herr, S. (1978) Persistent postcervical band as a cause of dystocia in a bitch (a case report). *Vet. Med./small Anim. Clin.*, **73**, 1533.
34. Hosek, J.J. (1972) Syntocinon – a treatment for agalactia in the dog. *Vet. Med./small Anim. Clin.*, **67**, 405.
35. Jiřina, K. (1974) Beitrag zur Therapie der Eklampsie des Hundes. *Berl. Münch. tierärztl. Wochenschr.*, **87**, 15–16.
36. Larsen, E. (1946) Obstetrics in the bitch. *Medlemsbl. danske Dyrlægeforen.*
37. Larsen, R.E. and Wilson, J.M. (1977) Acute metritis. In: *Current Veterinary Therapy. VI. Small Anim. Pract.*, 1227–1228. W.B. Saunders, Philadelphia, London, Toronto.
38. Leach, J.L.R. (1963) The technique of caesarean section in a bitch. *J. small Anim. Pract.*, **4**, 285–288.
39. Lederer, H.A. and Fisher, L.E. (1960) Ectopic pregnancy in the dog. *J. Amer. vet. med. Assoc.*, **137**, 61–62.
40. Lorin, D. (1975) Peripartale und postpartale Probleme bei der Hündin und den Welpen. *Wien. tierärztl. Monatsschr.*, **62**, 345–347.
41. Mather, G.W. (1971) Acute metritis. In: *Current Veterinary Therapy. IV. Small Anim. Pract.*, 760–762. W.B. Saunders, Philadelphia, London, Toronto.
42. Mitchell, B. (1966) Anaesthesia for caesarean section and factors influencing mortality rates of bitches and puppies. *Vet. Rec.*, **79**, 252–257.

43. Mudaliar, A.S.R. and Hussain, M.M. (1967) Puerperal tetany in a bitch. *Ind. vet. J.*, **44**, 804–805.
44. Naaktgeboren, C. (1967) Über Domestikationseinflüsse auf den Verlauf des Geburtsvorganges, mit besonderer Berücksichtigung der Canidae. *Arch. néerl. Zool.*, **17**, 278–280.
45. Nicholl, T.K. (1979) Extrauterine fetuses in a bitch. *Canine Pract.*, **6**(4), 16–22.
46. Niggli, H.B. (1963) Die Geburtshilfe beim Fleischfresser. *Schweiz. Arch. Tierheilk.*, **105**, 47–52.
47. North, A.F. and Somerville, N.J. (1940) Hernia, dystocia and complications. *Cornell Vet.*, **30**, 245–247.
48. Peck, G.K. and Badame, F.G. (1967) Extra-uterine pregnancy with fetal mummification and pyometra in a Pomeranian. *Canad. vet. J.*, **8**, 136–137.
49. Rasbech, N.O. (1973) *Husdyrenes Reproduktion I*. Copenhagen.
50. Reid, J.S. and Frank, R.J. (1973) Double contrast hysterogram in the diagnosis of retained placentae in the bitch. A case report. *J. Amer. Anim. Hosp. Assoc.*, **9**, 367–368.
51. Schlotthauer, C.F. and Wakim, K.G. (1955) Ectopic pregnancy in a dog. *J. Amer. vet. med. Assoc.*, **127**, 213.
52. Schulze, W. (1950) Geburtshilflich-gynäkologische Beobachtungen bei Hunden. *Mh. Vet.-Med.*, **5**, 239–247.
53. Schutte, A.P. (1967) Vaginal prolapse in the bitch. *J. S. Afr. vet. med. Assoc.*, **38**, 197–203.
54. Short, C.E. (1963) Abnormal pregnancy in the dog. *Small Anim. Clin.*, **3**, 390.
55. Siegel, E.T. (1977) *Endocrine Diseases of the Dog*. Lea and Febiger, Philadelphia.
56. Skydsgaard, J. (1960) Personal communication.
57. Smith, K.W. (1974) Female genital system. In: *Canine Surgery*, 2nd Archibald ed. 751–782.
58. Sokolowski, J.H. (1980) Prostaglandin F$_2$ alpha-THAM for medical treatment of endometritis, metritis, and pyometritis in the bitch. *J. Amer. Anim. Hosp. Assoc.*, **16**, 119–122.
59. Spira, H.R. (1960) Thiambutene for caesarean section in bitches and its antagonism by nalorphine. *Aust. vet. J.*, **36**, 232–234.
60. Stein, E. (1954) Beitrag zur Embryotomie bei Katzen und Zwerghunden. *Dtsch. tierärztl. Wochenschr.*, **61**, 75–79.
61. Stoye, M. (1973) Untersuchungen über die Möglichkeit pränataler und galaktogener Infektionen mit *Ancylostoma caninum* Ercolani 1859 (Anchylostomidae) beim Hund. *Zbl. Vet.-Med. B*, **20**, 1–39.
62. Tröger, C.P. (1968) Zum Problem des Prolapsus vaginae bei der Hündin. *Die Blauen Hefte*, **37**, 35–40.
63. Ullner, W. (1957) Dystokie bei den Carnivoren. *Dtsch. tierärztl. Wochenschr.*, **64**, 417–420.
64. Wright, J.G. (1934) Some aspects of canine obstetrics. 1. The diagnosis of pregnancy. *Vet. Rec.*, **14**, 563–585.
65. Wright, J.G. (1939) Caesarean hysterectomy – hysterectomy. *Vet. Rec.*, **51**, 1331–1346.

ADDITIONAL READING

Al-Bassam, M.A., Thomson, R.G. and O'Donnell, L. (1981) Normal postpartum involution of the uterus in the dog. *Canad. J. comp. Med.*, **45**, 217–232.
Baines, F.M. (1981) Milk substitutes and the hand rearing of orphan puppies and kittens. *J. small Anim. Pract.*, **22**, 555–578.
Barrett, E.P. (1949) A preliminary note on the treatment of uterine inertia in the bitch. *Vet. Rec.*, **61**, 783.

Dreier, H.-K. (1980) Physiologisches und pathologisches Puerperium bei der Hündin. *Tierärztl. Prax.*, **8**, 367–374.

Drew, R.A. (1974) Possible association between abnormal vertebral development and neonatal mortality in bulldogs. *Vet. Rec.*, **94**, 480–481.

Freak, M.J. (1975) Practitioners'–breeders' approach to canine parturition. *Vet. Rec.*, **96**, 303–308.

Günther, S. (1954) Ansichten, Beobachtungen und bisherige experimentelle Untersuchungsergebnisse über die Geburtsfunktion des Uterus bicornis unter den Bedingungen der Multiparität. *Arch. exp. Vet.-Med.*, **8**, 643–665.

Herzog, A. and Höhn, H. (1972) Chromosomenanomalien mit letaler Wirkung bei Welpen. *Kleintier-Prax.*, **17**, 176–179.

Irvine, C.H.G. (1964) Hypoglycaemia in the bitch. *N.Z. vet. J.*, **12**, 140–144.

Joshua, J.O. (1963) Absence of foetal fluids and dystocia in the bitch. *Vet. Rec.*, **75**, 956.

Lloyd, S., Amerasinghe, P.H. and Soulsby, E.J.L. (1983) Periparturient immunosuppression in the bitch and its influence on infection with *Toxocara canis*. *J. small Anim. Pract.*, **24**, 237–247.

Thiel, U. (1980) Myocarditis in puppies. *Vet. Rec.*, **106**, 43.

Thomée, A. (1978) *Zusammensetzung, Verdaulich- und Verträglichkeit von Hundemilch und Mischfutter bei Welpen unter besonderer Berücksichtigung der Fettkomponente*. Thesis, Hanover.

Waterman, A. (1975) Accidental hypothermia during anaesthesia in dogs and cats. *Vet. Rec.*, **96**, 308–313.

Wenger, J.B. (1967) How I perform cesarean section in the dog. *Vet. Med./small Anim. Clin.*, **62**, 1153–1154.

Zinnbauer, H. (1963) Ein radförmig eingerollter Fetus als Geburtshindernis bei einer Hündin. *Wien. tierärztl. Monatsschr.*, **50**, 601–604.

Part 2
Reproduction in the Cat

Chapter 11
Gynaecology of the Normal Female

ANATOMY AND PHYSIOLOGY

The anatomy of the genitalia is similar to that of the bitch apart from the dimensions and the fact that, in the queen, there are tubular cervical glands, major vestibular (Bartholin) glands and m. retractor clitoridis, which the bitch does not possess.

The ovaries

The ovaries are 8–9 mm long and situated beneath the 3rd–4th lumbar vertebra, partly covered by the bursa ovarica.

The oviducts or Fallopian tubes are 4–5 cm long, the uterine horns are 9–10 cm long with a diameter of 3–4 mm, and the body of the uterus (corpus uteri) is 2 cm long.

PUBERTY

The age of puberty, when the queen has her first oestrus and is said to be 'in season' or 'calling', is very variable, depending both upon the breed and the time of year. The majority of queen kittens have their first oestrus when they reach a bodyweight of 2.3–2.5 kg, that is at about 7 months, but in some cases maturity may be reached as early as 3 months,[20] and some purebred longhairs, such as persians, may not call until they are 12–18 months old.[38]

The onset of puberty is influenced more by the time of birth in relation to the breeding season than by age. Kittens born between October and December may not become sexually mature when the breeding season starts a few months later, but in the following season, thus being 12–16 months old at their first oestrus.[22,37]

Reproductive activity normally continues until the age of 14 years, and there have been observations of pregnancy in 20-year-old cats,[23] but with increasing age the litter size and the number of litters per year decrease.

BREEDING SEASON

The time of the year for, and the duration of, the breeding season depend on the geographical location (Table 11.1). In the northern

Table 11.1 Time and duration of the breeding season in female cats on various geographical locations. (After Hurni[15])

			Breeding season — Time of the year											
Country	Northern latitude	Duration in months	January	February	March	April	May	June	July	August	September	October	November	December
Algeria	37	10	▓	▓	▓	▓	▓	▓	▓	▓	▓	▓		▓
United States	40	7	▓	▓	▓	▓	▓	▓	▓					
United States	42	5		▓	▓	▓	▓	▓						
United States	43	7		▓	▓	▓	▓	▓	▓	▓				
England (Porton)	51	8		▓	▓	▓	▓	▓	▓	▓	▓			
England	51.5	6			▓	▓	▓	▓	▓	▓				
England (London)	52	5			▓	▓	▓	▓	▓					
Poland	52	6		▓	▓	▓	▓	▓	▓					

hemisphere the breeding season starts late in January and lasts till August–September, and is even reported to last until November–December, when the daylight is at its minimum; this is followed by the anoestrous period which lasts until the onset of the next breeding season.[1,32,37] Again, this only goes for the majority of cats, for the breeding season can still be affected by the breed of cat, or by environmental or even psychological factors. In some cases polyoestrous activity may continue all the year round especially in queens kept indoors and exposed to artificial light. This is more common in short-haired cats, particularly siamese, than in the long-haired breeds.[17] Psychological stress, such as that caused by moving house or entering a cattery, may also interrupt the anoestrous period.[34]

The length of daylight is an important factor in breeding control. Reports of the number of parturitions during the year in cats kept under natural daylight[17,30] differed from those of cats under laboratory conditions,[28,31,33] but it has been shown that daily exposure to 14 h of artificial light produces more constant results.[15] It is now recognized that a laboratory breeding programme can be controlled to some extent by changing the length of exposure to light.[16]

The first oestrus after a pregnancy usually occurs 8 days after the kittens are weaned, that is on average 8 weeks after parturition, with a variation from 1 week in non-lactating queens to 21 weeks in those fully lactating.[30,34] Further reports indicate that oestrus may occur as early as 7–10 days after parturition even in lactating queens[20] and certainly some lactating queens do mate.[33] Indeed fertile oestrus has been observed as early as 18–26 h after parturition. All this implies that a queen can be pregnant most of the year.[10,18]

OESTROUS CYCLE

The queen is polyoestrous and an induced ovulator – in other words, neither ovulation nor the formation of corpora lutea occur unless mating has taken place. The duration and the course of the cycle depend on whether mating occurs, whether ovulation follows, whether conception results, and whether pregnancy and delivery are followed by lactation (Figure 11.1).

An anovulatory cycle has a length of 2–3 weeks, but cycles of different lengths may occur, as may continuous oestrus. Joshua[20] found an oestrous cycle duration of 21 days in 74% of investigated cats, whereas 14% had cycles of different lengths, and continuous oestrus occurred in 12%. The oestrous cycle intervals are prolonged, averaging about 6 weeks (range 30–75 days), in cats who ovulate in connection with mating but fail to conceive.[11,29,39]

The total number of oestrous periods per year per queen in an indoor colony averages 13.00 ± 4.85 (SD) (range 4–25) cycles per year.[6]

Pro-oestrum

This period lasts 1–3 days, usually characterized by an increased desire to be petted and an increased frequency of urination and even urine spraying as in the tomcat.[38] Vulval oedema is either absent or not very pronounced, and there is no bloodstained vaginal discharge as in the bitch. During this period there is a selective enlargement of 3–7 ovarian follicles (2 mm in diameter), and atresia of others.[11]

Oestrus

The duration of oestrus is affected by the season and by the occurrence or non-occurrence of ovulation. In the spring the number of days in oestrus is increased (5–14 days/cycle), while in other seasons it is decreased (1–6 days/cycle).[6] In laboratory-maintained cats significantly longer periods of oestrus occur in March, April, August and December compared to the periods in June, September, October and November.[40] In the ovulating cat the oestrous period averages 5.7 days, with oestrous signs waning 24–48 h following mating,[5] whereas a duration of 8 days is normal in cats if ovulation does not occur.[29] Nevertheless, variations occur and there have been observations of

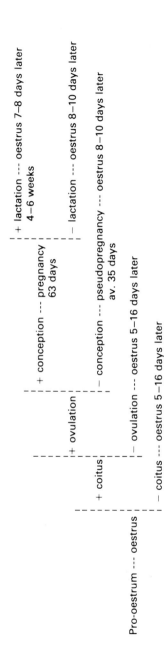

Figure 11.1 Potential courses of the reproductive cycle of the queen

the same duration of oestrus in both ovulating and non-ovulating queens,[41] as well as of a shorter duration in non-mated queens.[35]

During oestrus the queen alters her behaviour; she 'calls', displays lordosis, deviates the tail to one side and is willing to copulate. Spasmodic contractions of the perineal region may occur especially if the dorsal pelvic region is stroked gently. Anorexia and urine spraying commonly occur during oestrus. The follicles grow to a diameter of 2–3 mm.[11]

Ovulation

Copulation produces impulses in the vulval–vaginal region, causing release of releasing hormones from the hypothalamus. These lead to secretion of luteinizing hormone (LH) resulting in ovulation within 24–50 h after coitus.[12,24]

In some queens a single coital stimulus results in ovulation,[29] whereas in others several repeated coital stimuli are necessary.[12,24,33] In some cats even three copulations within 30 min are unsuccessful in causing ovulation,[37] presumably due to the fact that some queens allow copulation before maturation of the follicles.

Therefore, one single mating may induce ovulation, and repetition of matings increases the number of ovulations in some investigations,[8,33,42] though this has not been found to be the case in other instances.[21,30]

Metoestrum

The duration of metoestrum averages 21 days (range 14–28) in the non-mated female,[4,5,7] whereas a sterile mating or sham copulation is followed by a period of pseudopregnancy lasting on average 35 days (range 30–73).[11,29,39]

Anoestrum

In this period, which is characterized by sexual quiescence, the ovaries are small, and the size of the follicles averages 0.5 mm.[11]

VAGINAL CYTOLOGY

Examination of vaginal smears cannot, as in the bitch, be used for detection of the optimal time for breeding, since the queen is an induced ovulator, but it may be useful for verifying oestrus in combination with other methods of oestrous detection.[9,11,14,25–27,32,33]

Specimen collection should be performed very carefully, since deep introduction of any equipment for sample collection may elicit ovulation. A sterile cotton swab moistened with isotonic saline solution, sterile pipette or sterile short-tipped medicine dropper containing an isotonic saline solution may be used. The queen is restrained by a firm grasp of the scruff of the neck and the tail. After careful cleaning of the vulva the swab is inserted and a specimen withdrawn, or the tip of the pipette or medicine dropper is inserted about 1.5 cm into the vagina, the bulb squeezed several times and the recovered material placed upon a slide, stained and examined. To avoid contamination from the vestibule, the equipment should be inserted first dorsally and then cranially.

Fixation and staining can be performed as mentioned in Chapter 1 (page 21). If examination can be performed within 20–30 min, staining with methylene blue 1% solution can be used, but this preparation is of no use in a later examination.[27]

As in the bitch, vaginal smears reveal nucleated and cornified cells as well as leukocytes during the cycle, but erythrocytes do not appear in pro-oestrum. Progressive cornification, a criterion used in the bitch, is not quite so reliable in the female cat for indicating the onset of oestrus. Thus it has been found that queens in 32% of 168 cycles investigated showed oestrus before obvious cornification in the vaginal smear.[35] The characteristics of the epithelial cells in the vaginal smear and their normal frequency during the oestrous cycle are shown in Table 11.2. Basal cells are not present in the vaginal smear during oestrus and early metoestrum, though they may be present in limited number during other stages of the cycle. Parabasal cells have nearly completely disappeared in oestrus, and they reach a maximum during late metoestrum. Intermediate cells diminish in number during oestrus, increasing again rapidly during early metoestrum. Superficial cells without nuclei normally occur during the whole cycle, reaching a maximum during oestrus, whereas superficial cells with pyknotic nuclei decrease in number from pro-oestrum to metoestrum. By examining the vaginal smear it is possible to differentiate between ovulating and non-ovulating queens from 3 days after ovulation, but it has not yet been investigated whether it might be possible to differentiate between pregnant and pseudopregnant queens.[13]

The same indices as mentioned in Chapter 2 may be used in the queen. The superficial cell index (SCI) is high in pro-oestrum and oestrus, thus differentiating these from other stages in the cycle. For the queen a new index, the nucleus degeneration index (NDI), fluctuating similarly to SCI, has been introduced:[13]

$$NDI = \frac{\text{number of superficial cells} \times 100}{\text{number of intermediate cells}},$$

Table 11.2 Characteristics of epithelial cells in vaginal smears, the average percentage of cells and the superficial cell index (SCI) during the oestrous cycle of the queen. (After Haenisch[13])

	Basal cells	Parabasal cells	Intermediate cells	Superficial cells		
				With pyknotic nucleus	Without nucleus	SCI:
Cell characteristics						
size, average	12 μm	19 μm	<41 μm	55 μm	55 μm	
shape	round	round–oval	polygonal	polygonal	polygonal	
nucleus, average diameter	8–10 μm	8–10 μm	10 μm	6 μm		
Stage of the oestrous cycle			*Average percentage of cells:*			
pro-oestrum	few	7.9	35.6	36.3	20.3	127
oestrus		0.5	20.4	33.6	45.8	400
metoestrum early		7.8	54.8	28.4	9.0	59
metoestrum, late	few	23.0	56.7	15.9	4.9	25
anoestrum	few	20.5	39.3	21.4	18.6	67

A characteristic feature during oestrus is the clearing of the vaginal smear,[35] also described as the absence of cellular debris.[27] The clearing may begin before the onset of the follicular phase, and is evident in the majority of cycles during the follicular phase and in some cycles after the end of this phase.[35]

HORMONES

Only a limited number of publications are available concerning hormone concentrations in the cat's plasma during the oestrous cycle, pregnancy and pseudopregnancy.

Luteinizing hormone (LH)

The mating stimulus is necessary for LH release from the pituitary gland,[38] and multiple copulations produce elevated and prolonged serum LH levels.[40] This is evident during the first 90 min of stimulation, but if mating continues beyond this period, the stimulus becomes virtually ineffective at inducing further LH release.[19] LH release is induced within minutes after coitus, and one coital input only can cause LH release for as long as 16–20 h.[36]

Luteinizing hormone levels of 17 ± 2 ng/ml were found in ovulating cats and 8 ± 2 ng/ml in non-ovulating cats 10 min after mating.[8] After 1 h the concentration was 34 ± 9 ng/ml in ovulating cats once-mated versus 73 ± 11 ng/ml in multiple-mated cats. In ovulating cats maximum levels were reached at 1–4 h with a return to normal after 8 h, whereas in non-ovulating cats normal values were reached after 4 h.

Progesterone (P)

In the pregnant cat P is not detectable during the first 2–3 days after mating, but the concentration then rapidly increases to 22.9 ± 4.1 ng/ml on day 11.[39] After a slight decrease there is a further increase in concentration to 34.9 ± 6.2 ng/ml on day 21, and thereafter the concentration decreases to 4–5 ng/ml prior to parturition, and to below 1 ng/ml immediately post partum (Figure 11.2).

In the pseudopregnant cat the level of P has been found to be almost parallel to that of the pregnant cat with a peak of 24.6 ± 6 ng/ml on day 21,[39] or a few days earlier.[29] Thereafter the concentration decreases more rapidly than during pregnancy to 2 ng/ml on day 50, and it is below 2 ng/ml on the 62nd day after oestrus. In cats undergoing normal polyoestrous cycles, being neither pregnant nor pseudopregnant, the concentration of P varies very little (from 0.1 to 0.6 ng/ml).

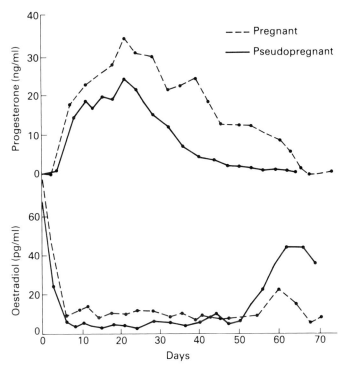

Figure 11.2   Average concentration of progesterone and oestradiol in pregnant and pseudopregnant queens. (After Verhage *et al.*[39])

Oestradiol

On the day of mating the concentration of oestradiol is approximately 60 pg/ml, and during the following 5 days the concentration decreases rapidly to 8–12 pg/ml. This level is maintained until days 58–62, when a slight increase occurs prior to parturition, after which the concentration decreases again (see Figure 11.2).[39]

In cases of pseudopregnancy the concentration, although lower, has a similar course to that during pregnancy for the first 40 days, but thereafter considerable individual variations occur. In all pseudopregnant cats an increase of oestradiol to 40 pg/ml after days 62–69 corresponds with an immediate decline of P to 1 ng/ml. The concentration varies during the cycle in polyoestrous queens. During cycles of 15.8 ± 3.8 days, peak concentrations of 59.5 ± 13.4 pg/ml occur between intervals of concentrations of 8.1 ± 3.8 pg/ml.

Prolactin (PRL)

The levels of PRL are elevated in the last trimester of pregnancy reaching a concentration of 31.2 ± 5.1 ng/ml throughout the last week, and 43.5 ± 5.4 ng/ml on the last 3 days of gestation.[3] The concentration is similarly elevated (40.6 ± 7.2 ng/ml) during the lactation period except for the last 2 weeks, when PRL levels decline significantly to 27.8 ± 3.1 ng/ml, and the PRL concentration reaches basal levels within 1–2 weeks following weaning. In non-lactating queens basal levels of PRL are achieved within 1 week postpartum. During pseudopregnancy there are no significant changes in PRL concentrations.

Major shifts in PRL secretion do not accompany the seasonal anoestrous period in the cat, but PRL secretion may be modified daily in response to photoperiod.[2]

REFERENCES

1. Asdell, S.A. (1964) *Patterns of Mammalian Reproduction.* 2nd ed. Comstock Publishing Associated, Cornell University Press, Ithaca, New York.
2. Banks, D.R. and Stabenfeldt, G.H. (1983) Prolactin in the cat: II. Diurnal patterns and photoperiod effects. *Biol. Reprod.,* **28**, 933–939.
3. Banks, D.R., Paape, S.R. and Stabenfeldt, G.H. (1983) Prolactin in the cat: II. Pseudopregnancy, pregnancy and lactation. *Biol. Reprod.,* **28**, 923–932.
4. Burke, T.J. (1975) Feline reproduction. *Feline Pract.,* **5**(6), 16–19.
5. Burke, T.J. (1976) Feline reproduction. *Vet. Clin. N. Amer.,* **6**, 317–331.
6. Cline, E.M., Jennings, L.L. and Sojka, N.J. (1980) Analysis of the feline vaginal epithelial cycle. *Feline Pract.,* **10**(2), 47–49.
7. Colby, E. (1975) Feline reproductive physiology. *Amer. Anim. Hosp. Assoc., 42nd Ann. Meet.,* **2**, 329–333.
8. Concannon, P., Hodgson, B. and Lein, D. (1980) Reflex LH release in estrous cats following single and multiple copulations. *Biol. Reprod.,* **23**, 111–117.
9. Dawson, A.B. (1950) The domestic cat. In: *Care and Breeding of Laboratory Animals,* ed. Farris, E.J., Chapter 8. John Wiley, New York.
10. Djerassi, C. (1973) Reproduction control in dogs and cats. *Svensk Vet.-T.,* **25**, 641–647.
11. Foster, M.A. and Hisaw, F.L. (1935) Experimental ovulation and the resulting pseudopregnancy in anoestrous cats. *Anat. Rec.,* **62**, 75–90.
12. Greulich, W.W. (1934) Artificially induced ovulation in the cat (*Felis domestica*). *Anat. Rec.,* **58**, 217–223.
13. Haenisch, V. (1980) *Vaginalzytologische Untersuchungen an der Hauskatze (Felis domestica) unter besonderer Berücksichtigung der Abgrenzung von anovulatorischem und gravidem Zyklus.* Thesis, Hanover.
14. Herron, M.A. (1977) Feline reproduction. *Vet. Clin. N. Amer.,* **7**, 715–722.
15. Hurni, H. (1975) Einfluss der Tageslänge auf die jahreszeitliche Verteilung der Würfe in einer Katzenzucht. *Z. Versuchstierk.,* **17**, 121–128.
16. Hurni, H. (1981) Daylength and breeding in the domestic cat. *Lab. Anim.,* **15**, 229–233.
17. Jemmet, J.E. and Evans, J.M. (1977) A survey of sexual behaviour and reproduction of female cats. *J. small Anim. Pract.,* **18**, 31–37.
18. Jöchle, W. (1974) Progress in small animal reproductive physiology, therapy of reproductive disorders, and pet population control. *Folia vet. lat.,* **4**, 706–731.
19. Johnson, L.M. and Gay, V.L. (1981) Luteinizing hormone in the cat. II. Mating-induced secretion. *Endocrinology,* **109**, 247–251.

20. Joshua, J.O. (1968) Abnormal behavior in cats. In: *Abnormal Behavior in Animals*, ed. Fox. M.W., Chapter 23. W.B. Saunders, Philadelphia, London, Toronto.
21. Kehrer, A. and Starke, P. (1975) Erfahrungen über die Zucht, Aufzucht und Haltung von Katzen für Versuchszwecke unter konventionellen Bedingungen. *Berl. Münch. tierärztl. Wochenschr.*, **88**, 101–107.
22. Klug, E. (1969) *Die Fortpflanzung der Hauskatzer (Felis domestica) unter besonderer Berücksichtigung der instrumentellen Samenübertragung*. Thesis, Hanover.
23. Lamm, A.M. & Settergren, I. (1971) Reproduction in the cat. *Almänt vet.-möte, Stockholm*, **II**, 11–20.
24. Longley, W.H. (1911) The maturation of the egg and ovulation in the domestic cat. *Amer. J. Anat.*, **12**, 139–172.
25. Michael, R.P. (1958) Sexual behaviour and the vaginal cycle in the cat. *Nature (Lond.)*, **181**, 567–568.
26. Mills, J.N., Valli, V.E. and Lumsden, J.H. (1979) Cyclical changes of vaginal cytology in the cat. *Canad. vet. J.*, **20**, 95–101.
27. Mowrer, R.T., Conti, P.A. and Rossow, C.F. (1975) Vaginal cytology. An approach to improvement of cat breeding. *Vet. Med./small Anim. Clin.* **70**, 691–696.
28. Nobunaga, T., Okamoto, M.T. and Takahaski, K.W. (1976) Establishment of a breeding colony of cats in Japan using small cages, and some breeding data. *Jap. J. Anim. Reprod.*, **22**, 82–88. (*Anim. Breed. Abstr.* (1979) **47**, p. 43, no. 429).
29. Paape, S.R., Shille, V.M., Seto, H. and Stabenfeldt, G.H. (1975) Luteal activity in the pseudopregnant cat. *Biol. Reprod.*, **13**, 470–474.
30. Prescott, C.W. (1973) Reproduction patterns in the domestic cat. *Aust. vet. J.*, **49**, 126–129.
31. Robinson, R. and Cox, H.W. (1970) Reproductive performance in a cat colony over a 10-years period. *Lab. Anim.*, **4**, 99–112.
32. Scott, P.P. (1970) Cats. In: *Reproduction and Breeding Techniques for Laboratory Animals*, ed. Hafez. E.S.E., Chapter 10. Lea and Febiger, Philadelphia.
33. Scott, P.P. and Lloyd-Jacob, M.A. (1955) Some interesting features in the reproductive cycle of the cat. In: *Studies on Fertility*, ed. Harrison, R.G., Chapter XII. Blackwell Scientific Publishers, Oxford.
34. Shille, V. and Edqvist, L.-E. (1978) Normal reproduction physiology in the cat. *XIII Nord. vet.-Congr., Åbo*, 111–114.
35. Shille, V.M., Lundström, K.E. and Stabenfeldt, G.H. (1979) Follicular function in the domestic cat as determined by estradiol-17β concentrations in plasma: Relation to estrous behaviour and cornification of exfoliated vaginal epithelium. *Biol. Reprod.*, **21**, 953–963.
36. Shille, V.M., Munro, C., Farmer, S.W., Papkoff, H. and Stabenfeldt, G.H. (1983) Ovarian and endocrine responses in the cat after coitus. *J. Reprod. Fertil.*, **68**, 29–39.
37. Stabenfeldt, G.H. and Shille, V.M. (1977) Reproduction in the dog and cat. In: *Reproduction in Domestic Animals*, eds. Cole, H.H. & Cupps, P.T., 3rd. ed., Chapter 19. Academic Press, New York, San Francisco, London.
38. Stein, B.S. (1975) The genital system. In: *Feline Medicine and Surgery*, ed. Catcott, E.J., 2nd ed., Chapter 13. Amer. Vet. Publ. Inc., California.
39. Verhage, H.G., Beamer, N.B. and Brenner, R.M. (1976) Plasma levels of estradiol and progesterone in the cat during polyestrus, pregnancy, and pseudopregnancy. *Biol. Reprod.*, **14**, 579–585.
40. Wildt, D.E., Guthrie, S.C. and Seager, S.W.J. (1978) Ovarian and behavioral cyclicity of the laboratory maintained cat. *Hormones Behav.*, **10**, 251–257.
41. Wildt, D.E., Seager, S.W.J., Dukelow, W.R. and Chakraborty, P.K. (1979) Ovulatory and LH response of the domestic cat. *Fed. Proc.*, **38**, p. 1031, Abstr. no. 4246.
42. Wildt, D.E., Seager, S.W.J. and Chakraborty, P.K. (1980) Effect of copulatory

stimuli on incidence of ovulation and on serum luteinizing hormone in the cat. *Endocrinology*, **107**, 1212–1217.

ADDITIONAL READING

van Aarde, R.J. (1978) Reproduction and population ecology in the feral house cat, *Felis catus*, on Marion Island. *Carnivore Genet. Newsletter*, **3**, 288–316.

Banks, D.H. and Stabenfeldt, G. (1982) Luteinizing hormone release in the cat in response to coitus on consecutive days of estrus. *Biol. Reprod.*, **26**, 603–611.

Burke, T.J. (1975) Feline reproduction. *Feline Pract.*, **5**(6), 16–19.

Cairoli, F., Colombo, G. and Zambetti, G. (1979) Vaginal cytology during oestrous cycle of the cat. *Clin. Vet.*, **102**, 661–668.

Fabian, G. and Preuss, F. (1966) Messungen am Uterus der Katze in den Zyklusphasen. *Zbl. Vet.-Med. A.*, **13**, 337–351.

Richkind, M. (1978) The reproductive endocrinology of the domestic cat. *Feline Pract.*, **8**(5), 28–31.

Shille, V.M. and Stabenfeldt, G.H. (1980) Current concepts in reproduction of the dog and cat. In: *Advanc. vet. Sci. comp. Med.*, ed. Brandly, C.A. and Cornelius, C.E. Vol. 24, pp. 211–243. Academic Press, New York, London, Toronto, Sydney and San Francisco.

Stover, D.G. and Sokolowski, J.H. (1978) Estrous behavior of the domestic cat. *Feline Pract.*, **8**(4), 54–58.

Verhage, H.G. and Brenner, R.M. (1975) Estradiol-induced differentiation of the oviductal epithelium in ovariectomized cats. *Biol. Reprod.*, **13**, 104–111.

Verhage, H.G., Akbar, M. and Jaffe, R.C. (1979) Cytosol and nuclear progesterone receptor in cat uterus and oviduct. *J. Steroid Biochem.*, **11**, 1121–1128.

West, N.B., Verhage, H.G. and Brenner, R.M. (1976) Suppression of the estradiol receptor system by progesterone in the oviduct and uterus of the cat. *Endocrinology*, **99**, 1010–1016.

West, N.B., Verhage, H.G. and Brenner, R.M. (1977) Changes in nuclear estradiol receptor and cell structure during estrous cycles and pregnancy in the oviduct and uterus of cats. *Biol. Reprod.*, **17**, 138–143.

Wildt, D.E. (1980) Effect of transportation on sexual behavior of cats. *Lab. Anim. Sci.*, **30**, 910–912.

Chapter 12
Breeding and Mating

Cat-lovers maintain that man has never managed to fully domesticate the cat, and nowhere is this more evident than in his attempt to control its breeding habits. The 'normal' mating behaviour described below is 'observed' behaviour, often under laboratory conditions, and this may explain the many variations and anomalies reported. The 'sexual orgy' that takes place 'on the tiles' every night may well be different and is certainly more prolific.

NORMAL MATING AND MATING BEHAVIOUR

The mating pattern of cats depends upon both sexual experience and hormonal state.[12] Maiden queens resist the attention of the tomcat until the first copulation occurs. After this the queen becomes progressively more receptive, allowing intercourse from the male with greater frequency and with less restraint. It has been shown that after 6 mating attempts and approximately 20 intromissions the female readily and completely accepts strange males.[12]

A tomcat shows no sexual interest in a queen in the anoestrous or metoestrous state, but during pro-oestrum and oestrus he is attracted both by her calling and general behaviour, as well as by the smell of her urine. He will mark his territory by spraying urine and rubbing against areas not already marked with urine, and may lick his partly erect penis as a sign of readiness to mate.[6]

Sexual pheromones
Cats appear to have at least two sexual pheromones.[2] One is valeric acid, which is in the vaginal secretion of an oestrous female. Apart from attracting the tom this is also a female–female pheromone which induces, or helps to induce, oestrus in other females, thus leading to the phenomenon of several cats in a cattery coming into season at the same time.

The other is tomcat smell, which when used for territorial demarcation will also bring some queens into oestrus.

Behaviour just prior to mating
During pro-oestrum the female will reject the tomcat (Table 12.1). At the start of oestrus she becomes more active and nervous and she will 'call'; this heat cry may be mimicked by the male, and then be

Table 12.1   Behaviour of the queen in relation to the vaginal smear. (After Mowrer
et al.[8], by permission of Merck and Co., Inc. proprietor)

Days from mating	Stage of the cycle	Behaviour	Cytology
−12	Anoestrum	No overt signs, no stance	Epithelial cells in clumps, little cellular debris
−4	Early pro-oestrum		Cornified cells and increasing cellular debris
−2	Late pro-oestrum	No heat symptoms and no stance, but increasing affection towards handler	Increasing number of cornified cells
0	Oestrus	Stance fairly good, rubbing against perches, some rolling	Uniform cornified cells, markedly less debris – 'clearing'
+2	Mid to late oestrus	Stance strong, some vocalization, being ridden by other queens	Mature cornified cells predominate other cell types
+4	Metoestrum	Regression of stance, no longer receptive to tomcats	Cellular debris
+8	Anoestrum	No stance, no overt signs	

repeated to and fro. She rubs her head and neck on various objects as
well as on the floor, and often rolls on the floor. These symptoms are
intensified when a tomcat approaches. By the presence of, or even
the mere smell of, a tomcat, the queen in oestrus will lower her back,
raise the pelvic region and bend the tail to one side while making
treading movements with the hindlegs. Some queens do not walk at
this stage but move forward on bent elbows with the pelvis raised
(Figure 12.1). These reactions may be induced when the queen is
stroked gently over the back and the pelvic region is touched. The
symptoms may become more pronounced if the skin of the neck is
grasped and the perineal region stroked at the same time. If the
queen is not yet fully receptive when mounted, she will turn round on
the tomcat (or a human handler), usually with claws bared, but will
often resume the oestrous behaviour when the tom has retreated. Not
all the above-mentioned oestrous symptoms are always present at the
first oestrus. Often only screaming, rolling and treading are present,
frequently leading to a false emergency call to the veterinarian to
attend kittens 'rolling in agony'.[4] There is also a considerable breed
variation and many long-haired breeds do not vocalize at all.

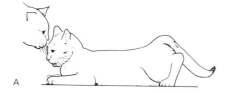

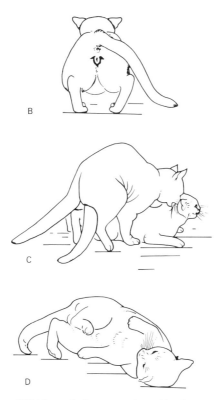

Figure 12.1 (A and B) The typical posture adopted by the queen in full oestrus. Note the flexion of the limbs, lordosis, and deflection of the tail. (C) Tomcat holding the queen during intromission. (D) Postcoital rolling by the queen. (After Scott[11])

Mating

During mating the tomcat bites firmly in the neck of the queen, grasping round her chest with the forelegs while treading with the hindlegs. An eager tomcat may start too far forward on the back of the queen, and will have to move further back while the queen will try to make the vulval region easier to reach, and during this manoeuvre both cats make treading movements with their hindlimbs. The penis, which normally points backwards, is at this stage partly erect, pointing forward and down at an angle of 20–30° with the horizontal plane.[9] The penis is introduced with a thrusting movement followed immediately by ejaculation. As the penis is withdrawn, the female moves forward while uttering the 'copulation cry' (thought to be due to the cornified papillae on the penis) and the male cat moves away to avoid her aggressiveness. The queen rolls aside and licks the vulval region, while the tomcat watches from a safe distance. After a pause lasting from a few minutes to an hour the mating is repeated. Mating continues until approximately 7 ejaculations have occurred. Although the frequency and intervals between matings depend on particular individuals, it is usually the female who eventually puts an end to mating.[13]

Table 12.2 shows the characteristic behaviour patterns of males

Table 12.2   Duration and characteristics of the premating, mating and post-mating periods. (Cited in Klug[6], after Rosenblatt and Aronson[10])

Duration	Tomcat	Queen
Premating		
10 s–5 min	Moves foward	Crouches
	Sniffs genitalia	Rubs its nose and mouth
	Circulates	Rolls
	Calls	Calls
	Moves forward	Stamps
Mating		
1–3 min	Neckbiting	Crouches
	Mounts with forelegs	
	Mounts with hindlegs	
	Rakes with forelegs	Stamps
	Stamps	
	Arches its back	
5–10 s	Coital movements	Bends the tail to one side
	Penis erection	
	Lowers pelvic region	Raises the pelvic region
	Intromission of penis	Dilation of vagina
	Ejaculation	
	Withdraws penis	Copulation cry
		Turns around towards tomcat
Postmating		
<1 min	Licks penis and forepaws	Licks genitalia
	Stays near the queen	Rolls
		Rubs nose and mouth
		Licks the paws
		Views the tomcat
		Re-encourages the tomcat

and females, as well as the duration of the premating, mating and post-mating periods. In an investigation on sexual exhaustion and recovery[13] it was found that the male normally grips the female neck after 16 s, mounts after 19 s, begins thrusting movements after 66 s and achieves the first mating after 107.5 s. Matings may be repeated for up to 4 days (average 2 days), and in this period the tomcat loses weight due to a limited food intake. The kidneys of the tomcat are large, and when a male with a progressive breeding season becomes thinner, the kidneys may be seen prominently beyond the ribs like tumour formations.[4]

Sometimes a tomcat shows no interest in a queen even though she is in obvious oestrus. This may be due to unfamiliar surroundings, and a period of 2 months may be necessary for acclimatization.[3] Some will not breed off their territory – this is frequently seen when the tomcat is taken to the queen's home rather than vice versa. The reverse is also seen; some queens, particularly those of exotic breed, refuse the attentions of stud tomcats but willingly mate with the local alley cat.

ABERRANT MATING AND MATING BEHAVIOUR

Sometimes a tomcat may try to mate with another tomcat. If a tomcat is mounted, he may show a crouched posture with a deviated tail. This behaviour frequently occurs in young cats approaching puberty, and normally stops without treatment when puberty is reached. The same behaviour may be seen under laboratory conditions, when a 'foreign' tomcat is brought to another male cat's accustomed surroundings, resulting in mounting on the 'foreign' cat, who passively permits the neck grip and the copulatory thrusts.[7] The same reaction may occur under natural conditions in a mature tomcat which has been deprived of mating and has thus built up a sexual tension.[1]

A male cat not socialized to other cats may try to mate with other species of animals. This problem does occur in cats normally weaned, but arises more often in orphan kittens.

Masturbation may occur in cats living in extremely restricted surroundings, such as laboratory conditions. Tomcats may use their forepaws, and females lick or rub the genital region against the floor.[5]

REFERENCES

1. Beaver, B.V. (1977) Mating behavior in the cat. *Vet. Clin. N. Amer.*, **7**, 729–733.
2. Bland, K.P. (1979) Tom-cat odour and other pheromones in feline reproduction. *Vet. Sci. Comm.*, **3**, 125–136.
3. Hart, B.L. and Voith, V.L. (1977) Feline behavior – sexual behavior and breeding problems in cats. *Feline Pract.*, **7**(1), 9–12.
4. Joshua, J.O. (1968) Abnormal behavior in cats. In: *Abnormal Behavior in Ani-*

mals, ed. Fox, M.W., Chapter 23. W.B. Saunders Co., Philadelphia, London, Toronto.

5. Kling, A., Kovach, J.K. and Tucker, T.J. (1969) The behaviour of cats. In: *The Behaviour of Domestic Animals*, ed. Hafez, E.S.E., 2nd ed., Chapter 15. Baillière, Tindall and Cassell, London.

6. Klug, E. (1969) *Die Fortpflanzung der Hauskatze (Felis domestica) unter besonderer Berücksichtigung der instrumentellen Samenübertragung*. Thesis, Hanover.

7. Michael, R.P. (1961) Observations upon the sexual behaviour of the domestic cat (*Felis catus L.*) under laboratory conditions. *Behaviour*, **18**, 1–24.

8. Mowrer, R.T., Conti, P.A. and Rossow, C.F. (1975) Vaginal cytology. An approach to improvement of cat breeding. *Vet. Med./small Anim. Clin.*, **70**, 691–696.

9. Redlich, G. (1963) Das Corpus penis des Katers und seine Erektionsveränderung eine funktionell-anatomische Studie. *Gegenbaurs morph. Jb.*, **104**, 561–584.

10. Rosenblatt, J.S. and Aronson, L.R. (1958) The decline of sexual behavior in male cats after castration with special reference to the role of prior sexual experience. *Behavior*, **12**, 285–338.

11. Scott, P.P. (1970) Cats. In: *Reproduction and Breeding Techniques for Laboratory Animals*, ed. Hafez, E.S.E., Chapter 10. Lea and Febiger, Philadelphia.

12. Whalen, R.E. (1963) The initiation of mating in naive female cats. *Anim. Behav.*, **11**, 461–463.

13. Whalen, R.E. (1963) Sexual behavior of cats. *Behavior*, **20**, 321–342.

Chapter 13
Infertility and Hormone Treatment
in the Female

INFERTILITY

Infertility in the queen may be the result of anatomical defects, such as hermaphroditism,[12,15,22] clinical or subclinal infections of the genital system, or dysfunction of the system. In many colony cats housed under conditions permitting oestrous cycles all the year round, some queens did not conceive.[5] It was found that 13 (39%) of 33 infertile cats had obvious functional and/or structural abnormalities of ovaries, uterus, oviducts or the uterotubal junction. The most common disorder was uterine cystic hyperplasia (7 of 33) together with either cystic or functioning ovaries.

The diagnosis of infertility may result from the history and the clinical and gynaecological examinations, if necessary supplied with a radiographic or a laparoscopic examination (see Chapter 1), with an examination for parasites, haematological examinations, urine analyses or serological tests for viral infections and toxoplasmosis. In the case of hereditary origin, treatment should be avoided, and in the case of a chronic inflammation the prognosis for normal fertility is rather limited.

Prolonged anoestrum

This may be a sign of the conditions mentioned above, but it is more directly associated with a systemic disease, parasite invasion or environmental factors such as poor nutrition, excessive exposure to cold, lack of sufficient light or overcrowded surroundings.[24] It is also often seen in queens used for shows, which even at the age of 2 years may show abnormal cycles and refuse to mate. Prolonged anoestrum in the domestic cat should always be considered abnormal[19] and treatment of the condition is not always successful; for example, the reason why a $2\frac{1}{2}$-year old burmese cat with primary anoestrum did not respond to gonadotropin stimulation was inactive germinal epithelium in the ovaries (lacking follicles and primordial germ cells) associated with X-chromosome monosomy (37, XO).[17]

Therapy
Once a full clinical examination has been made and anatomical defects and infectious conditions eliminated as causal agents, treat-

ment is aimed at the induction of oestrus. This is indicated in the case of queens that have never shown signs of oestrus; queens with prolonged anoestrum; and queens with irregular cycles. The induction of oestrus may also be useful where programmed pregnancies and parturitions are wanted.

As the queen is an induced ovulator, induction of oestrus has to be followed by induction of ovulation when the queen is to be artificially inseminated.

As has been described earlier, changes in the length of daylight influence a queen's oestrous cycle, and exposure to 14 h of light per day may be sufficient to bring a queen into oestrus. Furthermore it has been shown that exposure to 12 h of daylight per day, after a period of days of shorter daylight, induces oestrus, and that this can be repeated at least six times a year.[16]

As has also been described, oestrus may be stimulated by the presence of another queen in oestrus (see Chapter 11), and it is worth trying this, particularly in the case of family pets normally kept in virtual isolation.

Oestrus may be induced by the administration of pregnant mare serum gonadotropin (PMSG)[8] or of genuine follicle-stimulating hormone (FSH),[32] both of which may be followed by the administration of luteinizing hormone (LH) or human chorionic gonadotropin (HCG).

Pregnant mare serum gonadotropin (PMSG)

The treatment regimen depends on the queen being in or out of breeding season; thus the total dosage of PMSG used should be increased outside the breeding season and so should the period of treatment. PMSG has been used successfully with an initial dose of 100 IU followed by a daily administration of 50 or 25 IU as shown in Table 13.1. Oestrus is normally induced on days 6 or 7 and mating is performed a day later. As natural mating is accompanied by ovulation, no treatment for obtaining ovulation is necessary, but an

Table 13.1 Tentative dose schedules to induce oestrus in the domestic cat based on calendar months and days of injections. (After Colby[8])

Calendar months	PMSG (IU)							
December–January	100	50	50	50	50	50	50	50
February–August	100	50	50	25	25	25	25	25
September–November	100	100	50	50	50	50	50	50
Day	1	2	3	4	5	6	7	8

intramuscular injection of 250 IU HCG may be used if desired.[8,32] The total number of follicles rupturing and forming corpora lutea is significantly influenced by the dosage of HCG administered,[31] and the duration of oestrus is significantly longer in HCG-treated queens.[32]

Other workers have found that injection of PMSG 100 IU followed 7 days later by HCG 50 IU shows results comparable with natural breeding, whereas daily administration of PMSG, 300 IU in total, resulted in fewer pregnancies and the ability of the kittens to survive till weaning was adversely affected.[7] Again, injection of PMSG 100 IU to anoestrous cats followed by 50 IU 2 days later has resulted in oestrus 3 days after second injection.[25]

Many of these reports are from work on laboratory cats in connection with programmed breeding, and in the cases of prolonged anoestrum it may be necessary both to increase the dose and to prolong the treatment. Administration of PSMG (or FSH) in too high a dose or for too long a period may result in the formation of cysts due to a hypersensitive reaction of the ovarian tissue to the hormone. It has also been shown that the LH effect of PMSG may cause follicle rupture before mating.

The dosage advocated is independent of the queen's bodyweight, but should be related to her age. Queens less than 1 year old are often overstimulated regardless of size and stage of breeding season, resulting in either development of follicle cysts or superovulation with up to 60 ovulations or even more. Cats over 5 years of age ovulate fewer eggs resulting in smaller litter size.

Follicle-stimulating hormone (FSH)
A dose of 2 mg/day for a maximum of 5 days has resulted in oestrus in 4.6 ± 0.5 days after the first injection, lasting 6.2 ± 1.2 days, with an average production of 5.7 ± 1.3 follicles on the first day of oestrus.[32] When this treatment was followed by HCG 250 IU on the first or second day of oestrus the mean ovulation rate was 10.5 ± 2.1. These results vary both in the acceptance of the tomcat and in number of mature or ruptured follicles. There was less variation in mature queens, particularly those having had limited contact with tomcats.

It is possible to induce ovulation in 50% of the cats by administration of 0.075 IU of pig FSH, 1200 μg horse LH, 0.0234 RU of unfractionated sheep gonadotropin, 6 IU HCG or 4 μg synthetic luteinizing hormone releasing hormone (LHRH).

Prolonged pro-oestrum and oestrus

Variations may occur in the duration of pro-oestrum and oestrus, and as a length of oestrus of 40 days is not uncommon[19] it may be questionable whether this is abnormal. Prolonged pro-oestrum and oestrus are reported to be effectively treated with chlormadinone acetate (CAP) 1–2 mg daily for 7–10 days.[26]

Implantation failure, abortion and premature birth

These events are not infrequent in queens, and the history is one of no pregnancy in spite of normal matings, no parturition in spite of a previous positive pregnancy diagnosis, or expulsion of fetuses or vaginal discharge some time during the pregnancy period. Kittens born before day 56 rarely survive, whereas those born later may do, although delivery before day 60 is considered premature.

In some queens abortion may occur habitually and typically in the 4th–7th week of pregnancy, evident by a blood-stained vaginal discharge possibly containing small pieces of fetal tissue.[2,4] The queen appears normal, is afebrile, and the discharge may persist for 5–6 days, or may be intermittent. Even though the oestrous activity is resumed 4–6 weeks later, abortion rearrests pregnancy.

Fetal death is not always accompanied by immediate abortion; the fetus may undergo mummification, maceration or emphysematous changes before it is expelled if at all.

Fetal mummification, death of all the fetuses, and dehydration of the fetal tissue and membranes may be the reason for absence of parturition without any clinical symptoms. In other cases some fetuses may be delivered alive while a mummified fetus situated near the apex is retained. If this is the sole fetus, it may be expelled and eaten by the queen before term, or it may be retained and even reabsorbed.

The death and maceration of one fetus may not affect other fetuses. The clinical symptoms in such cases include a reddish vaginal discharge of varying consistency. Later there may be expulsion of fetal tissue and, depending on the pathogenicity of the microorganisms involved, there may be fever and other clinical symptoms of uterine disease. There is a pronounced depression if the fetus undergoes emphysematous changes.

Specific infections Various specific infections may cause interruption of pregnancy in queens.[10,11,13,20,21,27,30] Feline viral rhinotracheitis (FVR) and feline panleukopenia virus (FPV) may cause abortion, fetal mummification and delivery of stillborn kittens; FPV may also cause early neonatal deaths. Feline leukaemia virus (FELV) may cause fetal resorption and abortion as well as fading kitten syndrome. Toxoplasmosis may also be the reason for abortion in queens.

Non-specific infections Colibacteria, streptococci, staphylococci and salmonella bacteria may be responsible for premature termination of the pregnancy.

Diagnosis
This is based upon serological tests and viral or bacteriological isolation in combination with the clinical diagnosis.

Antibiotic therapy
Queens with reproduction failure caused by viral or toxoplasma infections should not be used for breeding, and in catteries the disease should be eradicated before new cats are taken in, after isolation and negative serological tests. Non-specific infections may be treated with antibiotics, in accordance with sensitivity tests, together with supportive therapy.

Hormone therapy
Habitual abortions, especially when occurring at the same time during a pregnancy, and where no other cause can be found, may be related to an insufficient luteal function. Progestogens may be used for therapy, but overdosing may lead to masculinization of female kittens and to prolonged pregnancy.[1]

Progesterone (P)
This may be used effectively in a dose of 3 mg intramuscularly at the first sign of abortion, and results in continuation of the pregnancy and delivery of full-term healthy kittens.

Medroxyprogesterone acetate (MAP)
MAP administration, 10 mg orally per week, starting 1 week before the corresponding time of the previous abortion and until 1 week before the anticipated parturition, is effective.[6]

Repositol progesterone
When administered weekly in intramuscular injections of 1–2 mg/kg for the same period as in MAP therapy above, this is also effective.[29]

HORMONE THERAPY

Sex hormones may be indicated for purposes other than controlling reproduction.

Pseudopregnancy

Sterile matings induce ovulations accompanied by similar endometrial changes as if the queen were pregnant. Unlike the bitch, in the queen there is rarely any mammary enlargement; the condition is considered a normal event in the queen and treatment is seldom required.

If indicated, mild tranquillization may be used. Attempts to cause regression of corpus luteum by administration of prostaglandins have failed.[28,33]

The cystic endometrial hyperplasia–pyometra complex

This condition most often occurs in middle-aged or older queens due to P produced from a retained corpus luteum developed after a sterile mating. When it occurs in queens that have never been mated, as it often does, it has been assumed that an affectionate pat on the hindquarters from the owner might be a sufficient stimulus to induce ovulation.[9] It may also be caused by the administration of exogenous hormones.

Symptoms
The symptoms are often mild anorexia and depression. If the cervix is dilated, there may be vaginal discharge of a watery or thick and viscous consistency, the colour varying from light brown-pink to dark red. In more advanced cases abdominal distension is present together with weight loss, polydipsia, polyuria and dehydration. The temperature is often normal, but fever may occur, and subnormal temperature accompanies toxaemia. If the cervix is closed the uterus may be extremely distended filling nearly the whole abdominal cavity, and rupture may follow.

Diagnosis
This is made by abdominal palpation, radiographic examination and haematological examination revealing neutrophilia. Normal pregnancy, uterine torsion, ascites, peritonitis, diabetes mellitus and malnutrition are differential diagnoses to bear in mind.

Therapy
The treatment of choice is ovariohysterectomy combined with antibiotics and other supportive therapy.

Medical treatment has been tried with antibiotics and two intramuscular injections of 5 mg repositol stilboestrol, 5 days apart, for dilation of the cervix and drainage of the uterus possibly via an inserted catheter.[29]

Treatment of feline pyometra with $PGF_{2\alpha}$ has been tried in combination with an antibiotic.[3] $PGF_{2\alpha}$ was administered between 6 and 22 times in a total dose of 420–3750 μg. The corpus luteum regressed in

8.8 ± 3.39 days, and 12 out of 16 treated queens became pregnant and delivered living kittens at normal term. As in the bitch, administration of prostaglandins to the queen is followed by side-effects such as ataxia, respiratory distress, muscular tremor, bowel movements and loose stool, and further investigations are needed before this drug can be of practical use in the treatment of pyometra in the queen.

Vaginitis

Vaginitis, caused by trauma during coitus or by a uterine infection, may be accompanied by intensive licking of the vulval region. Other signs include vaginal discharge, no acceptance of the tomcat, aberrant oestrous cycles or sterility.

Diagnosis
This is based upon the history and a gynaecological examination along with other examinations, as mentioned in Chapter 3.

Therapy
This comprises antibiotics, in accordance with a sensitivity test, administered locally and systemically.

Nymphomania

This condition is characterized by excessive sexual behaviour and screaming, presumably caused by prolonged or excessive production of oestrogen in non-ovulated follicles and cysts in the ovaries. Some queens, such as siamese, although not suffering from nymphomania, may scream excessively during oestrus, but this condition should not be treated medically, and the only solution is either to let them have kittens or to spay them.

Therapy

Medroxyprogesterone acetate (MAP)
MAP 2–3 mg orally daily for 1–8 days or 50 mg subcutaneously may be effective.[23]

Chlormadinone acetate (CAP)
CAP 1–2 mg daily for 7–10 days stops the screaming after 2 days' treatment,[26] and in many cases this is sufficient. In other cases the screaming starts again 2 days after the end of treatment, but repeated treatments may increase the intervals between attacks.

In queens not responding to medical treatment an ovariohysterectomy may be indicated.

Alopecia

Alopecia occurring after spaying can be treated by implantation of testosterone 25 mg.[18]

Undesirable sexual and social behaviour

These conditions are more common in the male (see Chapter 15, page 257). Treatment with MAP or megestrol acetate (MA) in the female is exactly the same, but there is an additional possible side-effect of mammary gland enlargement.[14]

Excessive grooming

In some cats this can be eliminated by administration of MAP or MA.

REFERENCES

1. Acland, G.M. (1975) Habitual abortion. *Feline Pract.,* **5**(1), 8.
2. Acland, G.M. and Butcher, D.R. (1974) Habitual abortion in cats. *Aust. vet. J.,* **50**
3. Amano, T. and Koi, Y. (1980) Treatment of feline pyometra with prostaglandin F₂. *J. Japan. vet. Assoc.,* **33**, 115–119.
4. Brown, C.V. (1974) Breeding problems in a cattery. *Feline Pract.,* **4**(3), 6.
5. Cline, E.M., Jennings, L.L. and Sojka, N.J. (1981) Feline reproductive failures. *Feline Pract.,* **11**(3), 10–13.
6. Christiansen, Ib J. (1980) Unpublished observations.
7. Cline, E.M., Jennings, L.L. and Sojka, N.J. (1980) Breeding laboratory cats during artificially induced estrus. *Lab. Anim. Sci.,* **30**, 1003–1005.
8. Colby, E.D. (1970) Induced estrus and timed pregnancies in cats. *Lab. Anim. Care.,* **20**, 1075–1080.
9. Dow, C. (1962) The cystic hyperplasia–pyometra complex in the cat. *Vet. Rec.,* **74**, 141–147.
10. Dubey, J.P. (1978) Toxoplasmosis. *Pract. Vet.,* **Winter**, 10–14.
11. Gilbride, A.P. (1972) Toxoplasmosis and the cat. *Feline Pract.,* **2**, 1, 10.
12. Gregson, N.M. and Ishmael, J. (1971) Diploid–triploid chimerism in three tortoiseshell cats. *Res. vet. Sci.,* **12**, 275–279.
13. Hansen, J.S. (1970) A case of fetal maceration in a cat. *Vet. Med./small Anim. Clin.,* **65**, 1077–1078.
14. Hart, B.L. (1979) Feline behavior – evaluation of progestin therapy for behavioral problems. *Feline Behav.,* **9**(3), 11–14.
15. Herron, M.A. and Boehringer, B.T. (1975) Male pseudohermaphroditism in a cat. *Feline Pract.,* **5**(4), 30–32.
16. Hurni, H. (1981) Day length and breeding in the domestic cat. *Lab. Anim. (GB)* **15**.
17. Johnston, S.D., Buoen, L.C., Madl, J.E., Weber, A.F. and Smith, F.O. (1983)

X-chromosome monosomy (37, XO) in a Burmese cat with gonadal dysgenesis. *J. Amer. vet. med. Assoc.*, **182**, 986–989.

18. Joshua, J.O. (1971) Some conditions seen in feline practice attributable to hormonal causes. *Vet. Rec.*, **88**, 511–514.

19. Joshua, J.O. (1975) Feline reproduction: the problem of infertility in purebred queens. *Feline Pract.*, **5**(5), 52–54.

20. Kilham, L., Margolis, G. and Colby, E.D. (1971) Cerebellar ataxia and its congenital transmission in cats by feline panleukopenia virus. *J. Amer. vet. med. Assoc.*, **158**, 888–901.

21. McKinney, H.R. (1973) A study of toxoplasma infections in cats as detected by indirect fluorescent antibody method. *Vet. Med./small Anim. Clin.*, **68**, 493–495.

22. McQuown, J.B. (1940) An unusual case of sexual excitement in a kitten. *J. Amer. vet. med. Assoc.*, **97**, 266.

23. Moltren, H. (1963) Hinausschiebung der Läufigkeit bei Hunden und Katzen mit Perlutex Leo. *Kleintier-Prax.*, **8**, 25–27.

24. Mosier, J.E. (1975) Common medical and behavioral problems in cats. *Mod. vet. Pract.*, **56**(10), 699–703.

25. Nobunaga, T., Takahashi, K.W. and Imamichi, T. (1976) Studies on the ovulatory response of Japanese cats (*Felis catus*) to gonadotropins. *Jap. J. Anim. Reprod.*, **22**, 89–94.

26. Oettel, M., Arnold, P. and Arnold, H.-I. (1969) Der Einsatz von Chlormadinonazetat bei Hund and Katze. *Fortpfl. Haustiere*, **5**, 358–364.

27. Scott, F.W. and Gillespie, J.H. (1973) Feline viral diseases. *Vet. Scope*, **17**, 2–11.

28. Shille, V.M. and Stabenfeldt, G.H. (1979) Luteal function in the domestic cat during pseudopregnancy and after treatment with prostaglandin $F_{2\alpha}$. *Biol Reprod.*, **23**, 1217–1223.

29. Stein, B.S. (1975) The genital system. In: *Feline Medicine and Surgery*, ed. Catcott. E.J., Chapter 13, 2nd ed. American Veterinary Publishers Inc., California.

30. Vainisi, S.J. and Campbell, L.H. (1969) Ocular toxoplasmosis in cats. *J. Amer. vet. med. Assoc.*, **154**, 141–152.

31. Wildt, D.E. and Seager, S.W.J. (1978) Ovarian response in the estrual cat receiving varying dosages of HCG. *Hormone Res.*, **9**, 144–150.

32. Wildt, D.E., Kinney, G.M. and Seager, S.W.J. (1978) Gonadotropin induced reproductive cyclicity in the domestic cat. *Lab. Anim. Sci.*, **28**, 301–307.

33. Wildt, D.E., Panko, W.B. and Seager, S.W.J. (1979) Effect of prostaglandin $F_{2\alpha}$ on endocrine-ovarian function in the domestic cat. *Prostaglandins*, **18**, 883–892.

ADDITIONAL READING

Chesney, C.J. (1976) The response to progestagen treatment of some diseases of cats. *J. small Anim. Pract.*, **17**, 35–44.

Daykin, P.W. (1971) Use and misuse of steroids in canine and feline dermatology. *J. small Anim. Pract.*, **12**, 425–430.

Fennell, C. (1975) Some demographic characteristics of the domestic cat population in Great Britain with particular reference to feeding habits and the incidence of the feline urological syndrome. *J. small Anim. Pract.*, **16**, 775–783.

Fowler, N.G. and Foster, S.J. (1971) The use of megestrol acetate in the treatment of miliary eczema in the cat. *Vet. Rec.*, **88**, 374.

Leyva, H. and Stabenfeldt, G. (1983) Manipulation of the estrous cycle of the cat with photoperiod and melatonin. *Biol. Reprod.*, **28**, Suppl. 1, 122.

Radecky, M. and Wolff, A. (1980) Anomaly of the reproductive organs in an infertile cat. *Vet. Med./small Anim. Clin.*, **75**, 434.

Turner, T. (1977) Eosinophilic granulomas in cats. *Vet. Rec.*, **100**, 327.

Chapter 14
Andrology of the Normal Male

ANATOMY AND PHYSIOLOGY

The testes of the mature tomcat measure 14×8 mm and are situated in the scrotum immediately ventral to the anus. The scrotum is densely covered with hair and is not pendulous, compared to the scrotum of the dog, which is pendulous and more sparsely covered with hair. The prostate gland, measuring 5×2 mm, is divided into a left and right lobe separated by a septum of connective tissue. These prostate lobes do not cover the ventral aspect of the urethra, whereas in the dog they completely surround it. Unlike the dog, the cat has a pair of bulbo-urethral glands, 4×3 mm, situated laterally and anteriorly to the base of the penis.

The penis, encased in a free prepuce, is located ventral to the scrotum and is directed backwards (Figure 14.1). The os penis is not deeply grooved as in the dog. The glans penis is undivided and conical. On the cranial two-thirds of the penis there are 100–200 cor-

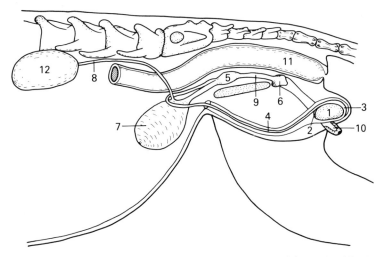

Figure 14.1 Reproductive structure, *in situ*, of the cat. (1) Testis. (2) Caput epididymis. (3) Cauda epididymis. (4) Ductus deferens. (5) Prostate gland. (6) Bulbo-urethral gland. (7) Bladder. (8) Ureter. (9) Urethra. (10) Penis. (11) Rectum. (12) Kidney. (After McKeever[6])

nified papillae, 0.75–1 mm long and directed towards the base of the penis. They are fully developed at the age of 6–7 months and are believed to be of importance during mating, as their rasping effect stimulates the female resulting in a greater chance of ovulation.[4]

The testes have normally descended into the scrotum at birth, but may move up and down the inguinal canal for some months, although they are usually in a permanent position in the scrotum by 12–14 weeks or at least within 6 months. The Leydig cells mature at 5 months and spermatozoa may be present in the tubules at 6–7 months.

Neck-biting, mounting, pelvic thrusts and spraying may occur at 4 months, but intromission and ejaculation are not common at this age as spermatogenesis does not begin until 5 months.

Sexual maturity is achieved at an average age of 9 months, at a bodyweight of approximately 3.5 kg. Some wild-living cats do not reach puberty until 12–18 months, whereas stud cats can, and often do, sire litters when mating queens at 7 months.[2] Reproductive activity may continue for up to 14 years, but if a tomcat is meant for breeding he should be used only for 4–6 years.[11]

ANDROLOGICAL EXAMINATION

This may be performed in the same way as described for the male dog, and the findings must be compared with the history (see Chapter 4).

Scrotum and testes

Traumatic injuries from territorial fights causing pain, increased scrotal temperature, swelling, haematoma, fever and depression are not uncommon. The treatment comprises antibiotics, cool moist applications and possibly castration.

Cryptorchidism

Unilateral cryptorchidism is more common than the bilateral form, and the condition occurs more frequently in purebred males. As in the male dog the condition may be of congenital origin and it implies atrophy of the inguinally or abdominally retained testicle, degeneration of the seminiferous tubules, aspermatogenesis and a risk of testicular neoplasia; castration is therefore advocated, possibly by using techniques described by Kirby[3] or Wolff.[13]

Orchitis

Orchitis with cool, painless, fluctuating distension of the scrotum may accompany feline infectious peritonitis (FIP) and *Brucella* infection.

Tumours
Tumours are very rare in cat testes, but Sertoli cell tumours may occur.

Penis and prepuce
Penile haematoma may accompany urethral obstruction, and should be corrected by gentle manipulation. A ring of hair may interfere with mating (see Chapter 15).

Prostate and the bulbo-urethral gland
These are very rarely the site of pathological conditions, but prostatic hyperplasia and adenocarcinoma may occur.

Semen collection

Semen is collected either to examine it or for artificial insemination. Collection is either by electroejaculation or by the use of an artificial vagina. A specimen of semen for examination can be obtained from ejaculated semen deposited within the vagina by the use of a cotton-tipped applicator stick moistened with warm sterile saline.

Artificial vagina
Up to 20% of tomcats can be trained to ejaculate in an artificial vagina,[5] and it is possible to obtain a good result within 2–3 weeks.[8]

An artificial vagina may be made of a 2 ml rubber bulb-pipette with the bulb end cut off and a 3 × 44 mm test tube fitted instead to make a watertight system.[12] It is placed in a 35 × 75 mm (60 ml) polyethylene bottle filled with water at 52 °C to maintain an internal working temperature of 44–46 °C. The rolled end of the bulb is stretched over the rim of the polyethylene bottle for fixation. A very small amount of a non-spermiotoxic lubricant is used, and often a teaser queen is needed. When the tomcat mounts and erection develops, the artificial vagina is slipped onto the penis, and within 1–4 min the semen is ejaculated and collected. A firm grip and pressure on the dorsal part of the pelvic region may stimulate the tomcat.

Electroejaculation
This method may be used for collection of semen, but, due to the accompanying pain, only during total anaesthesia.[7,9,11] A teflon rectal probe has been used, 10 × 12 cm, with three longitudinal stainless steel electrodes 5 cm each in length, and a stimulus of 2–8 V and 5–220 mA.[9]

The ejaculate

Using an artificial vagina, ejaculates with an average volume of 0.04

Table 14.1   Characteristics of cat semen collected by electroejaculation or by the use of an artificial vagina. (After Platz *et al.*[8])

Method of collection	Number of ejaculates	Total sperm count	Volume (ml)	Range of motility (%)
Electroejaculation	200	$15-130 \times 10^6$	0.1–0.5	60–95
Artificial vagina	34	$15-130 \times 10^6$	0.02–0.12	60–95

ml (range 0.01–0.12), a constant of 57×10^6 spermatozoa per ml (range $13-143 \times 10^6$) and a motility of 80–90% have been obtained.[12] Collection of semen by artificial vagina is possible two to three times weekly without any significant decrease in the concentration.

Using electroejaculation, collection of semen is possible once a week with an average volume of ejaculate of 0.233 ml a total number of spermatozoa of 28×10^6, and a motility of 60%.[9] The volume of the ejaculate is greater than when using an artificial vagina, probably due to the greater amount of secretion from the accessory glands.

The pH in the ejaculate averages 7.4 (range 7.0–7.9) and the motility of the spermatozoa is 80–90%.[12]

The normal dimension of the spermatozoa is 55–65 µm, and the length and width of the head are 6.5 and 3 µm, respectively.[10] Sperm abnormalities occur with a frequency of 10% without interference in the fertility, the most frequent ones being tail abnormalities and cytoplasmic droplets.

Examination of Semen

Examination of semen is carried out in the same ways as in the dog. The volume of the ejaculate and the concentration of spermatozoa depend on the method of collection (Table 14.1). Eosin B is recommended for determining the sperm viability.[1]

REFERENCES

1. Dooley, B.S., Auen, K.L. and Brown, G.G. (1980) A dye exclusion assay for the determination of cat (*Felis catus*) sperm viability. (Abstr.) *Fertil. Steril.*, **33**, 226.
2. Joshua, J.O. (1968) Abnormal behavior in cats. In: *Abnormal Behavior in Animals*, ed. Fox, M.W. Chapter 23. W.B. Saunders Co., Philadelphia, London, Toronto.
3. Kirby, F.D. (1980) A technique for castrating the cryptorchid dog or cat. *Vet. Med./small Anim. Clin.*, **75**, 632.
4. Klug, E. (1969) *Die Fortpflanzung der Hauskatze* (Felis domestica) *unter besonderer Berücksichtigung der instrumentellen Samenüber-tragung*. Thesis, Hanover.
5. Lamm, A.M. and Settergren, I. (1971) Reproduction in the cat. *Almänt vet.-möte, Stockholm*, **II**, 11–20.
6. McKeever, S. (1970) Male reproductive organs. In: *Reproduction and Breeding*

Techniques for Laboratory Animals, ed. Hafez, E.S.E., Chapter 2. Lea and Febiger, Philadelphia.

7. Platz, C.C. and Seager, S.W.J. (1978) Semen collection by electroejaculation in the domestic cat. *J. Amer. vet. med. Assoc.*, **173**, 1353–1355.
8. Platz, C., Follis, T., Demorest, N. and Seager, S. (1976) Semen collection, freezing and insemination in the domestic cat. *VIII Int. Congr. Anim. Reprod., Krakow*, 1053–1056, volume IV.
9. Platz, C.C., Wildt, D.E. and Seager, S.W.J. (1978) Pregnancy in the domestic cat after artificial insemination with previously frozen spermatozoa. *J. Reprod. Fertil.*, **52**, 279–282.
10. Schmaltz, R. (1921) Das Geschlectsleben der Haussäugetiere, 3rd ed. Richard Schoetz, Berlin (cited in Klug[4]).
11. Scott, P.P. (1970) Cats. In: *Reproduction and Breeding Techniques for Laboratory Animals*, ed. Hafez, E.S.E. Chapter 10. Lea and Febiger, Philadelphia.
12. Sojka, N.J., Jennings, L.L. and Hamner, C.E. (1970) Artificial insemination in the cat (*Felis catus L.*). *Lab. Anim. Care*, **20**, 198–204.
13. Wolff, A. (1981) Castration, cryptorchidism, and cryptorchidectomy in dogs and cats. *Vet. Med./small Anim. Clin.*, **76**, 1739–1741.

ADDITIONAL READING

Dooley, M.P., Murase, K. and Pineda, M.H. (1983). An electroejaculator for the collection of semen from the domestic cat. *Theriogenology*, **20**, 297–310.

König, J.E., Klawiter-Pommer, J. and Vollmerhaus, B. (1979) Korrosionsanatomische Untersuchungen an der Harnröhre und den Penisschwellkörpern des Katers. *Kleintier-Prax.*, **24**, 351–360.

Thuline, H.C. and Norby, D.E. (1961) Spontaneous occurrence of chromosome abnormality in cats. *Science*, **134**, 554–555.

Chapter 15
Infertility and Hormone Treatment in the Male

INFERTILITY

Only a limited amount of information is available on infertility in male cats. The diagnosis is based upon the history and a clinical examination. Infertility, reversible or irreversible, may be due to various circumstances both psychological and physical in origin.

A tomcat being placed in a new, unfamiliar environment may show arrest of sexual desire for a period of a month,[11] and an aggressive queen can dominate and discourage an inexperienced tomcat. A tomcat kept constantly in a cage may become sexually inactive, but he returns to normal after some time in freedom.[1,14]

The condition of the tomcat is of importance, so malnutrition or obesity may result in temporary infertility or lowered libido. Reversible degeneration of the testes and azoospermia are described in tomcats following a large intake of liver.[13]

Testicular hypotrophy may be of congenital origin or a result of fetal or neonatal panleukopenia.[2,10]

Unsuccessful matings without intromission in spite of mounting activity and pelvic thrusting may be due to accumulation of hair around the basis of the glans penis. If the male cat does not remove this hair ring during grooming it is easily removed by sliding it over the glans after retracting the preputial sheath, and then successful mating will be possible.

HORMONE TREATMENT

Various problems in male cats are treated with sex hormones. The most common are: aggressive behaviour, either to humans or other animals; spraying, that is, the emission of small jets of urine against walls, furniture, etc., originating in the need for territorial demarcation; problem urination, such as on the owner's bed or some other inappropriate site for no apparent reason; and alopecia after castration.

Therapy

Urine spraying and aggressive behaviour[5,6] are problems which are often solved by castration (see Chapter 17), but in about 10% of neutered cats the operation does not arrest this behaviour.[4,8] Administration of progestogens may be used instead of castration, or in cases where castration has proven ineffective.

Medroxyprogesterone acetate (MAP) or megestrol acetate (MA)
These may be used, one being more successful than the other in some instances, and vice versa. Both may produce side-effects in the form of weight gain, decreased appetite and depression.

Hart[7] using both MAP and MA found that 20% of cats with problem urination, 33% of cats treated for spraying, and 60% of cats with aggressive behaviour responded satisfactorily to this treatment.

Both parenteral and oral administration is usually effective within 7 days, but in many cases the effect wears off in variable periods of between 1 month and 1 year, and the treatment must be repeated.

The following dose regimen is advocated: parenteral injection, either subcutaneous or intramuscular, 100 mg; oral administration, 5 mg daily for 7–10 days, followed by 5 mg every 2nd day for 2 weeks, then 5 mg twice a week for 4 weeks, and thereafter 5 mg per week.

Delmadinone acetate (DMA)
This has been used successfully in the case of vagrant spraying tomcats,[3] but again the effect has only lasted for 3–12 months. The advocated dose rate is 0.25–1.0 mg orally daily for 2 weeks, or alternatively 10–20 mg/kg once or even twice within a 24 h period.

Alopecia

Alopecia occurring after castration may be treated effectively by implantation of 25 mg testosterone[9] or by administration of chlormadinone acetate (CAP).[12]

REFERENCES

1. Bloom, F. (1954) Pathology of the dog and cat. *Amer. Vet. Publ.*, Santa Barbara, California (cited in Stein[14]).
2. Csiza, C.K., Scott, F.W., de Lahunta, A. and Gillespie, J.H. (1971) Feline viruses. XIV. Transplacental infections in spontaneous panleukopenia of cats. *Cornell Vet.*, **61**, 423, 439.
3. Gerber, H.A., Jöchle, W. and Sulman, F.G. (1973) Control of reproduction and of undesirable social and sexual behaviour in dogs and cats. *J. small Anim. Pract.*, **14**, 151–158.
4. Hart, B.L. (1973) Behavioral effects of castration. *Feline Pract.*, **3**(2), 10–12.
5. Hart, B.L. (1975) Spraying behavior. *Feline Pract.*, **5**(4), 11–13.
6. Hart, B.L. (1976) Feline behavior – inappropriate urination and defecation. *Feline Pract.*, **6**(2), 6–7.
7. Hart, B.L. (1979) Feline behavior – evaluation of progestin therapy for behavioral problems. *Feline Pract.*, **9**(3), 11–14.

8. Hart, B.L. and Barrett, R.E. (1973) Effects of castration on fighting, roaming, and urine spraying in adult male cats. *J. Amer. vet. med. Assoc.*, **163**, 290–292.
9. Joshua. J.O. (1971) Some conditions seen in feline practice attributable to hormonal causes. *Vet. Rec.*, **88**, 511–514.
10. de Lahunta, A. (1971) Comments on cerebellar ataxia and its congenital transmission in cats by feline panleukopenia virus. *J. Amer. vet. med. Assoc.*, **158**, 901–906.
11. Michael, R.P. (1961) Observations upon the sexual behaviour of the domestic cat (*Felis catus L.*) under laboratory conditions. *Behaviour*, **18**, 1–5.
12. Oettel, M., Arnold, P. and Arnold, H.-I. (1969) Der Einsatz von Chlormadinonazetat bei Hund und Katze. *Fortpfl. Haustiere*, **5**, 358–364.
13. Seawright, A.A., English, P.B. and Gartner, R.J.W. (1970) Hypervitaminosis A of the cat. *Adv. Vet. Sci. comp. Med.*, **14**, 1–27.
14. Stein, B.S. (1975) The genital system. In: *Feline Medicine and Surgery*, ed. Catcott, E.J., 2nd ed., Chapter 13. American Veterinary Publications Inc., California.

ADDITIONAL READING

Beaver, B.V. (1981) The marking behaviors of cats. *Vet. Med./small Anim. Clin.*, **76**, 792–793.
Gysler, K. (1981) Behandling av uønsket urinering hos katt med Perlutex® Leo. *Norsk Vet.-T.*, **93**, 535–537.
Hagamen, W.D., Zitzmann, E.K. and Reeves, A.G. (1963) Sexual mounting of diverse objects in a group of randomly selected, unoperated male cats. *J. comp. physiol. Psychol.*, **56**, 298–302.
Hart, B.L. (1975) Feline behavior. Behavioral patterns related to territoriality and social communication. *Feline Pract.*, **5**(1), 12–14.
Hart, B.L. (1975) Feline behavior. A quiz on feline behavior. *Feline Pract.*, **5**(3), 12–14.
Hart, B.L. (1975) Feline behavior. Learning ability in cats. *Feline Pract.*, **5**(5), 10–12.
Hart, B.L. (1976) Feline behavior. Quiz on feline behavior. *Feline Pract.*, **6**(3), 10–14.
Hart, B.L. (1976) Feline behavior. The role of grooming activity. *Feline Pract.*, **6**(4), 14–16.
Hart, B.L. (1977) Feline behavior. Aggression in cats. *Feline Pract.*, **7**(3), 22–28.
Hart, B.L. (1978) Feline behavior. Psychosomatic aspects of feline medicine. *Feline Pract.*, **8**(4), 8–12.
Hart, B.L. (1978) Feline behavior. Water sprayer therapy. *Feline Pract.*, **8**(6), 13–16.
Hart, B.L. (1979) Feline behavior. Problem solving. *Feline Pract.*, **9**(1), 8–10.
Hart, B.L. (1980) Feline behavior. Prescribing cats. *Feline Pract.*, **10**(1), 8–12.
Michael, R.P. (1961) 'Hypersexuality' in male cats without brain damage. *Science*, **134**, 553–554.

Chapter 16
Artificial Breeding and Embryo Transfer

ARTIFICIAL BREEDING

Artificial insemination in the cat is possible with fresh as well as deep frozen semen. It is a prerequisite of success that the queen is in oestrus, which is verified partly by examination of a vaginal smear and partly by the behaviour when touching the neck and pelvic region.

Induction of ovulation may be achieved by mating with a vasectomized tomcat or by injection of human chorionic gonadotropin (HCG) on days 1 and 2 of oestrus.[4] Ovulation occurs within 25–26.6 h of intravenous or intramuscular injection of HCG.[5,6]

Fresh semen

Fresh semen for artificial insemination should be diluted immediately after collection with an isotonic saline solution to a volume of 1 ml, and 0.1 ml of this dilution is used for insemination. Even though pregnancy may be achieved after insemination with 1.25×10^6 spermatozoa, the number of spermatozoa per insemination dose should be at least 5×10^6 for routine work (Table 16.1). The semen is deposited deep into the vagina or the cervix by a catheter made of a 9 cm long 20-gauge cannula tipped with polyethylene tubing. The insemination should by synchronized with hormonal treatment for induction of ovulation or within 27 h of its administration. A conception rate of 50% has been achieved after synchronized insemination,

Table 16.1 Fertilization rates in relation to different numbers of spermatozoa in fresh semen inseminated into oestrous queens. (After Sojka, Jennings and Hamner[6])

| Number of spermatozoa inseminated | Queens | | | |
	Number inseminated	Number fertilized	Percentage fertilized	Average litter size
0.5×10^6	3	0	0	—
1.25×10^6	3	1	33	3
5.0×10^6	7	3	43	4
10.0×10^6	7	4	57	3
25.0×10^6	6	3	50	2
50.0×10^6	6	4	67	2

but the rate increases to 75% if the insemination is repeated 24 h later simultaneously with a supplementary injection of HCG 50 IU.[6] Fertilization can take place when the spermatozoa are introduced into the vagina up to 49 h after an ovulation-inducing injection of HCG, whereas no fertilization results after matings that occur 50 h or more after this injection, that is approximately 24 h after ovulation.[1]

Deepfrozen semen

The diluent used for deepfreezing semen consists of deionized water containing 20% (v/v) egg yolk, 11% (w/v) lactose and 4% (v/v) glycerol, together with streptomycin sulphate 1000 μg and penicillin G 1000 IU/ml.[4]

Immediately after collection the semen is mixed with 200 μl sterile 0.9% saline and 200 μl diluent at room temperature (22–23 °C). Motility and sperm content must be judged before further dilution to a proportion of at least 1:1 (semen to diluent, v/v), and after an equilibration period of 20 min at 5 °C a further 200 μl diluent plus an amount equivalent to the collected ejaculate volume are added. Then the sample is frozen in pellets, transferred to, and stored in vials in liquid nitrogen.

The highest recovery rate results from rapid thawing in 0.154 mol NaCl at 37 °C.

For artificial insemination with deepfrozen semen, the number of spermatozoa should be 50–100 × 10[6] motile spermatozoa, so 2 or more ejaculates are required. As the insemination dose is only 0.1 ml, concentration is needed by centrifugation of the thawed semen, and removal of some of the diluent. The female cat is anaesthetized and placed in dorsal recumbency with the hindquarters elevated during, and until 20 min after, the insemination.

The freezing and thawing procedures do not seem to lower significantly the viability and motility of the spermatozoa. In ejaculates collected into an artificial vagina, and by the use of electroejaculation, the motility was found to be 83 and 70% respectively immediately after collection, and after deepfreezing and thawing 71 and 54%.[4] In spite of these rates only 6 pregnancies resulted after artificial insemination of 56 cats, probably due to lack of exact synchronization of insemination and ovulation. The litter sizes in these pregnancies were 1, 1, 2, 2, 2, 2 and 4, and all kittens were normally developed with normal birthweights.

EMBRYO TRANSFER

Embryo transfer, surgically performed in the natural cycle, has succeeded in cats.[2,3] The synchronization was within ± 1 day and ovulation was induced by either injection of HCG or by mating with a vasectomized tomcat. 47 embryos were recovered in 17 collections. In nine transfer experiments all recovered embryos were transferred, resulting in four pregnancies diagnosed and four offspring. Attempts to fertilize eggs *in vitro* have only succeeded after 2–24 h incubation of the spermatozoa in uterus; but the spermatozoa were unable to fertilize eggs after incubation in follicular fluid and cumulus cells.

REFERENCES

1. Hamner, C.E., Jennings, L.L. and Sojka, N.J. (1970) Cat (*Felis catus L.*) spermatozoa require capacitation. *J. Reprod. Fertil.,* **23**, 477–480.
2. Kraemer, D.C., Flow, B.L., Schriver, M.D., Kinney, G.M. and Pennycook, J.W. (1979) Embryo transfer in the nonhuman primate, feline and canine. *Theriogenology,* **11**, 51–62.
3. Kraemer, D.C. Kinney, G.M. and Schriver, M.D. (1980) Embryo transfer in the domestic canine and feline. *Arch. Androl.,* **5**, 111.
4. Platz, C.C., Wildt, D.E. and Seager, S.W.J. (1978) Pregnancy in the domestic cat after artificial insemination with previously frozen spermatozoa. *J. Reprod. Fertil.,* **52**, 279–282.
5. Scott, P.P. (1970) Cats. In: *Reproduction and Breeding Techniques for Laboratory Animals,* ed. Hafez, E.S.E., Chapter 10. Lea and Febiger, Philadelphia.
6. Sojka, N.J., Jennings, L.L. and Hamner, C.E. (1970) Artifical insemination in the cat (*Felis catus L.*). *Lab. Anim. Care,* **20**, 198–204.

Chapter 17
Limitation of Fertility in the Female and the Male

THE FEMALE

Limitation of fertility in the queen is desired for much the same reasons as in the bitch, but with greater urgency because of the queen's reproductive capabilities. Limitation may be effected either by contraception or termination of pregnancy (Table 17.1).

Table 17.1 Methods for reproduction control in the queen.

Contraception	Surgical	Ovariohysterectomy Ovariectomy Salpingectomy
	Physical	Sham mating
	Chemical Prevention of oestrus	Progestogens 　Medroxyprogesterone acetate (MAP) 　Megestrol acetate (MA) 　Norethisterone acetate (NET) 　Chlormadinone acetate (CAP) 　Delmadinone acetate (DMA) Androgens 　Mibolerone
	Suppression of oestrus	Progestogens 　Medroxyprogesterone acetate (MA) 　Megestrol acetate (MA) 　Norethisterone acetate (NET) 　Delmadinone acetate (DMA)
	Prevention of implantation	Stilboestrol Oestrogens 　Oestradiol cypionate 　Oestradiol benzoate Progestogens 　Megestrol acetate (MA) 　Chlormadinone acetate (CAP) 　Delmadinone acetate (DMA)
	Various	Irradiation Immunization Vaccination Asexuality
Termination of pregnancy	Surgical	Ovariohysterectomy
	Chemical	Prostaglandins

Contraception

Contraception can be established surgically, physically or chemically.

Surgical contraception – spaying

This comprises ovariohysterectomy, ovariectomy or salpingectomy.

Ovariohysterectomy or ovariectomy

Ovariohysterectomy is the method of choice as, by its very nature, it effectively prevents any further uterine disorder. The operation must be carefully performed when ligating the ovarian ligaments, if necessary by the use of haemostatic clips[47] and care must be taken to ensure the removal of intact ovaries, since crushing the ovarian tissue may result in bits dropping into the abdominal cavity, and consequently result in oestrous activity after vascularization.[18,34] A case of abdominal pregnancy in an ovariohysterectomized cat has also been described.[5] This was presumably due to dislodgement of fertilized ova into the abdominal cavity at the time of ovariohysterectomy followed by establishment of a blood supply via the omentum.

Pyometra may develop after ovariohysterectomy if some uterine tissue is left. This has been reported to occur in the uterine stump following treatment of a spayed queen with megestrol acetate (MA).[25,33,44,50]

The resulting deficiency of sex hormones after ovariohysterectomy or ovariectomy contributes to an increase in the bodyweight because of an increasing amount of subcutaneous fat. Furthermore alopecia may develop with sparsity of hair on the posterior abdomen and medial surface of the thigh, and in some cases loss of hair on the ventral surface of the body, the lateral and posterior thigh region and even the forelimbs and the trunk. Implantation of 25 mg testosterone may be effective in such cases.[26]

Salpingectomy

Salpingectomy, or tubal ligation, may be used for prevention of pregnancy, and the operation performed by laparotomy with removal of a part of the oviduct after ligation[31] followed by histological examination to ensure removal of the right organ. Alternatively the oviduct may be cauterized.[48]

This intervention prevents pregnancy but it does not alter the behaviour of the queen; oestrous symptoms will develop with normal intervals and there is still a risk of uterine diseases as in an entire queen.

Physical contraception

Sham mating
Because the female cat is an induced ovulator, oestrus may be stopped for a short period by an artificial copulatory stimulus resulting in a period of pseudopregnancy. The stimulus can be made by the use of a sterile cotton swab without lubrication or a sterile mating with a tomcat.[3,27] Elimination of oestrus several times in this way is not recommended because of a possibility of inducing the cystic endometrial hyperplasia–pyometra complex.

Chemical contraception
This type of contraception may be established as oestrous control (prevention or suppression) or pregnancy control (prevention of implantation, see page 267). Much work has been done on the use of progestogens and androgens for the prevention or suppression of oestrus in the cat (the results are given below), but one must seriously consider whether oestrous control is to be advocated in queens meant for later breeding, even though pregnancies are reported to occur in queens bred after cessation of the treatment.

Prevention of oestrus by progestogens
To minimize the risk of inducing uterine disorders the treatment should be initiated in real anoestrum, that is, during the short period of sexual inactivity. In the case of a queen having just given birth, the treatment should be started during the first week after parturition in an effort to avoid the first postpartum oestrus.

Prolonged administration of progestogens can result in side-effects such as cystic endometrial hyperplasia, with or without pyometra, and development of mammary hyperplasia. There is a single report of non-neoplastic mammary hypertrophy in the cat associated with oral therapy.[21]

Medroxyprogesterone acetate (MAP)   A dose of 0.01 mg/kg daily should be sufficient to maintain anoestrum in most queens, and 0.05 mg/kg is effective in maintaining an anoestrous stage in all queens; queens given this dose may return to oestrus and litter kittens 83–184 days after cessation of the treatment.[12]

It is reported that oral administration of 2.5 mg MAP per queen daily for 6–14 days prevents oestrus for 1–3 months, and, if the treatment is repeated, the intervals between treatments may be increased.[29]

It is found that a dose of 5 mg MAP orally per queen per week is effective in the prevention of oestrus for the duration of the treatment.[28]

Injection of 25 mg MAP subcutaneously at 6 month intervals is effective for permanent prevention of oestrus, and this can also be obtained by oral administration of 2.5–5.0 mg per queen per week.[42]

Megestrol acetate (MA)   A normal dose for prevention of oestrus is 2.5 mg per queen per day for 8 weeks, or 2.5 mg per queen per week for up to 18 months.[3,7,32]

Even a high dose seems to be without deleterious side-effects; a dose of 10 mg MA per queen per week is found to have no side-effects, except for a slight tendency to increased bodyweight.[9] A rest period of 2–3 months has proved beneficial, and mating can be recommended at the second call after cessation of treatment, with pregnancy and average litter size as a result.

Norethisterone acetate (NET)   Prevention of oestrus will result after a daily oral dose of 0.2 mg/kg; in some siamese queens, though, 0.4 mg/kg may be necessary.[38] After cessation of the treatment oestrus will reappear from 2–4 weeks to 6 months later.

Chlormadinone acetate (CAP)   This can be used for prevention of oestrus if injected in a dose of 20–30 mg with an interval of 3 weeks between the first 2 injections, and thereafter intervals of 3 months.[1]

Delmadinone acetate (DMA)   Administered orally in a dose of 0.25–0.7 mg/kg once a week DMA is effective for prevention of oestrus after 2 weeks, and if treatment is continued once a week, permanent prevention is achieved.[10] Subcutaneous administration of 2.5–5.0 mg/kg is effective after 2–3 days and for up to 12 months, but treatment should be repeated twice a year. Queens treated early in the season may require the double dose for 1 or 2 weeks.

Prevention of oestrus by androgens

Mibolerone   Mibolerone, an androgenic–anabolic steroid, may prevent oestrus in a high percentage of queens when given orally on a continuous daily basis starting 30 days prior to the expected onset of oestrus.[4,41] A dose of 50 μg per day of mibolerone given to adult queens prevents oestrus as long as the treatment continues without any apparent effect on the subsequent oestrus, mating, conception, litter size or parturition.

Treatment with mibolerone is followed by side-effects such as a 2- or 3-fold increase in the size of glans clitoris and thickening of the cervical dermis.[48] Mibolerone does not arrest pro-oestrum or oestrus and therefore cannot be used for suppression of oestrus.

Suppression of oestrus by progestogens

Medroxyprogesterone acetate (MAP)   Interruption and suppression of oestrus may result if MAP is administered in a dose of 10–20 mg daily for 3–4 days followed by daily administration of 5–10 mg for 12–14 days. Very rarely cystic endometrial hyperplasia is seen in connection with MAP treatment.[6,45]

Megestrol acetate (MA)   This can effectively be used for suppression of oestrus.[35,40] Houdeshell and Hennessey[22] used 5 mg MA daily for 3 days and continued treatment with 2.5–5.0 mg once a week for 10 weeks. The queen should be kept away from tomcats for the first few days, since the treatment may not prevent conception if mating occurs in this period. This treatment is effective for rapid remission of behavioural signs of oestrus; vocalization, rolling, awkward body contortions and seeking males disappeared in 41% of cats after 3 days of treatment, 47% were anoestrous on day 5, and 92% by the end of the 1st week.

A single dose of 2 or 3 mg/kg can successfully prevent ovulation in cats.[7] Tsakalou and Vlachos[46] found that 2 mg/kg MA for 7 consecutive days postponed oestrus for 1–4 months.

Norethisterone acetate (NET)   Suppression of oestrus is possible if NET is administered orally in a dose of 1 mg/kg for 5 days, followed by 0.5 mg/kg daily for 5 days and then daily administration of 0.2 mg/kg as long as suppression is wanted. The effect is observed within 72 h after initiation of treatment,[38] but 8 days after the cessation of treatment a new oestrous period may start.

Delmadinone acetate (DMA)   Given orally in a daily dose of 0.5–1.0 mg/kg for 6 days, DMA is effective for suppression of heat symptoms, and attraction of males diminishes within 2–3 days; this effect is achieved within 1 day after subcutaneous injection of 2.5–6.75 mg/kg once or twice within 24 h. The effect lasts on average 2–4 months after oral and 6–9 months after subcutaneous administration.

Prevention of implantation

This is less frequently requested than in the bitch, partly because many pet owners do not realize that their cats have been mated. When it is requested, considerable care should be taken to confirm that mating has occurred. As the queen is an induced ovulator and signs of oestrus disappear rapidly after mating, this confirmation must be based partly on the history.

Examination of a vaginal smear may give information of the stage in the cycle, and during the first hours after mating sperm may be present in the smear. A clinical examination may reveal signs of mating in the form of patchy hair loss, and scratches and perforations of the skin on the back of the neck from the tomcat's mating grip.

Normally the ova take 4–5 days to complete their passage down the uterine tubes, but administration of oestrogen results in retarded transport to at least the sixth postcoital day and causes degeneration of the retarded ovum.[20]

Stilboestrol
Doses of 2 mg stilboestrol intramuscularly 2, 3, 7 and 10 days after mating should prevent implantation,[43] but the owner should be informed of the risks of prolonged oestrus and subsequent anoestrum,[43] chronic endometritis and pyometra.[11]

Stilboestrol dipropionate, 0.5 ml intramuscularly 5 h after straying, has failed to prevent pregnancy, and in one account a dead deformed fetus was delivered at term.[8]

Oestrogens
These can be used with the same risks as mentioned for stilboestrol treatment.

Oestradiol cypionate   Implantation can be prevented by 0.25 mg injected intramuscularly within 2 days of mating.[2]

Oestradiol benzoate   Implantation can also be prevented by intramuscular injection of 0.5–1 mg 2–3 days after coitus.

Progestogens

Megestrol acetate (MA)   Given as a single oral dose of 2.0 mg during oestrus this is effective for prevention of implantation.[7]

Chlormadinone acetate (CAP)   This may be used for prevention of implantation when given in an oral dose of 5 mg within 24 h after mating, and the next oestrus will follow within 20–30 days.[23,24]

Delmadinone acetate (DMA)   Implantation can be prevented if 2.5 mg is given orally after the mating, and the next oestrus will occur within 20–30 days.[23,24]

Other methods
Reproduction control by X-ray, immunological intervention, vaccination or induction of asexuality has not been reported in the queen, but there may be other possibilities as in the bitch, and several investigations are underway.

Termination of pregnancy
A pregnancy may be interrupted surgically or chemically.

Surgical termination

Ovariohysterectomy
Ovariohysterectomy is an effective solution for pregnancy following mismating and will of course prevent future pregnancies.

Chemical termination

Prostaglandins
$PGF_{2\alpha}$-THAM administered subcutaneously twice in doses of 0.5–1.0 mg/kg 24 h apart results in abortion, or parturition if treatment is instituted after the 40th day of pregnancy.[30] Treatment after the 55th day of pregnancy results in delivery of clinically normal kittens that survive, and lactation is normal.

THE MALE

Limitation of fertility can be established surgically, but chemical sterilization methods have also been tried.

Surgical methods
The most frequently used method is castration, but vasectomy may also be used.

Castration
This operation is used for the control of reproduction and as a means of correcting certain unwanted behavioural patterns. It may be performed at any age once the testicles have descended into the scrotum, but the age of 4–5 months is ideal for most kittens. There are some who prefer postpubertal castration citing such possible complications as urethral obstruction due to failure of normal development of the urethra,[17] and preputial adhesions.[15,16] However, it has been shown that the urethral circumference is not affected by the age at castration.[17]

The effect of castration upon the mating behaviour depends upon the age at, and the sexual experience prior to, castration. Sexually inexperienced tomcats exhibit a less pronounced mounting behaviour, and the ability to achieve intromission is considerably reduced and persists only for a short period following castration.[36,37] In the adult male castration results in a decrease of the mounting behaviour over a certain period of time, and in some tomcats the ability to achieve intromission may persist for several years.

As regards behavioural changes a rapid decline has been obtained in fighting (53%), roaming (56%), and urine spraying (78%), followed by a further gradual decline in 35%, 38%, and 9% respectively.[13,14] Most often a rapid change is seen in one of these patterns, whereas only a gradual decline or even no change at all is noticed in one or both of the other behavioural patterns. Even a castrated tomcat may start spraying urine at an advanced age due to an environmental change, such as for example the introduction of a strange cat to the house or neighbourhood.

Vasectomy
This operation is performed under general anaesthesia.[19,31] The owner should be informed that the cat may remain fertile for a short time after the vasectomy, and that a vasectomized male will copulate routinely with a willing female, continue to maintain libido and yet be sterile. Live spermatozoa have been detected in the ejaculate for as long as 120 h after vasectomy.[49] After vasectomy the testicular volumes are increased presumably due to increased epididymal fluid retention.

Chemical sterilization
Alphachlorohydrin affects the motility of the spermatozoa and results in sterility.[39]

REFERENCES

1. Brandt, H.-P. (1978) Die hormonelle Unterdrückung der Brunst. *Prakt. Tierarzt*, **59**, 92–94.
2. Burke, T.J. (1977) Pregnancy prevention and termination. *Current Veterinary Therapy. VI. Small Anim. Pract.* pp. 1241–1252. W.B. Saunders, Philadelphia, London, Toronto.
3. Burke, T.J. (1977) Fertility control in the cat. *Vet. Clin. N. Amer.*, **7**, 699–703.
4. Burke, T.J., Reynolds, H.A. and Sokolowski, J.H. (1977) A 180-day tolerance efficacy study with mibolerone for suppression of estrus in the cat. *Amer. J. vet. Res.*, **38**, 469–477.
5. Carrig, C.B., Gourley, I.M. and Philbrick, A.L. (1972) Primary abdominal pregnancy in a cat subsequent to ovariohysterectomy. *J. Amer. vet. med. Assoc.*, **160**, 308–310.
6. Christiansen, Ib J. (1980) Unpublished observation.
7. David, A., Edwards, K., Fellowes, K.P. and Plummer, J.M. (1963) Anti-ovulatory and other biological properties of megestrol acetate. *J. Reprod. Fertil.*, **5**, 331–346.
8. Davies, M. (1978) Treatment of feline misalliances. *Vet. Rec.*, **103**, 453.
9. Findlay, M.A. (1975) Oral progestagens in cats. *Vet. Rec.*, **96**, 413.
10. Gerber, H.A., Jöchle, W. and Sulman, F.G. (1973) Control of reproduction and of undesirable social and sexual behaviour in dogs and cats. *J. small Anim. Pract.*, **14**, 151–158.
11. Gruffydd-Jones, T.J. (1978) Treatment of feline misalliances. *Vet. Rec.*, **103**, 498.
12. Harris, T.W. and Wolchuk, N. (1963) The suppression of estrus in the dog and cat with long-term administration of synthetic progestational steroids. *Amer. J. vet. Res.*, **24**, 1003–1006.

13. Hart, B.L. (1973) Behavioral effects of castration. *Feline Pract.*, **3**(2), 10–12.
14. Hart, B.L. and Barrett, R.E. (1973) Effects of castration on fighting, roaming, and urine spraying in adult male cats. *J. Amer. vet. med. Assoc.*, **163**, 290–292.
15. Herron, M.A. (1971) A potential consequence of prepubertal feline castration. *Feline Pract.*, **1**(1), 17–19.
16. Herron, M.A. (1971) Prepubertal adhesions in prepubertally castrated cats. *Feline Pract.*, **1**(1).
17. Herron, M.A. (1971) The effect of prepubertal castration on the penile urethra of the cat. *J. Amer. vet. med. Assoc.*, **160**, 208.
18. Herron, M.A. (1976) Estrus after ovariohysterectomy. *Feline Pract.*, **6**(5), 28.
19. Herron, M.A. and Herron, M.R. (1972) Vasectomy in the cat. *Mod. vet. Pract.*, **53**(6), 41–43.
20. Herron, M.A. and Sis, R.F. (1974) Ovum transport in the cat and the effect of estrogen administration. *Amer. J. vet. Res.*, **35**, 1277–1279.
21. Hinton, M. and Gaskell, C.J. (1977) Non-neoplastic mammary hypertrophy in the cat associated either with pregnancy or with oral progestagen therapy. *Vet. Rec.*, **100**, 277–280.
22. Houdeshell, J.W. and Hennessey, P.W. (1977) Megestrol acetate for control of estrus in the cat. *Vet. Med./small Anim. Clin.*, **77**, 1013–1017.
23. Jöchle, W. and Jöchle, M. Hormonal influences on sexual and social behavior in male and female cats. Unpublished observations 1966–1973. (Jöchle, W. (1974) Progress in small animal reproductive physiology, therapy of reproductive disorders, and pet population control. *Folia vet. lat.*, **4**, 706–731).
24. Jöchle, W. and Jöchle, M. Hormonal influences on sexual and social behavior in male and female cats. Unpublished observations 1966–1973. (Jöchle, W. (1974) Pet population control: chemical methods. *Canine Pract.*, **1**, 8–18).
25. Jones, A.K. (1975) Pyometra in the cat. *Vet. Rec.*, **97**, 100.
26. Joshua, J.O. (1971) Some conditions seen in feline practice attributable to hormonal causes. *Vet. Rec.*, **88**, 511–514.
27. Kier, A. (1976) Solution to perpetual pregnancies. *Feline Pract.*, **6**(1), 4.
28. Linnet, A. (LEOpharma) (1980) Personal communication.
29. Moltzen, H. (1963) Hinausschiebung der Läufigkeit bei Hunden und Katzen mit Perlutex Leo. *Kleintier-Prax.*, **8**, 25–27.
30. Nachreiner, R.F. and Marple, D.N. (1974) Termination of pregnancy in cats with prostaglandin $F_{2\alpha}$. *Prostaglandins*, **7**, 303–308.
31. Norsworthy, G.D. (1975) Alternative surgical procedures for feline birth control: tubal ligation – vasectomy. *Feline Pract.*, **5**(1), 24–27.
32. Öen, E.O. (1977) The oral administration of megestrol acetate to postpone oestrus in cats. *Nord. Vet.-Med.*, **29**, 287–291.
33. Orhan, U.A. (1972) Pyometritis in spayed cats *Vet. Rec.*, **91**, 77.
34. Putnam, R.W. (1966) Occurrence of estrus in an ovariohysterectomized cat. *Canad. vet. J.*, **7**, 155.
35. Remfry, J. (1978) Control of feral cat populations by long-term administration of megestrol acetate. *Vet. Rec.*, **103**, 403.
36. Rosenblatt, J.S. and Aronson, L.R. (1958) The influence of experience on the behavioural effects of androgen in prepubertally castrated male cats. *Anim. Behav.*, **6**, 171–182.
37. Rosenblatt, J.S. and Aronson, L.R. (1958) The decline of sexual behavior in male cats after castration with special reference to the role of prior sexual experience, *Behaviour*, **12**, 258–338.
38. Rüsse, M. and Jöchle, W. (1963) Uber die sexuelle Ruhigstellung weiblicher Hunde und Katzen, bei normalem und gestörtem Zyklusgeschehen mit einem peroral wirksamen Gestagen. *Kleintier-Prax.*, **8**, 87–89.
39. Shille, V.M. (1974) Clinical approach to small animal reproductive problems. *Amer. Anim. Hosp. Assoc.*, 41st Ann. Meet., 501–509.

40. Skerritt, G.C. (1975) Oral progestagens and pyometra in the cat. *Vet. Rec.*, **96**, 573.
41. Sokolowski, J.H. (1976) Androgens as contraceptives for pet animals with specific reference to the use of mibolerone in the bitch. In: *Pharmacology in the Animal Health Sector*. Colorado State University Press. Fort Collin. pp. 164–175, eds. Davis, L.E. and Falkner, L.C. (cited in Burke[3]).
42. Stabenfeldt, G.H. (1974) Physiologic, pathologic and therapeutic roles of progestins in domestic animals. *J. Amer. vet. med. Assoc.*, **164**, 311–316.
43. Stein, B.S. (1975) The genital system. In: *Feline Medicine and Surgery*, ed. Catcott, E.J., 2nd ed., Chapter 13. American Veterinary Publications Inc., California.
44. Teale, M.L. (1972) Pyometritis in spayed cats. *Vet. Rec.*, **91**, 129.
45. Thornton, D.A.K. (1967) Uterine cystic hyperplasia in a Siamese cat following treatment with medroxyprogesterone. *Vet. Rec.*, **80**, 380–381.
46. Tsakalou, P. and Vlachos, N. (1976) The suppression of oestrus in cat by means of the progestagen megestrol acetate. *Ellenikes Kteniatrikes*, **19**(1), 15–23.
47. Wallace, L.J. and Stevens, G.E. (1975) The use of hemostatic clips in feline ovariohysterectomy and orchiectomy. *Feline Pract.*, **5**(6), 20–27.
48. Wildt, D.E., Kinney, G.M. and Seager, S.W.J. (1977) Reproduction control in the dog and cat: an examination and evaluation of current and proposed methods. *J. Amer. Anim. Hosp. Assoc.*, **13**, 223–231.
49. Wildt, D.E., Seager, S.W.J. and Bridges, C.H. (1981) Sterilization of the male dog and cat by laparoscopic occlusion of the ductus deferens. *Amer. J. vet. Res.*, **42**, 1888–1897.
50. Wilkins, D.B. (1972) Pyometritis in a spayed cat. *Vet. Rec.*, **91**, 24.

ADDITIONAL READING

Bareither, M.L. and Verhage, H.G. (1980) Effect of estrogen and progesterone on secretory granule formation and release in the endometrium of the ovariectomized cat. *Biol. Reprod.*, **22**, 635–643.

Dunbar, I.F. (1975) Behaviour of castrated animals. *Vet. Rec.*, **96**, 92–93.

Dürr, U.M. (1979) Kastrierwinkel: ein Hilfsmittel für die Katerkastration. *Kleintier-Prax.*, **24**, 222–223.

Edney, A.T.B. (1980) Vasectomy and the behavioural effects of neutering in the cat. *Pedigree Digest*, **6**(5), 7–14.

Fischer, R. (1962) Einseitige Hämatometra nach Ovarektomie. *Wien. tierärztl. Monatsschr.*, **49**, 156–157.

Goldston, R.T. and Seybold, I.M. (1981) Feline urologic syndrome: incidence, diagnosis, and treatment. *Vet. Med./small Anim. Clin.*, **76**, 1430–1431.

Johnson, M.S. and Gourley, I.M. (1980) Perineal hernia in a cat: a possible complication of perineal urethrostomy. *Vet. Med./small Anim. Clin.*, **75**, 241–243.

Mailhac, J.M., Barraud, F., Valon, F. and Chaffaux, S. (1980) Maîtrise de la reproduction chez les carnivores. Les interventions chirurgicales. *Le Points Vét.*, **10**, 17–39.

Sheppard, M. (1951) Some observations on cat practice. *Vet. Rec.*, **68**, 685–689.

Taylor, R.A. (1980) Reconstructive surgery after use of a urethral prosthesis in a castrated cat. *Vet. Med./small Anim. Clin.*, **75**, 437–440.

Tucker, H.Q., Jr. and Keating, L.K. (1980) Use of a copper intracystic device to treat urinary obstruction in male cats. *Vet. Med./small Anim. Clin.*, **75**, 435–436.

Wolff, A. (1981) Castration, cryptorchidism, and cryptorchidectomy in dogs and cats. *Vet. Med./small Anim. Clin.*, **76**, 1739–1741.

Zagraniski, M.J. (1980) Ovariohysterectomy in the pregnant cat utilizing a nylon cable tie band. *Feline Pract.*, **10**(4), 41–44.

Chapter 18
Pregnancy

INTRODUCTION

The ova pass through the oviduct within 2–4 days, and they enter the uterus 4–5 days after the mating,[15] and 10 days later implantation takes place. Obvious signs of pregnancy occur in the 6th–7th week. The size of the mammary glands increases and the nipples become firm. The distension of the abdomen increases (Figure 18.1), and the queen rests more than usual. During the last 2–3 weeks of pregnancy fetal movements are visible through the abdominal wall.

The duration of pregnancy averages 63–66 days (range 52–71).[18,27] When calculating the time for parturition, Table 18.1 may be of use. In Figure 18.2 the growth curve and representative developmental stages of the cat embryo are shown. Fetal dimensions during the pregnancy expressed by the crown–rump length are shown in Table 18.2. There is a relationship between the three parameters, gestational age, fetal weight and fetal crown–rump length. The gestational age in days can be calculated: $15.335 + 3.9805 \times$ (crown–rump length in cm) $- 0.0675 \times$ (weight in g). The accuracy is found to be ±3 days.[25] The pregnancy period is shortened in cases with large litters,[29] but these changes cannot be used for pregnancy diagnosis.

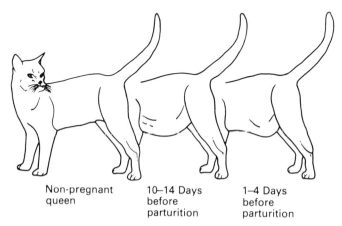

| Non-pregnant queen | 10–14 Days before parturition | 1–4 Days before parturition |

Figure 18.1 Estimation of time for parturition. (After Haemmerli and Hurni[12])

273

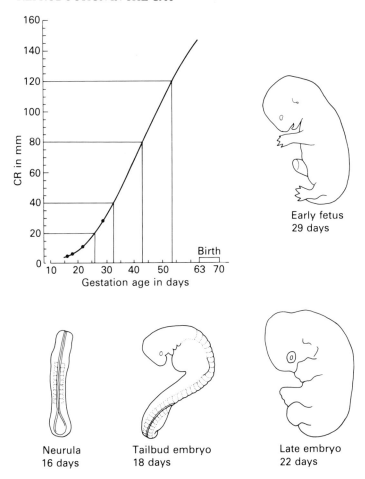

Figure 18.2 Growth curve and representative development stages of the cat. (After Evans and Sack[9])

Days of gestation	
13	Somite formation begins
14	Optic vesicle and otic placode present; neural tube forming
15	Three branchial arches present; torsion of caudal end of embryo
17	Fourth branchial arch present; otic vesicle present
18	*Fore-limb bud* present; olfactory placode formed
19	*Hind-limb bud* formed
20	Head touches heart bulge; olfactory pits formed; *hand plate* present
21	Acoustic meatus forming; eyes well formed and pigmented; shallow grooves between digits; *intestines herniated* into umbilical cord

Table 18.1   The dimension of the smallest diameter of the uterine swellings measured on radiographs and the true short width measured on sacrificed cats. (After Tiedemann and Henschel[30])

	Short width in mm	
Days after mating	X-ray	True
17	12–18	11–13
18	15–22	12–16
19	13–23	
20	15–23	18–22
21	18–25	19–22
22	19–27	
23	22–32	23–26
24	22–30	20–25
25	25–33	27–29
26	26–33	25–28
27	31–34	
28	30–38	37
29	30–37	
30	30–38	35–37

Superfecundation may result from multiple matings by several tomcats during the same oestrous period, resulting in fertilization of separate ova by spermatozoa from the various tomcats.

Superfetation occurs, since 10% of pregnant cats have oestrus and will accept male cats. Oestrus during pregnancy most often occurs from the 21st to the 24th day of pregnancy[2,6,14,16,17,19,23], but may also occur around the 6th week.[28]

Extrauterine pregnancy occurs in the queen presumably secondary to uterine rupture.[1,7,10,24,26,31]

22	Mammary ridge present; forelimb hand plate notched; pinna present as ridge
24	Forelimb digits separating distally; *tactile hair follicles* on lips, snout and above eyes; mammary primordia present; pinna triangular and projecting rostrally; eyelids forming
26	Hind-limb digits separating distally
27	All digits widely spread; tongue visible; pinna almost covers acoustic meatus; hair follicles present on body
28	Eyelids closing; claws forming
30	Eyelids almost closed; pinna covers acoustic meatus
31	*Eyelids fused*
32	*Palate fused*
37	Tactile hairs present on face
46	Fine hairs appearing on body; nose pigmented; claws hardening at the tips
50	Fine hairs cover body; claws white and hard; skin pigmented
60–63	Birth

Table 18.2 Fetal growth: crown–rump measurements in fetuses at various stages of pregnancy. (After Dawson[5] and Boyd[4])

Days after mating	Crown–rump, length (mm)	Days after mating	Crown–rump, length (mm)
17	6	36	50
19	7	38	58
20	8–9	40	68
21	10	42	80
22	12	44	84–85
23	14	45	86
24	15	47	90
25	19–20	48	95
26	21–22	49	100–102
27	23	50	102–103
28	26	52	106
29	26–28	53	109–110
30	27–31	54	112–115
31	35–36	56	120–122
32	37–38	57	125
33	37–38	58	130
34	43	60	136
35	45	At parturition	145–150

ANTENATAL EXAMINATIONS

Examinations may be indicated as in the bitch, and the first, prior to mating, should include a full history of previous breeding activity, physical examination of the genitalia and the mammary glands, and any laboratory examinations indicated. Owners should be kept fully informed after each examination, and the majority will find a straightforward explanation of the feline reproductive process helpful. Guidance on correct feeding should also be given.

Examinations during pregnancy comprise pregnancy diagnosis and necessary physical examinations.

Examination just prior to parturition, including the last checkup, is carried out in the queen's home, and any indicated abdominal palpation and vaginal exploration should be done gently and quietly to avoid disturbances of a parturition in progress.

DIAGNOSIS OF PREGNANCY

As in other animals a positive pregnancy diagnosis does not necessarily mean that live kittens will be produced. Abortion may occur, but most often this is not observed, as the fetuses are ingested by the queen. Cessation of pregnancy may be evident only from frequent checks of the queen's bodyweight.

A diagnosis of pregnancy can be made either by abdominal palpation, by ultrasonic or X-ray examinations, or by laparoscopy.

Diagnostic methods

Abdominal palpation
This is easy to perform due to the thin and usually relaxed abdominal wall. The optimal time for this examination is the period from days 20–30, during which the fetal swellings are distinctly separated (Figure 18.3), but great care must be taken as heavy palpation will cause

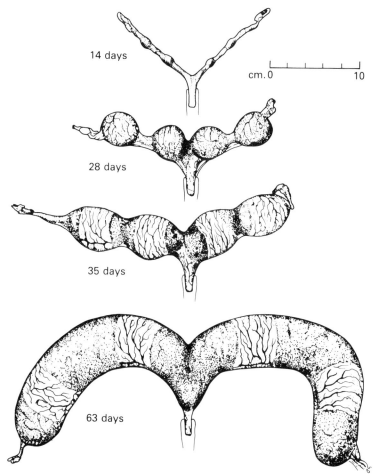

14 days

cm. 0   10

28 days

35 days

63 days

Figure 18.3   Note the growth of the uterus and its contents during pregnancy, with the formation of a zonary (girdle) placenta for each fetus. Individual conceptuses can be identified up to the thirtieth day. After this, 'conversion' occurs from round to ovoid and individuals cannot be distinguished. (After Scott[28])

abortion. Later, the swellings become more confluent, resulting in a more uniform diameter of the uterus. Late in the pregnancy the fetuses may be palpated through the abdominal wall.

Ultrasound

Ultrasonic examination after the Doppler principle, can be used from about day 30 by determining the pulsation in the navel artery, in the fetal heart or the uterine artery. B-Scanning examination can be used for pregnancy determination from the 19th day after mating.[21,22]

X-ray

This may be used as early as day 17 for determining uterine swellings. From the size of the short width of the swellings it is possible to estimate the approximate stage of pregnancy.[30] The voltage of the apparatus should be maintained at 50 kV, the mA at 50 for thin cats and at 60 for fat ones, and the focal distance (focus to film) kept at 100 cm.

Table 18.1 compares the short width of the uterine swellings measured directly from radiographs with swellings measured on uteri taken from queens sacrificed between day 17 and 30 of pregnancy.

From day 39–40 of pregnancy X-ray examination can reveal ossification centres, and the stage of pregnancy may be estimated from their location (Table 18.3).

Laparoscopy

This may be used for pregnancy diagnosis in the queen and the investigation is performed in the same way as described for the bitch (see Chapter 8, page 166). After examination 1 suture is performed in the peritoneum and 1 in the skin.

Table 18.3 Survey of the radiographic appearance of fetuses from the 17th day of pregnancy until full term. (After Boyd[4] and Tiedemann and Henschel[30])

Days after mating	X-ray appearance
17–30	Uterine swellings
25–35	Uterus is enlarged with a more anterior position in the abdomen
38	Ossification of the fetal mandible, the frontal maxillary and parietal bones of the skull and the scapula, humerus, femur, the ribs and the bodies of the vertebrae
41	Radius and ulna
43	Tibia, fibula, ileum, ischium and the occipital bones of the skull
47	Interparietal bones
49	Metacarpal and metatarsal bones
52	Phalanges of the digits and the sternum
56	Molar teeth

Other examinations

Vaginal smears

These can be used for the identification of an ovulatory cycle from the third day after ovulation, but whether it is possible to differentiate between pregnancy and pseudopregnancy in vaginal smears has not been investigated.[13]

Haematological examination

As in the bitch the erythrocyte count, the haemoglobin concentration and the packed cell volume decrease during the later part of the pregnancy, but normalization is evident within 1 week after the parturition and, as in the bitch, this examination cannot be used alone for pregnancy diagnosis. The leukocyte count and the plasma total protein are constant during the period of pregnancy.[3] It must be pointed out, though, that the number of leukocytes may be increased if the blood sample is taken from excited cats.[8]

FEEDING

The energy intake is 1465 kJ (350 kcal) per day for pregnant cats with a bodyweight of 3.5 kg, and 2510 kJ (600 kcal) for lactating cats weighing 2.5 kg.[20] Restricted maternal protein intake during pregnancy and lactation results in kittens that exhibit increased emotional responsiveness in stressful situations, and retardation in the development of maternal attachment formation.[11]

Owners should be encouraged to feed their pregnant queens several small meals a day during the last 3 weeks of pregnancy, as pressure on the stomach from the uterus discourages the intake of the usual single large meal.

REFERENCES

1. Barke, H., Sekeles, E. and Marcus, R. (1980) Extra-uterine mummified fetus in the cat. *Feline Pract.*, **10**(3), 44–47.
2. Beaver, B.G. (1973) Supernumerary fetation in the cat. *Feline Pract.*, **3**(3), 24–25.
3. Berman, E. (1974) Hemogram of the cat during pregnancy and lactation and after lactation. *Amer. J. vet. Res.*, **35**, 457–460.
4. Boyd, J.S. (1971) The radiographic identification of the various stages of pregnancy in the domestic cat. *J. small Anim. Pract.*, **12**, 501–506.
5. Dawson, A.B. (1950) The domestic cat. In: *The Care and Breeding of Laboratory Animals*, ed. Farris, E.F. John Wiley and Sons, Inc., New York and Chapman and Hall, Ltd., London.
6. Doak, J.B. (1962) A case of superfetation in a cat. *Vet. Med.*, **57**, 242.
7. Dörrie, H. and Dörrie, S. (1976) Abdominalgravidität bei der Katze. *Prakt. Tierarzt*, **57**, 25–30.
8. Doxey, D.L. (1966) Cellular changes in the blood as an aid to diagnosis. *J. small Anim. Pract.*, **7**, 77–89.

9. Evans, H.E. and W.O. Sack (1973) Prenatal development of domestic and laboratory mammals: growth curves, external features and selected references. *Zbl. Vet.-Med., Reihe C*, **2**, 11–45.
10. Fry, P.D. and Jones, S.C. (1973) A case of ectopic pregnancy in a cat. *J. small Anim. Pract.*, **14**, 361–365.
11. Gallo, P.V., Werboff, J. and Knox, K. (1980) Protein restriction during gestation and lactation: Development of attachment behaviour in cats. *Behav. Neurol. Biol.*, **29**, 216–223.
12. Haemmerli, M. and Hurni, H. (1980) Hysterektomie und Aufzucht keimfreier Katzen. *Jahrestagung der G.V., Lausanne*.
13. Haenisch, V. (1980) *Vaginalzytologische Untersuchungen an der Hauskatze* (Felis domestica) *unter besonderer Berücksichtigung der Abgrenzung von anovulatorischem und gravidem Zyklus*. Thesis, Hanover.
14. Harman, M.T. (1917) A case of superfetation in the cat. *Anat. Rec.*, **13**, 145–153.
15. Herron, M.A. and Sis, R.F. (1974) Ovum transport in the cat and the effect of estrogen administration. *Amer. J. vet. Res.*, **35**, 1277–1279.
16. Hoogeweg, J.H. and Folkers, Jr., E.R. (1970) Superfetation in a cat. *J. Amer. vet. med. Assoc.*, **156**, 73–75.
17. Hunt, H.R. (1919) Birth of two unequally developed cat fetuses (*Felis domestica*). *Anat. Rec.*, **16**, 371–377.
18. Jemmet, J.E. and Evans, J.M. (1977) A survey of sexual behaviour and reproduction of female cats. *J. small Anim. Pract.*, **18**, 31–37.
19. Jepson, S.L. (1883) A case of superfetation in a cat. *Amer. J. Obstet. Gynec.*, **16**, 1056–1057.
20. Kronfeld, D.S. (1976) Canine and feline nutrition. *Mod. vet. Pract.*, **57**(1), 23–26.
21. Laiblin, Ch., Schmidt, S. and Dudenhausen, J.W. (1982) Erste Erfahrungen mit dem ADR-Real-Time-Scanner zur Trächtigkeitsdiagnose bei Schaf, Schwein, Hund und Katze. *Berl. Münch. tierärztl. Wochenschr.*, **95**, 473–476.
22. Mailhac, J.M., Chaffaux, St., Legrand, J.J., Carlier, B. and Heitz, F. (1980) Diagnostic de la gestation chez la chatte: utilization de l'echographie. *Rec. Méd. vét.*, **156**, 899–907.
23. Markee, J.E. and Hinsey, J.C. (1935) A case of probable superfetation in the cat. *Anat. Rec.*, **6**, 241.
24. Morgan, A.F. (1976) Extrauterine mummified fetus in a Burmese. *Feline Pract.*, **6**(3), 55–56.
25. Nelson, N.S. and Cooper, J. (1975) The growing conceptus of the domestic cat. *Growth*, **39**, 435–451.
26. De Nooy, P.P. (1979) Extrauterine pregnancy and severe ascites in a cat. *Vet. Med./small Anim. Clin.*, **74**, 349–350.
27. Prescott, C.W. (1973) Reproduction patterns in the domestic cat. *Aust. vet. J.*, **49**, 126–129.
28. Scott, P. P. (1970) Cats. In: *Reproduction and Breeding Techniques for Laboratory Animals*, ed. Hafez, E.S.E., Chapter 10. Lea and Febiger, Philadelphia.
29. Stein, B.S. (1975) The genital system. In: *Feline Medicine and Surgery*, ed. Catcott, E.J., 2nd ed., Chapter 13. American Veterinary Publications Inc., California.
30. Tiedemann, K. and Henschel, E. (1973) Early radiographic diagnosis of pregnancy in the cat. *J. small Anim. Pract.*, **14**, 567–572.
31. Tomlinson, J., Jackson, M.L. and Pharr, J.W. (1980) Extrauterine pregnancy in a cat. *Feline Pract.*, **10**(5), 18–24.

ADDITIONAL READING

Arbeiter, K. (1981) Trächtigkeitsdiagnose bei Hund und Katze. *Tierärztl. Prax.*, **9**, 367–373.

McIntire, J.W. and Waugh, S.L. (1981) Uterine torsion in a cat. *Feline Pract.*, **11**(3), 41–42.

Windle, W.F., Orr, D.W. and Minear, W.L. (1934) The origin and development of reflexes in the cat during the third fetal week. *Physiol. Zoöl.*, **7**, 600–617.

Chapter 19
Parturition and Newborn Kittens

PARTURITION

A kittening box, 45 × 30 × 25 cm, should be placed at a suitable location some time before parturition. The activity of the queen decreases in the ninth week of pregnancy, and she seeks a solitary, quiet and warm dark place in which to have her kittens. The queen may display restlessness due to abdominal distension, she may frequently change sides or prefer a sitting position, and fetal movements may be seen and result in a quick turn of her head towards the flank. The muscular tone of the perineal region becomes flaccid and relaxed, and the nipples of the swollen mammary glands become prominent and deep pink as parturition approaches.

As far as is known, the birth mechanism is the same as described for the bitch (see Chapter 9). The first stage of labour is characterized by increasing restlessness, pacing and vocalization up to 24 h before parturition, and just prior to parturition nesting behaviour and antagonism towards strangers may be visible and intensified. In some queens the rectal temperature is lowered. The appetite may be decreased or lost, but some queens continue to eat until the kittens are born, and even during the intervals between expulsion of the kittens.

After rupture of the fetal membranes the expulsion of the first fetus often follows very rapidly by three to four contractions within as many minutes, though the birth of the first kitten in some cases may take up to 30–60 min. The expulsion may be accompanied by a loud scream. The queen most often licks the kitten vigorously and severs the navel cord. Some may lick and cleanse themselves before giving any attention to the newborn kitten, and others only sever the umbilical cord with no further attention to the newborn kitten. In spite of the normally quite vigorous licking, cannibalism of newborn kittens is rare.[5] In cases of unequal distribution of fetuses in the two horns, the first fetus to be born comes from the horn containing most fetuses, and the first two kittens are delivered from each horn.[3,4]

The intervals between the delivery of the individual fetuses vary from 5 min to 1 h; often two kittens are born immediately after one another, and then a period of 10–90 min may elapse before the next expulsion. This second stage of labour is normally accomplished within 2–6 h. Birth of a litter of 12 kittens has succeeded without complications within 6 h. In some queens a physiological interruption of the parturition may, for unknown reasons, occur after the birth of the first kitten(s) and may last for 12–24 h. Labour activity and contractions cease, while the queen nurses the kitten(s), eats, rests, lactates and acts as if the parturition were already complete.[8,9] The remaining kittens are delivered alive and without problems when labour starts again. This phenomenon is not to be regarded as an abnormal occurrence in the cat and must be differentiated from dystocia.

The fetal membranes are often expelled shortly after the delivery of each kitten, but if two kittens are born within a few minutes, both placental membranes may be expelled simultaneously shortly afterwards, or all membranes may be extruded at the same time after delivery of the last kitten. Many queens ingest the fetal membranes soon after they are delivered. There is no evidence that this matters, but to avoid postpartum diarrhoea if they are ingested the membranes should be gently removed from the queening box.

KITTENS

Litter size

The litter size is influenced by various factors such as the breed, the number of litters already produced, the feeding condition and the health of the queen. The average size of a litter is four (range one to eight). It is characteristic that the first litter is relatively small, and the litter size increases markedly until the fourth litter, then decreases until the seventh. Thereafter a further increase follows with the maximum number of kittens in the ninth litter.[12] The maximum litter size of alive and weaned kittens registered is 14, but the birth of 15 kittens, of which 11 were weaned, has been recorded.[2]

The litter size of siamese cats is relatively large, but the weight of the kittens is relatively small, the average litter comprising six kittens, each weighing on average 100 g.

In litters of one to eight kittens there are only small variations in the frequency of stillbirth and postnatal deaths until weaning, whereas the frequency of postnatal deaths increases in litters comprising more than eight kittens; this litter size is rare though.[12] As the queen gets older the frequency of stillbirths and the number of postnatal deaths increase. There seem to be some breed-related differences in the frequency of stillbirths (Table 19.1), which may be associated with dystocia. More female than male kittens are stillborn, and there is a greater mortality rate among male kittens in the suckling period.[12] In an investigation comprising 790 litters with a total number of 3468 kittens born in catteries, the rate of stillbirth was

Table 19.1 Mortality among siamese and persian kittens. (After Prescott[11])

Breed	Number of kittens born	Born dead (%)	Died in 1st week (%)	Died between week 1 and weaning (%)	Weaning Number	%
Siamese	388	7.2	7.5	0	331	85.3
Persian	266	22.1	3.9	3.1	160	70.8

found to be 10.2%. Of live-born kittens 6.7% died within 24 h, a further 3.8% during the following 24 h, and within 1 week after birth 15.2% of the kittens had died.[15] The neonatal kitten is subjected to similar hazards as the neonatal puppy, and survival in each case depends on maternal behaviour.

The newborn kitten

At birth the kittens are deaf and the eyelids are closed, the senses of smell and taste are fully functional. Although kittens are resilient due to a glycogen depot in their liver at birth, they should be fed within 12 h of birth to avoid hypoglycaemia and hypothermia.[14] The queen stays near the kittens during the first 24–48 h after parturition, and the kittens ingest 2–3 ml of milk every 3 h.

The eyelids open during the second week of life. At an age of 2 weeks 5–7 ml of milk is ingested per meal, and the queen may leave the kittens for longer periods of up to several hours. After each suckling period she licks the kittens round the perineal region, thus promoting urination and defecation, and she will ingest the urine and the faeces. When 3 weeks old, the kittens are more active and start to play, and they learn to urinate and defecate some distance away from the nest. At 4 weeks old kittens may be offered some supplementary food such as small pieces of meat, canned food or babyfood.

Weaning is usually around the 6th–7th week of age, but sometimes a much longer period of lactation is seen. The queen may permit a kitten to suckle long past normal weaning age and may even persist throughout an ensuing pregnancy. Thus an 8 or 9-month-old kitten from a previous litter may sometimes be seen suckling with kittens from a recent litter.[8]

The weight of a newborn kitten is 110–120 g, and is nearly doubled within a week. By 3 weeks of age the weight is 300–350 g and by 6–7 weeks 700–800 g.[13]

Kehrer and Starcke[10] found an average weight gain per week of 91 ± 14 g from birth to the 8th week inclusive, and thereafter the weekly weight gain averaged 114 g until the 12th week of life.

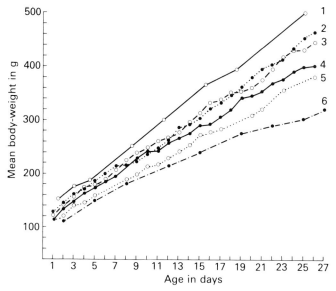

Figure 19.1 Average weight gain in kittens during the first month of life. The kittens originate from litters of different sizes, 1 = kittens from litters of two (*n* = 4); 2 = kittens from litters of three (*n* = 21); 3 = kittens from litters of four (*n* = 44); 4 = kittens from litters of five (*n* = 55); 5 = kittens from litters of six (*n* = 18); 6 = kittens from litters of seven (*n* = 7). (After Hurni[6])

The weight gain depends upon the litter size (Figure 19.1). Without supplementary food a kitten from a litter of two reaches a body weight of 350 g in 12 days, whereas 27 days are needed for a kitten from a litter of seven. Furthermore, kittens from large litters reach the adult weight 1–3 months later than kittens from small litters.[6]

The queen's milk

The composition of the queen's milk varies during the lactation period (Table 19.2). During the first 9 weeks of lactation the lactose content is almost constant, whereas considerable alterations occur in the fat and the protein contents. The protein content is found to increase from 6.5% to more than 10%; the fat content decreases during the first 4 weeks of lactation, then it increases and in the 8th week of lactation the content is at its initial level.[7]

Artificial feeding

Artificial feeding may be necessary under various circumstances as

Table 19.2   The contents of fat, protein and lactose in cat milk. (After Hurni and Montalta[7])

	Days after parturition	
	1–10	11–60
Fat (%)	3.7 ± 1.2	3.3 ± 1.3
Protein (%)	6.5 ± 0.9	8.7 ± 1.2
Lactose (%)	3.6 ± 0.3	4.0 ± 0.3

for instance in large litters and in the event of illness or death of the queen.

For artificial feeding a dilution of evaporated cows' milk contains adequate amounts of lactose and fat, but the protein content is insufficient. It is better to use 20 g skimmed-milk powder, 90 ml water and 10 ml olive oil.[14] For the first couple of days this mixture should contain 80 μg vitamin A and thereafter 50 μg vitamin A; it should be administered at 37 °C as indicated in Table 19.3. The substitute may be given by the use of an eyedropper or a 2 ml plastic syringe; most kittens learn to lap from a saucer at an age of 3 weeks. After feeding, the perineal region should be massaged to stimulate defecation and urination. The kittens should be weighed daily and their feeding adjusted accordingly. A weight gain of 10 g or more daily is ideal.[14] Care must be taken to prevent hypothermia, as this is one of the primary causes of neonatal death.

Table 19.3   Dosage and frequency of meals to kittens. (After Scott[14])

Weeks post partum	Feed	Number of meals in 24 h	Volume of milk per meal (ml)	Caloric requirement for 24 h	Expected bodyweight (g)
1	Milk mixture in bottle	12–9	2–7	40–80	100–200
2	Milk mixture in bottle	9	7–9	80–100	200–300
3	Milk mixture in bottle	9	10	112	300–360
4	Milk mixture in bottle	7	10	115	350–420
Introduce solids					
5	Reduce bottle, increase solids	7		120	400–500
6	Milk in bowl + solids	6		125	450–600
7	Weaning	3		130	550–700

Sex differentiation

Even though the testes have normally descended into the scrotum at birth, they may move up and down in the inguinal canal, and they do not descend permanently until the 10th–14th week of age. Sex differentiation is not difficult if one remembers that the distance from the anus to the preputial orifice is greater than the distance from the anus to the vulva (Table 19.4), and in the queen this area has only a few hairs (Figure 19.2).

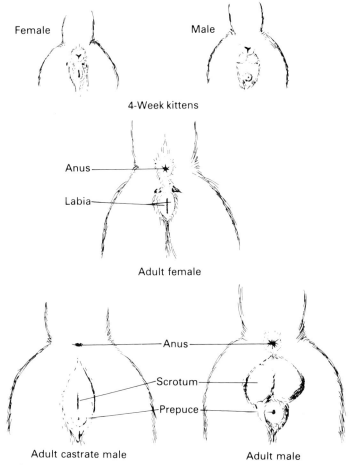

Figure 19.2   The distance between the anal and urethral apertures is always greater in the male than in the female cat. This is apparent at all ages. Scrotal sacs are covered by short fur, but the female labia are relatively free of hair. The firm round testes can be felt in the scrotal sacs of the intact male. In the castrated or cryptorchid cat, the sacs are empty. (After Scorr[14])

Table 19.4   Mean anogenital distance measured from birth to 42 days post partum in male and female kittens. (After Burke[1])

Age	Anogenital distance (mm)	
	Male	Female
At birth	12.9 ± 1.5	7.6 ± 1.0
21 days	17.8 ± 2.1	10.1 ± 1.1
42 days	20.3 ± 1.8	11.5 ± 1.6

REFERENCES

1. Burke, T.J. (1975) Determining sex of neonatal kittens. *Feline Pract.*, **5**(1), 44–50.
2. *Guinness Book of Records* (1978) 24th ed. *Guinness Superlatives Ltd.* Enfield.
3. Günther, S. (1954) Klinische Beobachtungen über die Geburtsfunktion bei der Katze. *Arch. exp. Vet.-Med.*, **8**, 739–743.
4. Günther, S. (1954) Experimentelle Untersuchungen über die Geburtsfunktion bei der Katze. *Arch. exp. Vet.-Med.*, **8**, 744–747.
5. Hart, B.L. (1972) Feline behaviour. *Feline Pract.*, **2**(5), 6–7.
6. Hurni, H. (1979) Fütterungsversuche mit SPF-Katzen. *Dtsch. tierärztl. Wochenschr.*, **86**, 356–360.
7. Hurni, H. and Montalta, J. (1980) Katzenmilch und ihr Ersatz für die künstliche Aufzucht. *Z. Versuchstierk.*, **22**, 32–35.
8. Joshua, J.O. (1968) Abnormal behavior in cats. In: *Abnormal Behavior in Animals*, ed. Fox, M.W., Chapter 23. W.B. Saunders, Philadelphia, London, Toronto.
9. Joshua, J.O. (1971) Some conditions seen in feline practice attributable to hormonal causes. *Vet. Rec.*, **88**, 511–514.
10. Kehrer, A. & Starcke, P. (1975) Erfahrungen über die Zucht, Aufzucht und Haltung von Katzen für Versuchszwecke unter konventionellen Bedingungen. *Berl. Münch. tierärztl. Wochenschr.*, **88**, 101–107.
11. Prescott, C.W. (1973) Reproduction patterns in the domestic cat. *Aust. vet. J.*, **49**, 126–129.
12. Robinson, R. and Cox, H.W. (1970) Reproductive performance in a cat colony over a 10-years period. *Lab. Anim.*, **4**, 99–112.
13. Rosenstein, L. and Berman, E. (1973) Postnatal body weight changes of domestic cats maintained in an outdoor colony. *Amer. J. vet. Res.*, **34**, 575–577.
14. Scott, P.P. (1970) Cats. In: *Reproduction and Breeding Techniques for Laboratory Animals*, ed. Hafez, E.S.E., Chapter 10. Lea and Febiger, Philadelphia.
15. Scott, F.W., Geissinger, C. and Peltz, R. (1978) Kitten mortality survey. *Feline Pract.*, **8**(6) 31–34.

ADDITIONAL READING

Baines, F.M. (1981) Milk substitutes and the hand rearing of orphan puppies and kittens. *J. small Anim. Pract.*, **22**, 555–578.
Benirschke, K., Edwards, R. and Low, R.J. (1974) Trisomy in a feline fetus. *Amer. J. vet. Res.*, **35**, 257–259.
Ledford, L.M. (1981) Urolithiasis in an 11-week-old kitten. *Vet. Med./small Anim. Clin.*, **76**, 516.

Patterson, D.F. (1980) Genetik in der Kleintiermedizin. *Kleintier-Prax.*, **25**, 104–106.
Saperstein, G., Harris, S. and Leipold, H.W. (1976) Congenital defects in domestic cats. *Feline Pract.*, **6**(4), 18–43.
Taylor, T.J. and Graham, D.L. (1971) A compilation of data on feline nutrient requirements. *Iowa State Univ. Vet.*, **33**, 144–148.
Wardrip, S.J. (1982) Cleft palate repair in a kitten. *Vet. Med./small Anim. Clin.*, **77**, 227–230.
Watson, A.D.J., Church, D.B., Middleton, D.J. and Rothwell, T.L.W. (1981) Weight loss in cats which eat well. *J. small Anim. Pract.*, **22**, 473–482.

Chapter 20
Dystocia, Obstetric and Postparturient Problems

DYSTOCIA

Dystocia is rare in the cat, but it may be suspected if strong non-productive contractions last for more than $\frac{1}{2}$–1 h after rupture of the fetal membranes, or if there are continuous contractions during 2–3 h after expulsion of the first kitten. Dystocia is obvious if labour ceases after partial delivery of a kitten, and also if labour is preceded or replaced by vaginal discharge of a bright to dark red colour.

Fetal dystocia

In fetal dystocia there may be, among other things, deviation of the head, an oversized fetus in a litter comprising only one or two kittens, and monstrosities such as congenital hydrocephalus, cyclopia and partial duplication.

Vertex presentation does not necessarily result in a difficult delivery; breech presentation occurs in 40–50% of kittens without causing dystocia, and simultaneous presentation is rare.[14]

Maternal dystocia

In maternal dystocia there may be among other things a narrow pelvic outlet caused by juvenile osteodystrophies or healed pelvic fractures, torsion of the uterus (usually only one horn is twisted 90–360° on its longitudinal axis), inguinal hernia including the pregnant uterus, and various tissue abnormalities. Dystocia due to a narrow pelvis may also occur in immature queens, particularly if they are mated in their first oestrus at an age of 4–5 months.

Uterine inertia can be anticipated in primiparous queens over 5 years of age and in multiparous queens over 8 years of age.[14] In obese queens primary inertia complicated with intrapelvic or perivulval fat deposits may interfere with the passage of the fetus. Secondary inertia may develop in multiparous queens if parturition is difficult and prolonged. If uterine inertia has developed, subcutaneous or intramuscular injection of 3–5 IU oxytocin may be used for initiating uterine contractions, and repeated after 30–45 min, if indicated. The uterus may be sensitized by a simultaneous intravenous injection of 1–2 ml calcium borogluconate solution (1.4% calcium)[13] or 5 ml 10%

calcium gluconate solution.[14] If the second injection of oxytocin does not result in uterine contractions, caesarean section should be performed within 30 min.[14]

OBSTETRIC PROBLEMS

Obstetric aid

This may be indicated in the case of fetal or maternal dystocia and may be established digitally, instrumentally or by surgical intervention, such as caesarean section or ovariohysterectomy.

As in cases of dystocia in other animal species, it is of the greatest importance to ensure sufficient amounts of lubricant, for example 5–10 ml of light mineral oil, as this may facilitate correction and extraction of the fetus.

Digital manipulation
Digital manipulation through the queen's rectum may be used for delivery of fetus which has reached the pelvis.

Instrumental extraction
Instrumental extraction of a fetus may be established by the use of an ovariectomy hook or of obstetric forceps as described for the bitch (see Chapter 10, page 208).

Caesarean section
This obstetric procedure is seldom indicated in the queen, but some breeds such as persian and siamese may need this more often than other breeds.

Anaesthesia   This may be instituted by inhalation of halothane and nitrous oxide, or ketamine hydrochloride may be used in a dose of 5–10 mg administered intravenously in combination with intubation and local analgesia.

Surgical technique   The operation can be performed through a flank or midline incision, the latter made as a 6–8 cm incision, a little anterior to the umbilicus and caudated. The uterus is gently exposed from the peritoneal cavity and incised in the corpus, and the fetuses with their placentas are gently manipulated towards and through the incision. The umbilical cord is clamped about 2 cm from the abdominal wall of the kitten. The uterus and laparotomy incisions are closed as in the bitch, and the kittens treated like puppies delivered by caesarean section.

Ovariohysterectomy

This may be indicated, for example, in the case of a complicated uterine torsion, though in valuable breeding queens only the affected horn and ovary may be removed.

POSTPARTURIENT PROBLEMS

Several pathological conditions may occur post partum which complicate the postpartum period and the survival of the kittens.

A primiparous queen who appears unwilling to nurse, shows restlessness and stays only a few minutes with her kittens or even shows aggression against neonatal kittens should be placed with the kittens in a quiet environment. In some queens oral administration of a small amount of a tranquillizer, such as acepromazine, 0.4 mg/kg orally, may be necessary. In others, P may be effective for treatment of these problems.[14]

Slow milk secretion may be treated with oxytocin in the form of a nasal spray or intramuscular injection of 2–5 IU.

Postpartum haemorrhage

Postpartum haemorrhage may follow dystocia and requires prompt treatment with one or two injections of 0.1 mg ergonovine maleate 20 min apart, or 5–10 IU oxytocin. If this does not control the haemorrhage, ovariohysterectomy should be performed.

Retention of placenta

The diagnosis here is made either by abdominal palpation or by radiographic determination of an enlargement in part of the uterus.[2,12] There may be no clinical symptoms, even for several weeks, if only a small amount of placental tissue is retained, but more often there is fever, anorexia, depression and arrest of lactation.

Treatment comprises oral administration of ergometrine, ergonovine maleate or another preparation of ergot, and antibiotics, and if this does not result in expulsion of the placental tissue, hysterotomy or ovariohysterectomy is advocated.

Prolapse of the uterus

Uterine prolapse[1,4,5,8,9,15] may occur during parturition or within the following 48 h. Usually all the kittens are expelled before the prolapse develops, but one empty horn may prolapse while the other still contains fetuses. The condition may occur at all ages due to over-relaxation and stretching of the pelvic musculature, uterine atony, trauma of the uterus, extreme prolonged tenesmus in cases of dystocia, or during lactation and oxytocin release.

Therapy

The therapy is manual reduction and replacement of the uterus under general or epidural anaesthesia if the condition has been present only a short time. In other cases replacement may require laparotomy and a combination of internal or external manipulations, but no vulval sutures are indicated, since there is no recurrence of prolapse. If the prolapse persists for several hours, and if the uterus is in a bad condition, ovariohysterectomy is indicated.

Acute metritis

Acute metritis produces clinical symptoms 12–96 h after parturition in the form of fever, polydipsia, vomiting, diarrhoea, depression, straining and cessation of lactation; often a dark red or light brown offensive vaginal discharge is present.

Therapy

The therapy comprises antibiotics, administration of fluids and electrolytes, and uterine evacuation, possibly supported by intramuscular or oral administration of ergonovine maleate. Ovariohysterectomy may be indicated if there is no response to medical treatment after 24 h. Supplementary feeding of the kittens should be established.

Lactation tetany

Lactation tetany[3,6,7,10,11] does not occur often but is found most frequently in multiparous queens with a large litter 2–4 weeks post partum. Hypocalcaemia may result in eclampsia causing tonoclonic convulsions and even death, if treatment is not instituted swiftly.

Initially the condition is characterized by restlessness, anorexia, vomiting, polydipsia, hyperpnoea and ataxia. Later there are muscular spasms, tonoclonic convulsions, increased body temperature and finally collapse and coma.

Therapy

The therapy is intravenous administration of 2–5 ml (or up to 10 ml) of a 10% calcium gluconate solution, but the condition will recur unless the kittens receive supplementary food. If the kittens are more than 3 weeks old, early weaning is advised.

Mastitis

Acute mastitis may occur in connection with parturition, most often with one or two posterior glands affected, due to infection with staphylococci and streptococci. The glands are warm and painful, the infection is accompanied by fever, depression and anorexia, and the kittens are often neglected, cold and hungry. The treatment is antibiotics for 7–10 days, supportive care and artificial feeding of the kittens.

Acute mastitis may also develop 1–2 weeks after parturition due to trauma from the sharp claw nails of the kittens. The condition at this stage is treated with antibiotics, the kittens being treated similarly. The kittens' claw nails should be trimmed, and in some cases the kittens should be removed altogether.

REFERENCES

1. Arnall, L. (1961) Prolapse of the uterus in the cat. *Vet. Rec.,* **73**, 750.
2. Cobb, L.M. (1959) The radiographic outline of the genital system of the bitch. *Vet. Rec.,* **71**, 66–68.
3. Edney, A.T.B. (1969) Lactational tetany in the cat. *J. small Anim. Pract.,* **10**, 231–236.
4. Egger, E.L. (1978) Uterine prolapse in a cat. *Feline Pract.,* **8**(1), 34–37.
5. Herbert, C.R. (1979) Prolapsed uterus in the cat. *Vet. Rec.,* **104**, 42.
6. James-Ashburner, P.W. (1961) A case of eclampsia in the cat. *Vet. Rec.,* **73**, 884–885.
7. Lawler, D.C. (1963) A case of lactational tetany in the cat and a review of the literature. *Vet. Rec.,* **75**, 811–812.
8. Luckhurst, J. (1961) Prolapse of the uterus in the cat. *Vet. Rec.,* **73**, 728.
9. Maxson, F.B. and Krausnick, K.E. (1969) Dystocia with uterine prolapse in a Siamese cat. *Vet. Med./small Anim. Clin.,* **64**, 1065–1066.
10. Michael, S.J. (1960) Suspected hypocalcemia in a Siamese cat. *J. Amer. vet. med. Assoc.,* **137**, 645.
11. Panel report (1973) Lactational tetany in cats. *Mod. vet. Pract.,* **54**(9), 95–97.
12. Reid, J.S. and Frank, R.J. (1973) Double contrast hysterogram in the diagnosis of retained placentae in the bitch. A case report. *J. Amer. Anim. Hosp. Assoc.,* **9**, 367–368.
13. Scott, P.P. (1970) Cats. In: *Reproduction and Breeding Techniques for Laboratory Animals,* ed. Hafez, E.S.E., Chapter 10. Lea and Febiger, Philadelphia.
14. Stein, B.S. (1975) The genital system. In: *Feline Medicine and Surgery,* ed. Catcott, E.J., 2nd ed. Chapter 13. American Veterinary Publications Inc., California.
15. Wallace, L.J., Henry, J.D. and Clifford, J.H. (1970) Manual reduction of uterine prolapse in a domestic cat. *Vet. Med./small Anim. Clin.,* **65**, 595–596.

ADDITIONAL READING

Bruinsma, D.L. (1981) Feline uterine prolapse (a case report). *Vet. Med./small Anim. Clin.,* **76**, 60.
Evans, I. (1980) Use of an allogeneic bone graft to enlarge the pelvic outlet in a cat. *Vet. Med./small Anim. Clin.,* **75**, 218–220.
Gruffydd-Jones, T.J. (1980) Acute mastitis in a cat. *Feline Pract.,* **10**(6), 41–42.

Haemmerli, M. and Hurni, H. (1981) Hysterektomie und Aufzucht keimfreier Katzen. *Z. Versuchstierk.*, **23**, 196–197.

Hurni, H. (1981) SPF-cat breeding. *Z. Versuchstierk.*, **23**, 102–121.

McAfee, L.T. (1979) Recurring feline dystocia. *Feline Pract.*, **9**(6), 32–36.

Part 3
Appendices and Index

Appendix I
Glossary of Abbreviations

**The following abbreviations are used in the text
(synonyms are given in brackets)**

CAP	Chlormadinone acetate
CI	Cornification index
CO_2	Carbon dioxide
DES	Diethylstilboestrol
DMA	Delmadinone acetate
DMSO	Dimethyl sulphoxide
ECP	Oestradiol 17-cyclopentylpropionate
EI	Eosinophilic index
FELV	Feline leukaemia virus
FIP	Feline infectious peritonitis
FPV	Feline panleucopenia virus
FSH	Follicle-stimulating hormone – pituitary origin
FVR	Feline viral rhinotracheitis
GH	Growth hormone (= STP, somatotropin)
GnRH	Gonadotropin-releasing hormone
HCG	Human chorionic gonadotropin – predominantly LH and some FSH activity
HP	Hydroxyprogesterone acetate
HPC	Hydroxyprogesterone caprionate
ICT	Interstitial cell tumour
IgG	Immunoglobulin, serum gammaglobulin
IVT	Illinois variable temperature extender
KPI	Karyopyknotic index
LH	Luteinizing hormone – pituitary origin
LHRH	Luteinizing hormone releasing hormone
MA	Megestrol acetate
MAP	Medroxyprogesterone acetate (or MPA, AMP)*
MGA	Melengestrol acetate*
MOE	Methyloestrenolone
NaCl	Sodium chloride
NDI	Nucleus degeneration index
NDTC	N-Desacetyl-thiocolchicine

*Abbreviations are widely used for the various hormones and hormonal compounds, but some confusion exists. Thus medroxyprogesterone acetate is normally abbreviated to MAP, but in French literature AMP and in some German literature MPA is used, and furthermore MAP has been used for melengestrol acetate instead of the normally used abbreviation MA.

NET	Norethisterone
P	Progesterone
PGF$_{2\alpha}$	Prostaglandin F$_{2\alpha}$ (or PG, PGF, PGF$_2$, PGF$_{2\alpha}$-THAM)
PIF	Prolactin-inhibiting factor (or PIH, prolactin-inhibiting hormone)
PMSG	Pregnant mare serum gonadotropin – predominantly FSH and some LH activity
PRF	Prolactin-releasing factor (or PRH, prolactin-releasing hormone)
PRL	Prolactin
SCI	Superficial cell index
SCT	Sertoli cell tumour
SEM	Seminoma
TCP	Depo-testosterone cyclopentylpropionate
TP	Testosterone propionate
TPT	A synthetic prostaglandin analogue
TRH	Thyrotropic releasing hormone (= TSH-RH)
TSH	Thyrotropic hormone

Appendix II
Units and Symbols

kg	kilogram	min	minute
g	gram, 10^{-3} kilogram	s	second
mg	milligram, 10^{-3} gram		
μg	microgram, 10^{-6} gram	mEq	milliequivalent
ng	nanogram, 10^{-9} gram	U	unit
pg	picogram, 10^{-12} gram	IU	international unit
		MU	mouse unit
l	litre	RU	rat unit
dl	decilitre, 10^{-1} litre	V	voltage
ml	millilitre, 10^{-3} litre	kV	kilovolt
μl	microlitre, 10^{-6} litre	mA	milliampere
		v/v	volume/volume
		w/v	weight/volume
cm	centimetre, 10^{-2} metre	w/cm³	waves/square centimetre
mm	millimetre, 10^{-1} centimetre		
μm	micrometre, micron, 10^{-3} millimetre	°C	degrees Celsius (Centigrade)
		kJ	kilojoule
w	week	kcal	kilocalorie
d	day		
h	hour	SD	standard deviation

Index